AF443084

Membrane Engineering for the Treatment of Gases

Volume 1: Gas-separation Problems with Membranes

Membrane Engineering for the Treatment of Gases

Volume 1: Gas-separation Problems with Membranes

Edited by

Enrico Drioli
*Department of Chemical Engineering and Materials,
The University of Calabria, Italy*

and

Giuseppe Barbieri
*National Research Council - Institute for Membrane Technology,
Italy*

RSCPublishing

ISBN: 978-1-84973-171-3

A catalogue record for this book is available from the British Library

Published by The Royal Society of Chemistry,
Thomas Graham House, Science Park, Milton Road,
Cambridge CB4 0WF, UK

Registered Charity Number 207890

For further information see our web site at www.rsc.org

Preface

Separations of downstreams and treatment of upstreams in any type of process require most of the energy necessary in the production cycle. The concentration and/or purification of the downstreams of a production unit on the basis also of the market specifications, the ratio adjustment of reactor feed streams to fit the best operating condition, the conditioning or pre-treatment of the feed streams for contaminants removal, *etc.* are some examples of typical separations that occur in the process industry every day.

Membrane engineering has been growing significantly in the last few years and membrane operations are the dominant technology in various areas today, *e.g.* in seawater desalination, in waste-water treatment and reuse, in artificial organs, in food juice treatment, *etc.* The intrinsic properties of membranes, such as molecular separations, the possibility of coupling reaction and separation in the same unit, *etc.* help to confirm membrane engineering as a powerful tool for realizing the process intensification strategy, which is the best answer today to sustainable industrial growth. For instance, reverse osmosis was demonstrated as requiring an energy load about 10 times lower than that of a thermal process. Therefore, in such fields of application, membrane separations are recognized today, among the different technologies, as the 'best available technology'. The use of membranes in the separation of gases is also a fast-growing field and in various cases membrane technology competes with traditional operations. The separation of air components or oxygen enrichment by means of membranes has been growing substantially during the past 10 years. The oxygen-enriched air produced has been used in various fields, including chemical and related industries, medical fields, food packaging, *etc.* In industrial furnaces and burners, for example, the injection of oxygen-enriched air (25–35% oxygen) leads to higher flame temperatures and reduces the volume of 'parasite' nitrogen to be heated. Mixtures containing more than 40% of O_2 or 95% by volume of N_2 from the air can be obtained together with

Membrane Engineering for the Treatment of Gases, Volume 1:
Gas-separation Problems with Membranes
Edited by Enrico Drioli and Giuseppe Barbieri
© Royal Society of Chemistry 2011
Published by the Royal Society of Chemistry, www.rsc.org

membrane systems which dominate the fraction of the nitrogen market today for applications of less than 50 tonnes day^{-1}. Membrane gas separation, however, is far from covering its real potential and further progress will be a challenge in the coming years in many applications, such as CO_2 capture and utilization, H_2 separation and purification, dehydration of gaseous streams, *etc.*

Currently, problems related to the pre-treatment of the streams, membrane life-time and their selectivity and permeability, still slow down the growth of large scale industrial applications. Membrane engineering will pursue the design and development of new polymeric, inorganic and hybrid materials with tailored and improved mass transport properties and which are able to withstand more aggressive environments and a wider range of operating conditions. The development of new knowledge for the better utilization of these unit operations in integrated membrane systems, combing various membrane operations in the industrial process will also be part of the work for a sustainable industrial growth.

The goal of this book is to present the main aspects and challenges related to membrane engineering for the treatment of gases starting from the fundamentals to the industrial application, focusing on polymeric, metallic and other inorganic membranes such as zeolites, perovskites and also carbon membranes. It is intended to provide a wide and critical state of the art on membrane technology for the treatment of gases, with emphasis for the application of membrane engineering in various fields of gas separation.

Analysis by molecular design and of aging phenomena show how to improve our understanding of the fundamentals of mass transport of gas molecules through thin and ultra-thin glassy polymer membranes. A state of the art of macro-scale simulation studies is provided, particularly for CO_2 post-combustion capture by means of membrane gas separation to identify the most relevant and efficient processes which fit the separation targets, including also an analysis of the target materials and cost performances. A very critical state of the art on current available polymers and recent progress on high performance polymers for next-generation gas separation applications is also presented. Membranes have to be assembled in modules before their utilization; their design depends also on the membrane configuration, such as flat sheet or hollow fiber. All the aspects concerning different module types and packing density, manufacturing costs and range of application are also analyzed and presented. An overview is provided about the applications of membranes in natural and biogas treatment, petroleum refining and petrochemicals production, presenting the technical challenges and the market for which the membranes are intended, giving a detailed presentation of existing commercial membrane technologies and considering the future trends for research in each addressed application. The potential of simulations in membrane engineering are shown in a techno-economic analysis of multi-stage membrane processes. This provides an insight into economics and related energy consumption and recovery avoidance costs with reference to the application of membranes CO_2 capture from coal-fired power plants. The commercial applications of membranes in gas separations are also widely described and the critical needs in the development of this technology and some of the factors which are impeding the growth of gas separation membranes are addressed together with a perspective on further commercial growth. The use of hybrid process solutions involving synergistic membranes and fully established non-permeating technologies, such as adsorption and absorption, for improving the separation performance with marginal costs is described in depth, highlighting the cooperative way of operating.

The main trends in Pd-based membrane development and criteria for their scale-up are presented in Chapters 10–14, specifically for very thin composite membranes. The basic features of membrane reactors are widely discussed for this application of this innovative technology, also considering membrane costs analysis on different membrane reactor architectures, by means of case studies. The packed bed and fluidized bed configurations most often used for membrane reactors for hydrogen production and purification studies are proposed, discussing the performances of the membrane reactor by means of modeling and simulations. An interesting strategy for the redesigning of more compact and efficient processes for pure hydrogen production than conventional ones is proposed, considering also the effect of the permeation reduction owing to concentration polarization and inhibition. This reduction is quantified by means of the concentration polarization and inhibition coefficients already included in the Sievert's law equation that can be used, in its integrated form, to evaluate the hydrogen permeating flux.

Relevant applications of carbon molecular sieve membranes are discussed for selected industrial separations such as CO_2/CH_4 in biogas upgrading, H_2/CH_4 wherever relevant, CO_2 capture from flue gases, air separation, petrochemical and high-temperature applications. The advantages and drawbacks offered by perovskite membranes are analyzed for high temperature oxygen separation, also by introducing engineering and scale-up issues as well as an economic evaluation. An overview on the current and potential applications of zeolite membranes in the treatment of gases is provided showing the possibility of zeolite use for separating light gases such as CO_2/N_2 and CO_2/CH_4 as well as in the deep purification of H_2 rich streams when these membranes are catalytically active. The application of mixed ionic–electronic conduction for oxygen separation is discussed, highlighting all the advantages over other methods, such as the large membrane area per unit packing volume, the reduced resistance to oxygen permeation, and the easy assembly into membrane modules offered by the hollow fiber configuration as well as the new strategies for improving the membrane properties, thereby making them ideal also for practical application in oxygen production.

The last chapter gives a more comprehensive approach and discusses the role of membrane gas separation and membrane engineering in the re-designing of industrial applications in terms of new, recently introduced metrics. It provides an analysis of some processes for hydrogen production/separation that can be easily extended in other separation processes. This is a useful tool for the evaluation of pros and cons during the design phase of a new plant, where the membrane operations would replace traditional ones to pursue the strategy of process intensification.

We wish to thank Dr Adele Brunetti for her collaboration in the preparation of this book, and for giving us the benefit of her knowledge in the field of gas separation and membrane reactors; she has been very useful for coordinating our activities during the various aspects of the final editing.

Enrico Drioli and Giuseppe Barbieri
The University of Calabria and
National Research Council -
Institute of Membrane Technology, Italy

Contents

Membrane Engineering for the Treatment of Gases, Volume 1:
Gas-separation Problems with Membranes
Edited by Enrico Drioli and Giuseppe Barbieri
© Royal Society of Chemistry 2011
Published by the Royal Society of Chemistry, www.rsc.org

Volume 2

Multi-scale Molecular Modeling Approaches for Designing/Selecting Polymers used for Developing Novel Membranes

ELENA TOCCI[a] AND PLUTON PULLUMBI[*b]

[a] Research Institute for Membrane Technology ITM-CNR, Via P. Bucci, Cubo 17/C, c/o Università della Calabria, I-87030 Rende (CS), Italy; [b] Air Liquide, Centre de Recherche Claude-Delorme, B.P. 126, Les-Loges-en-Josas, 78354, Jouy-en-Josas Cedex, France

1.1 Introduction

During the last decade computational chemistry and numerical simulations have had a favorable impact in almost all branches of materials research ranging from phase determination to structural characterization and property prediction.[1–7] An important effort has been focused on developing simulation tools to describe thermodynamic and transport properties of confined fluids.[7–14] The present contribution illustrates the benefit of coupling experiment to molecular modeling for selecting novel membrane materials with better separation properties for given gas mixtures as well as the limitations of the existent computational methodologies. New modeling and simulation tools based on multi-scale hierarchical modeling are needed to cope with the complexity of materials and associated phenomena at different length and time scales.[2–4]

Membrane Engineering for the Treatment of Gases, Volume 1:
Gas-separation Problems with Membranes
Edited by Enrico Drioli and Giuseppe Barbieri
© Royal Society of Chemistry 2011
Published by the Royal Society of Chemistry, www.rsc.org

Transport properties of small molecules in amorphous polymer matrices play an important role in many industrial applications such as gas separation of mixtures, packaging applications ranging from food conservation to controlled drug and cosmetics release, to special coatings for protecting specific substrates from gases.

Different aspects of technology and industrial application of polymer membranes from materials research to permeator design to their optimal configuration to enhance processes performance have been discussed in detail in a previous lecture and recently reviewed in the literature.[15–17] The potential application of a polymer as a separation membrane depends upon the selectivity towards the gas to be separated and the permeate flux. The selectivity determines the product purity and recovery whereas the permeability is related to the productivity of the membranes. This means that both the permeability and the selectivity should be as large as possible. The control of gas permeability and permselectivity of polymer membranes has become a subject of active research with worldwide participation in both industrial and academic laboratories.[17] The design and optimization of polymer membranes used in gas separation applications would be possible if reliable predictions of transport properties could be made rapidly in advance of synthesis and experiment. The actual status of available commercial software for modeling transport phenomena in polymer membranes,[2–4,18–20] does not allow the development of *de novo* material design approach. This is due not only to formidable time and length scales involved, but also to lack of detailed information on time evolution of the free volume and its distribution as a function of processing history during the manufacturing process. Rapid progress in computational methodology and validation of new simulation tools is improving the understanding of different facets of gas transport in polymer membranes and building the necessary tools for their effective use in materials design.[1–4,21–23] The possibility to predict transport properties of small molecules through polymer matrices permits the rational selection of polymer materials used in these applications and their optimal design. Although there has been reported an increasing number of studies on this subject over the last years,[24–43] the prediction of transport properties of gas molecules through glassy polymer membranes remains a difficult target. In many of the recent studies reporting molecular simulation predictions of diffusion and solubility of small gas molecules in several membrane models of the same glassy polymer a great scatter of the predicted values is observed. These results clearly indicate that the quality of the packing of the polymer chain into an amorphous cell membrane model strongly impacts the predicted gas transport properties.

The potential application of a polymer as a separation membrane depends upon the possible throughput and the purity of product.[15–17,44,45] This means that both the permeability of the gas that is transported more rapidly and the selectivity should be as large as possible. The permeability coefficient, *Pe*, of a small molecule through a polymer membrane is defined as:

$$Pe = D \cdot S \tag{1.1}$$

the product of the diffusion coefficient, D (kinetic parameter), and of the solubility coefficient, S (thermodynamic parameter). The estimation of these coefficients can be done, either by molecular dynamics (MD) and grand canonical Monte Carlo simulations, or by applying the transition state theory (TST) approach provided that the quality of the membrane amorphous cells used in the calculation represent the real distribution of torsion angles, of the free volume and its distribution, as well as the structural, conformational and volumetric properties of polymer membranes. The selectivity of a polymer membrane for a pair (i, j) of gas molecules is characterized by the ideal separation factor α_{ij} defined in eqn (1.2):

$$\alpha_{ij} = \frac{Pe_i}{Pe_j} = \frac{D_i}{D_j} \cdot \frac{S_i}{S_j} \tag{1.2}$$

Following this definition the selectivity of a membrane is the product of diffusion selectivity (D_i/D_j) and of solubility selectivity (S_i/S_j). In the case of glassy polymer membranes the overall selectivity is mainly controlled by diffusion selectivity. Two types of membranes are used commercially in gas separation technology. Glassy polymer membranes are made from stiff chain polymers and operate below their glass transition temperature. These membranes have moderately high free volume and separate gases predominantly based on differences in the sizes of the gas molecules. The smaller molecules (H_2, He, O_2) permeate more easily through the membrane than the larger ones (CH_4, C_2H_6, N_2). The second class of membranes is made either from highly flexible rubbery polymers or ultra-high free volume, glassy substituted polyacetylenes.[16,42] These membranes separate gases principally by differences in the solubility of gas molecules in these polymers. Larger and more soluble penetrants permeate faster than smaller and less soluble ones.

In the past 20 years, the control of gas permeability and permselectivity of polymer membranes has become a subject of active research with worldwide participation in both industrial and academic laboratories.[15–17,44–51] However, it has been found that simple structural modifications, which usually lead to an increase in polymer permeability, cause loss in permselectivity and *vice versa*. This so-called 'trade-off' relationship has been well described in the literature.[17,44,45,48–51] Here the log of the separation factor α versus the log of the higher permeability gas Pe yielded a limit for achieving the desired results. The upper bound limit is not fixed in the α–Pe space but moves with time as new polymers with optimized structures become available.

Recent studies of glassy polymer membranes indicate that in addition to the free volume content, gas transport parameters depend upon the backbone chain rigidity, its segmental mobility, the inter-chain distance and the chain interactions. For example, the introduction of bulky alkyl substituents opens up the polymer matrix resulting in a greater permeability. Also the reported introduction of *n*-alkyl side groups on a polymer backbone increases the side-chain flexibility as well as the membrane free volume with an overall increase of the permeability.[17,44,50,51] The free volume of a polymer, which corresponds to the unoccupied regions accessible to segmental motions, is an important parameter for understanding and predicting many of its characteristic properties. The free volume and the free volume distribution influence the molecular mobility and the transport properties of low molecular substances and gases in polymers.[52–55] Atomistic simulations allow a detailed investigation of these geometric characteristics. A complete description of definition of the free volume of membrane atomistic model configurations as well as detailed approaches for analyzing it has been reviewed recently.[18] Voronoi polyhedra and Delaunay simplices are used for analyzing the local arrangement of free volume in glassy polymer membranes cells. The calculated free volume and its distribution are used to describe the gas solubility and diffusion in amorphous polymer cells through a QSPR correlation as well as to control the quality of the constructed amorphous cells.

Many literature examples that use the QSPR approach associated with polymer permeability are based on the use of group contribution methods to establish a correlation between the structure of the repeat unit and some physical property of the polymer membrane such as the free volume, the mean segment distance or dielectric constant polarizability, which in its turn is used to predict permeation properties. As a result of such studies some practical criteria have emerged to guide synthetic researchers in improving the permselectivity of membranes which have evolved through extensive experimentation: (i) inhibition of inter-segmental packing meanwhile simultaneously inhibiting intra-segmental (backbone) mobility; and (ii) weakening of inter-chain interactions (reduction of charge transfer complexes). These design rules are based on phenomenological paradigms that provide guidelines for polymer selection. The QSPR approach that uses appropriate descriptors for representing at the same time the repeat unit, the unperturbed polymer chain and the packed amorphous cells, integrating data at different length scales including the some data from the processing history of the polymer membrane would lead to new criteria for materials improvement.

This contribution focuses on the need to shorten the research time of novel polymer materials used for membrane fabrication by combining several computational approaches and experimental techniques. It does not pretend to be a review or cover all aspects that might be considered traits of the computational materials design. Various examples are used to illustrate the use of existing numerical simulation and modeling tools for complementing experimental work. After a brief discussion of some computational approaches used to cover different aspects of polymer membrane simulation, methodologies used to characterize gas transport through polymer membranes as well as to identify factors that control thermodynamic and kinetic properties of confined fluids in membranes will be detailed. It is noticed that the successful application of modeling approaches to gas separation by membrane technology needs the development of models dealing with multi-component gas mixture transport through model membranes. Moreover, for a given polymer membrane, both gas diffusivity and gas solubility depend strongly on process parameters such as pressure difference, feed composition, and temperature. More information on the effects of process parameters on selectivity contribution should be thoroughly considered for identifying membrane materials suitable for each application.

1.2 Computational Methods

Progress in numerical simulation techniques has followed the developments in chemical engineering science and recently has received tremendous attention in both scientific and industrial communities due to the possibility to integrate micro-scale information in investigations that can be carried out at various levels of resolution on different scales of time and length. Several multi-scale modeling methodologies that are based on the information transfer between different scales starting from the molecular level and ending up at the industrial scale have been recently reviewed in the literature[3] indicate a clear trend towards coupling of the design of chemical engineering equipment/units with nanoscale modeling. It is generally expected that multi-scale modeling can lead to optimal unit design as well as to cost optimization of the final products. Fundamental research in materials and structure design now considers numerical simulation, as a complement to theory and experimentation. The integration of physical testing, advanced computations and system simulation would dramatically reduce the design and development time and costs.

In this contribution we do not aim to provide a detailed description of each of the numerical techniques contributing to the multi-scale modeling methodology but will only outline its basic principles, strengths and weaknesses, and potential applications. Interesting readers can refer to relevant books, reviews and research articles for details.[56–58]

The terminology used for characterizing multi-scale methods often varies with the application domain. Two general approaches have been developed for integrating different models at disparate spatial and temporal scales. The hierarchical modeling extracts information from lower-scale models and transfers it as parameters to the upper-scale (coarser) models of overlapping domains. The coarse-scale model is used over the entire computational domain but a higher-resolution modeling is applied to zoom into a particular sub-domain to obtain updated parameters within the corresponding grids of the coarse model.

Mesoscale modeling uses a basic unit just above the molecular scale, and is particularly useful for studying the behavior of polymers and soft materials. It can model even larger molecular systems, but with the commensurate trade-off in accuracy.[59–61]

Two methods dissipative particle dynamics (DPD) was initially devised by Hoogerbrugge and Koelman[59] as a particle-based off-lattice simulation method for the flow of complex fluids and to tackle hydrodynamic time and space scales beyond those available with MD. Since DPD is a coarse-grained model and individual atoms or molecules are not represented directly by the particles but they are grouped together into beads, these beads represent local 'fluid packages' able to move independently.

The fundamentals of the DPD method are now fairly well-established,[62–67] as are the technical subtleties[68,69] and the coarse-grained parametrization of the DPD particles, with improvements consistently being introduced.[70]

Furthermore, it is expected that DPD will play an ever-increasing role in multi-scale modeling approaches through bridging of the atomistic and continuum scales. In such approaches, atomistic simulations are performed to build the DPD models, followed by DPD simulations which provide the necessary input to the continuum codes.

DDFT, which was developed by Fraaije *et al.* in 1997,[62] is a field-based theoretical method for studying complex fluids, their kinetics and their equilibrium structures at micrometer length and microsecond time scales.[71–73] DDFT has been applied to the study of the self-assembly of block copolymers in bulk, under shear and in confinement,[74–76] and to study polymer blend compatibility.[77] Compared to the DPD method, DDFT is computationally extremely fast since larger elements can be modelled. Moreover, since the fluid elements can freely penetrate, larger time steps can be used, and furthermore it is less likely to become trapped in a local minimum. Since DPD is a particle-based method, it can provide somewhat more detailed structural information. Nonetheless, they are both powerful tools in simulating phase separated phenomena that occurs at the mesoscale and the consistency of results from the two methods for the same coarse-grain model is evaluated in this work.

The alternative to the hierarchical modeling is known as concurrent modeling and consists of combining different numerical models that simultaneously describe different sub-domains. Each numerical model runs in its sub-domain and exchanges information with its parent/child sub-domain at defined boundaries. The critical issue here is to define the criteria and protocols in order to automate the application of more detailed numerical models in a simulation domain in time and realizing the so-called temporal multi-scale calculations or/ and space referred as spatial multi-scale ones.

Spatial multi-scale methods are based on the paradigm that in many real situations the atomic description is only required within small parts of the simulation domain whereas for the majority the continuum model is still valid. This allows one to apply concurrent continuum and molecular simulations for the respective parts of the simulation domain using a coupling scheme that permits to connect between the two domains. The majority of the spatial domain is calculated by continuum solvers (computational fluid dynamics) which are very fast and only the 'active part' is calculated using molecular simulation methods. In some cases several other coarser-grained (mesoscale) methods than the atomic simulations ones are used as interfaces between the molecular simulation and the continuum domains. Such approaches are called hybrid molecular–continuum methods and allow the simulation of problems that are not accessible either by continuum or by pure molecular simulation methods.

Successful examples of application of hybrid molecular–continuum approaches are the fluid slippage past surfaces,[78] fracture propagation,[79] problems involving phase transitions,[80] fouling at surfaces, wetting, moving contact line between two immiscible liquids,[81] mass transport through membranes, systems with locally high stresses or strong gradients, flows in micro- or nanochannels with specific surface features, flows along surfaces with no ideal non-slip or slip condition or where the liquid molecules interact with wall molecules.

Modeling of gas separation process using membranes implies, as schematically reported in Figure 1.1, using different numerical simulation techniques going from the atomistic representation of the amorphous polymer matrix representing the thin layer of the active membrane to the hydrodynamics of fluid moves upon the membrane module. The computational fluid dynamics (CFD) simulations can reveal details of flow patterns, velocity distributions and resistances in an industrial module made of hundred thousands hollow fibers and correlate the pressure distribution upon membrane surface with the fluid flow pattern on the hollow fibers. However, in more realistic cases such as existence of the polarization layer the CFD calculations need to include details from lower scales in order to realistically predict permeate flux through the membrane module by integrating of modified boundary condition in the simulation and predicting the concentration polarization profile that is developed at the membrane fluid interface. The above described schematic representation of the hybrid atomistic-continuum is only one of the aspects of the numerical simulations applied to the R&D activity in membrane technology. The numerical simulation approach for predicting the morphology of the membrane is mainly based on a combination of atomistic and mesoscale methods. The final morphology of the membrane depends on the properties of the polymer materials as well as on the processing conditions. Most commercial membranes are prepared by the immersion precipitation process. Recently, it has been showed experimentally that the process of membrane formation

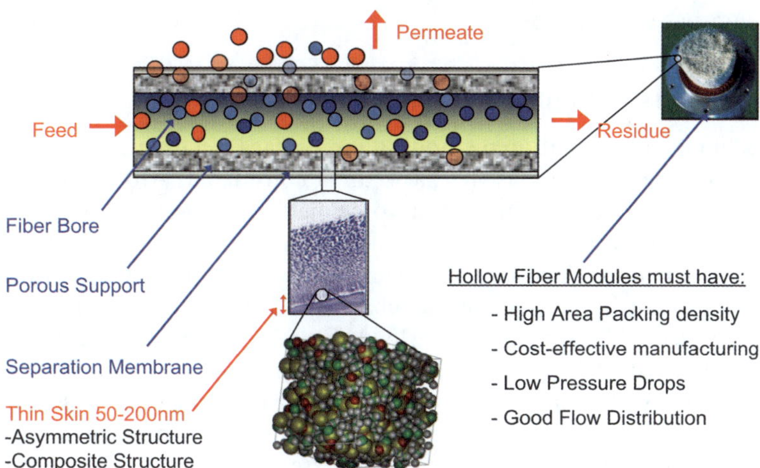

Figure 1.1   Schematic of different time- and length-scale computational domains in gas separation by membranes.

proceeds *via* spinodal decomposition to form the asymmetric membrane.[82] Lattice Boltzmann simulation of membrane formation in two dimensions[83] was able to capture the morphology of the interface between coagulation bath and polymer solution as well as the asymmetric morphology of the membrane. In a recent paper[84] the simulation has been revisited using a phase field model of the immersion precipitation process and membrane formation in three dimensions. The simulation approach looks very promising in terms of providing a deeper understanding of the immersion precipitation process and theoretical guidance on experiments design membrane formation.

Another class of spatial multi-scale methods concerns the quantum chemistry community where efforts have been focussed on the combination of quantum mechanics (QM) methods with continuum electrostatic theories in order to realistically represent the solvation free energy in a polar environment. These methods have been refined over the years and can now give a reasonable description of solvation properties of an isotropic and homogeneous medium.[85–87] However, these continuum models are not appropriate to represent the electrostatic and steric interactions of the structured environment with the active site. This is particularly true in the descriptions of complex systems like enzymes or catalysts. An appropriate description of such systems has been developed using a hybrid quantum mechanical/molecular mechanical (QM/MM) approaches[88–91] where the QM methods are used to describe the active site where chemical reactions or electronic excitations occur, and MM methods are employed to capture the effect of the environment on the active site.

In this contribution we will focus mainly on the methodology used for atomistic simulation of gas transport through models of amorphous together with the generation of atomistic models of polymer membranes and the prediction of gas transport properties of gas molecules through the membrane and briefly mention the recent contributions on coupling such calculations to coarser or continuum-based methods.

1.2.1 Atomistic Simulation Methods

The modeling and simulation methods at molecular level usually employ atoms, molecules or their clusters as the basic units considered. Atoms or molecules interact with each other through a force field, or intermolecular potential energy, and the accuracy of the force field directly determines the accuracy of the resulting calculations. The common simulation methods dealing with many-body systems can be divided into stochastic and deterministic ones. The first class is represented by the Monte Carlo (MC) method whilst the second one by the molecular dynamics (MD) method. Modeling of polymer membranes at the atomistic scale is predominantly directed toward the realistic representation of atomistic models of amorphous cells representing the membrane layer. However, the applicability and effectiveness of these methods often depends on the ability to fulfil large-scale computations; and despite the availability of high-performance computers, these methods are still restricted to solving systems that are too small for even nanoscale problems. Computer-aided molecular design of polymer membrane models at a detailed atomistic level has been reported in the literature[24–26,29–43,92–99] for investigating the sorption and diffusion of small gas molecules. MD simulation is used for exploring the structure and properties of bulk amorphous polymers and the diffusion of the penetrant molecules to be followed by exploring the trajectories generated during the simulations. MC simulations probe the configuration space of the model membranes by trial moves of particles either during the phase of amorphous cell building or during the simulation of sorption properties of gas molecules inside the amorphous cell using the so-called Metropolis algorithm to monitor the energy change from in successive steps as a trigger for accepting or rejecting a new configuration. Configurations resulting with a lower energy are accepted whereas those resulting with a higher energy are accepted with a probability governed by Boltzmann statistics. The algorithm ensures the correct limiting distribution and properties of a given system can be calculated by averaging over all MC moves within a given statistical ensemble.

1.2.1.1 Molecular Dynamics

Molecular dynamics (MD) is an atomistic simulation method for studying a wide class of materials, such as polymers, metals, ceramics and biomolecules under ambient as well as extreme conditions. MD allows one to predict the time evolution of a system of interacting particles (*e.g.* atoms, molecules, granules, *etc.*) and estimate the relevant physical properties.[100–102] Specifically, it generates such information as atomic positions, velocities and forces from which the macroscopic properties (*e.g.* pressure, energy, heat capacities) can be derived by means of statistical mechanics. MD simulation usually consists of three constituents: (i) a set of initial conditions (*e.g.* initial positions and velocities of all particles in the system); (ii) the interaction potentials to represent the forces among all the particles; (iii) the evolution of the system in time by solving a set of classical equations of motion for all particles in the system. MD methods are governed by the system Hamiltonian and the Hamilton's equations of motion are integrated to move particles to new positions and to assign new velocities at these new positions.

Given a force field for the potential energy, the Hamiltonian of a system of N atoms can be written as:

$$H(r^N, p^N) = \phi(r^N) + K(p^N) \tag{1.3}$$

where it is assumed that the kinetic energy, K, depends only on the momenta (p) and it is separable from the potential energy, ϕ, that depends only on atomic positions.

Particles in MD move naturally under their own intermolecular forces and follow Newton's second law:

$$\mathbf{F}_i = m_i \ddot{r}_i = -\frac{\partial \phi}{\partial r_i} \tag{1.4}$$

where m_i, $\ddot{r}_i$ and r_i are the mass, acceleration and position of particle i, respectively. During the simulation the configuration space as well as the phase space is explored allowing extracting information on dynamics of the system. In order to simulate the gas diffusion in a polymer membrane a force field representing the interactions between all the atoms of the system composed of the polymer amorphous cell and penetrant molecules is required. The force field has to be tested against experimental results and theoretical constraints.

The gas diffusion coefficients can be estimated either from MD simulations by using the Einstein (eqn (1.5)) or by means of the Green–Kubo (eqn (1.6)) formulations:

$$D = \lim_{t \to \infty} \frac{\left\langle [r(0) - r(t)]^2 \right\rangle}{6t} \tag{1.5}$$

$$D = \frac{1}{3} \int_0^{\infty} \langle v(0) - v(t) \rangle \tag{1.6}$$

The mechanism of diffusion in rubbery polymers is different from the glassy ones. The diffusion coefficients for small gas molecules in rubbery polymer membranes do not depend on concentration whilst in glassy polymer membranes they do depend and reach a constant value at relatively high concentrations. This is mainly due to the fact that glassy polymers are not in a thermodynamic equilibrium state. For these polymers the final 'metastable' chain configuration depends on the processing history of the membrane. This detail makes even more difficult the modeling of glassy polymer membranes due to lack of experimental structural data for validating computational approaches. The prediction of self-diffusion coefficients for non-polar small gas molecules in amorphous rubber polymer matrices is normally done through MD calculations.[27,103–105]

After construction of the amorphous cell using the method of Theodorou-Suter[106] and geometric free volume analysis of the cell several (four to six) penetrant molecules to improve sampling are inserted at the free volume positions. The cell is further relaxed by 100 ps of a NPT-MD (constant particle number, temperature and pressure) simulation at 1 bar and room temperature before starting a longer (nanoseconds) NVT dynamics. The recorded trajectories of each penetrant gas molecule are analyzed and the diffusion coefficient is determined by means of relation (eqn (1.5)). In Figures 1.2 and 1.3 the packed cell model of polydimethylsiloxane (PDMS) and the trajectory of N_2 molecules in the PDMS matrix are reported. The MD simulations show two types of motions of the N_2 molecules: jumps between cavities and local motion inside cavities.

The predicted self-diffusion coefficients depend principally on the quality of the force fields used to model the interactions not only between the penetrant and polymer matrix, but also intramolecular interactions between polymer chains. These last ones, strongly affect the quality of the amorphous polymer cell and in particular the total free volume its distribution and dynamics which in their turn affect the predicted values of diffusion coefficients. The role of chain relaxation and matrix fluctuations in explaining the diffusion mechanism of small gas penetrants as N_2 in rubber polymer membranes has been clearly

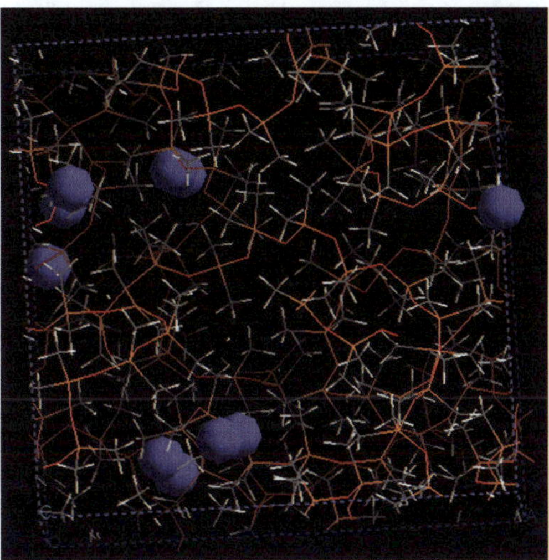

Figure 1.2   An amorphous PDMS cell.

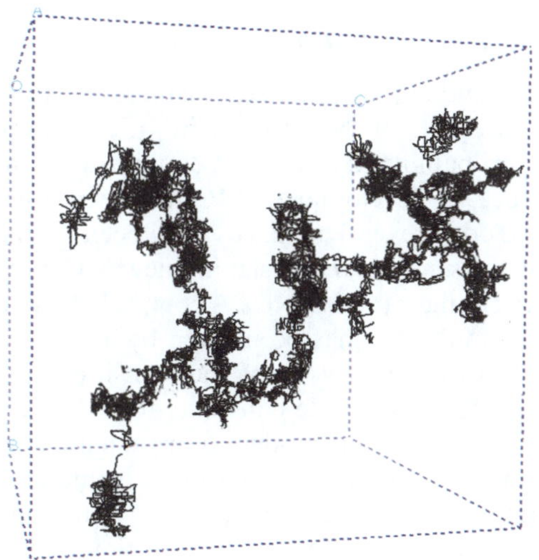

Figure 1.3   Trajectories of the N_2 molecule during MD simulations.

demonstrated through MD calculations in which the polymer matrix has been kept fixed.[27,103–105] MD simulation of gas diffusion in polymer membranes generates a wealth of information on the mechanism of gas transport but its use is limited to high free-volume rubber matrices and small gas molecules due to prohibitive CPU times for diffusion coefficients smaller than 4×10^{-7} cm^2 s^{-1}.[27] For this reason the technique is not adequate for systematic screening of a large number of polymer candidates and generating data to be used in a materials design approach.

The slowness of MD is an obstacle to the study of polymer properties, but not a barrier. This is especially true when the properties of interest are localized, as in the case of glass transition or diffusivity of small molecules through a polymer matrix. At the far end of what can be achieved, MD can also be applied to study polymer dynamics on the time and length scales of polymer entanglement. For current applications, we focus on methods that have been illustrated with several polymers and are on the verge of being competitive with existing empirical methods.

Predictions of polymer properties underscore the difficulty of developing MD methods that are competitive with existing empirical methods. Simulations of polymers require greater computational effort. In the case of gas penetrants, for example, the equilibrium concentrations are very small. This necessitates simulating a large number of polymer atoms just to simulate a few gas penetrant molecules in order to get a reasonable average. Long simulations of large systems are extremely expensive. On the other hand, academic motivation is very weak for developing MD methods that would be conclusively superior to empirical methods, as demonstrated with similarly large databases. The computational expense for polymer systems exacerbates this problem.

1.2.1.2 *Monte Carlo Simulations*

The Monte Carlo technique (MC) is a stochastic simulation method designed to generate a long sequence, or 'Markov chain' of configurations that asymptotically sample the probability density of an equilibrium ensemble of statistical mechanics.[107,108] Since its development, MC has been used to test statistical mechanics theories. Today several advances have been reached in devising new statistical mechanical ensembles and designing new MC moves for the efficient sampling of complex configuration spaces. A comprehensive review on progress and outlook in MC simulations has been given by Theodorou.[109]

MC implemented in the framework of the transition state theory (Section 1.2.1.3) can provide estimates of rate constants for infrequent events. For the construction of polymer membrane models, filling a basic cubic volume element under periodic boundary condition, a rotational isomeric state Monte Carlo technique incorporating long-range interactions[110] can be used.

The solubility coefficient can be calculated *via* simulations in the canonical ensemble in which the chemical potential is calculated using the Widom particle insertion method. The interaction energy of a gas particle inserted within the accessible free volume of the polymer matrix is calculated and the excess thermodynamic potential μ_{excess} can be estimated from eqn (1.7):

$$\mu_{\text{excess}} = RT \ln \langle \exp(-E_{\text{int}}/kT) \rangle \tag{1.7}$$

The solubility, S, is then obtained from eqn (1.8):

$$S = \exp(-\mu_{\text{excess}}/RT) \tag{1.8}$$

The simulation of sorption properties of gas molecules in the amorphous cells of glassy polymer can also be estimated using grand canonical Monte Carlo calculations. This needs as input the structural model of the amorphous cell and the force field describing sorbate/sorbent and sorbate/sorbate interactions. For the prediction of gas sorption in the generated amorphous cells, the interaction potential is the most important ingredient. A simplified interaction potential including only a dispersive-repulsive short-range potential, represented by a Lennard-Jones 6–12 potential combined with electrostatic interactions between partial charges on the adsorbent and guest atoms is used. The multipole–multipole interactions are calculated according to:

$$E_{\text{structure(polym-gas)}} = \sum_i \sum_j \left[\frac{A_{ij}}{r_{ij}^{12}} - \frac{B_{ij}}{r_{ij}^6} + \frac{q_i q_j}{r_{ij}} \right] \tag{1.9}$$

where A_{ij} is the repulsion constant and B_{ij} is the dispersion constant and q_i the point partial charges located at the atomic positions of the adsorbent and sorbate molecules.

1.2.1.3   Transition State Theory

The prediction of diffusivity in polymer glasses and low-temperature rubbery polymers *via* direct MD simulations would require extremely long simulation times because the penetrant diffusion becomes too slow to be predictable by MD.

The TST is a well-established methodology for the calculation of the kinetics of infrequent events in numerous physical systems. According to the TST method the gas transport mechanism through a dense polymer system is described as a series of activated jumps. For each transition, a 'reaction trajectory' leading from a local energy minimum to another through a saddle point in the configuration space is tracked, and the transition rate constant is evaluated.

The general approach followed to obtain the diffusivity, based on atomistic TST-based determination of rate constants for individual jumps executed by the penetrant in the polymer matrix and subsequent use of these rate constants within a kinetic MC simulation to track displacement at long times, is another good example of hierarchical modeling.

Three different approaches have been applied for describing the coupling between the jumping of a penetrant and the motions of nearby polymer chains.

In the original Gusev–Suter TST method[111] a frozen polymer method has been used. All polymer chains are considered fixed in place, and TST-based rate constants are calculated from the energy barriers found for a penetrant to pass from one local potential energy minimum to another. This method is the most straightforward; however, in polymers it yields rate constants that lead to diffusion coefficients much lower (by factors of 10^3–10^6) than experimental values, because neglecting chain fluctuation contributions is physically unrealistic in a polymeric material. Gusev and Suter implemented the frozen polymer method taking into account the thermal vibrations of the polymer matrix[27,112] with the assumption that the polymer atoms, in a sorption site, execute uncorrelated harmonic vibrations around their equilibrium positions to accommodate the guest molecules. The magnitude of the fluctuations is controlled by a parameter similar to the Debye–Waller factor in X-ray scattering. In the most recent method,[20,113] referred as the explicit polymer method, the dimensionality of the jump path was increased to include explicitly both penetrant displacements and polymer chain motions.

Each resulting rate constant then captures the details of the particular chain motions that accompany the jump, rather than mean-field average motions, but at the expense of a much higher computational cost. This method has been applied to a few penetrants in polypropylene,[113,114] with reasonable agreement compared to experiment and correlation.

The TST approach permits the calculation of D by using Einstein's relation (eqn (1.4)) only and the solubility through eqn (1.10):

$$S = \frac{1}{kTV} \int_V \rho(r)\,\mathrm{d}V \tag{1.10}$$

1.2.1.4 *Quantitative Structure Property Relationships*

Computational techniques intended to automate generation and mining of virtual libraries of compounds have successfully been developed and used in recent years for molecular and materials discovery and optimization. This activity has been extensively used in the field of drug discovery where quantitative structure property relationships and quantitative structure activity relationships (QSPR/QSAR) have been used to build correlations between structural molecular features and the experimentally measured properties or activities of the molecules. QSPR/QSAR approaches have been reported in the literature to predict many physicochemical properties, such as vapor pressures,[115] aqueous solubility[115–118] and boiling points.[119–121] They have recently been applied also to polymer materials for predicting properties in a number of applications.[122–128] The generation of practical design rules for polymers used in membrane technology would profit from the development of reliable QSAR/QSPR methodologies applicable to this field.

Most of the examples from the literature that use the QSAR approach for predicting gas permeability in polymer membranes use of group contribution methods to establish a correlation between the structure of the repeat unit and some physical property of the polymer membrane such as the free volume, the mean segment distance or dielectric constant polarizability, which in its turn is used to predict permeation properties.[129] Some practical criteria have emerged to guide synthetic researchers in improving the permselectivity of membranes going through extensive experimentation steps: (i) inhibition of inter-segmental packing whilst simultaneously inhibiting intra-segmental (backbone) mobility; (ii) weakening of inter-chain interactions (reduction of charge transfer complexes). These design rules are based on phenomenological paradigms that provide guidelines for polymer selection correlate the target property with other physical or chemical properties of the polymers, for example using group contribution properties.[130,131] Building a QSPR/QSAR approach that would integrate data at different length scales including processing conditions of the polymer membrane in order to capture information at a multi-scale level would lead to new criteria for polymer materials discovery. The first step in a QSPR/QSAR study is to collect data for a set of polymers that are used for making the membrane. Such data could be either measured experimentally or obtained by numerical simulations. The next step consists in calculating molecular descriptors that mirror fundamental physicochemical factors that in some way relate to gas solubility and diffusion are needed. It is desirable that the description is reversible, so that the model interpretation leads forward to an understanding of how the modification of chemical structure influences gas transport and equilibrium properties. In recent QSPR/QSAR studies[132–134] the following conditions are of fundamental importance for building reliable correlations: (i) the selection of the training set, (ii) the selection of adequate descriptors, (iii) the accuracy of the initial activity-data set used to generate the correlation, (iv) the algorithm used to reduce the number of descriptors to the best set and develop the property model by appropriate regression.

In Figure 1.4 a schematic diagram of the QSAR methodology is reported. The generation of the molecular models for a large number of polymer membranes as well as the collection of reliable simulated values of S and D for several gas molecules in these membranes is the first step towards building a useful QSAR. Generation of appropriate physically meaningful descriptors based on molecular information not only of the repeat unit but also of

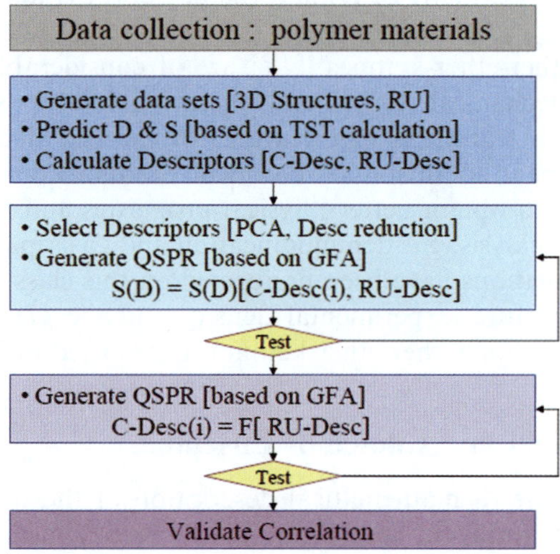

Figure 1.4 Schematic diagram of the applied QSAR methodology.

amorphous cells is key to developing and searching for associations between apparently disparate or disjointed datasets. Generation of different kind of descriptors (topological, geometrical and electronic) is followed by principal component analysis which is a multivariate statistical factor analysis technique. This leads to data reduction since it points to the possibility that only a limited number of properties have to be measured or calculated in order to explain the major part of the information concerning gas permeation through the membranes. Next genetic forms of statistical regression methods such as genetic function approximation or genetic partial least squares are used to generate multiple QSAR models for the collected experimental gas permeation data. The use of this methodology for building reliable correlations between the structure of the polymer, the characteristics of the amorphous cell and the properties of the penetrant molecule with gas transport properties through the polymer membrane is under progress. The development of rapid algorithms that would permit rapid construction and estimation of S and D for different gas molecules paralleled by accurate experimental characterization of well-defined polymer membranes is needed to accelerate this process.

1.3 Numerical Simulation of Polymer Membranes

Here we detail through a specific example the results of MD modeling, *i.e.* the numerical approach used for simulating gas transport properties and the use and limits of detailed atomistic description of polymer membrane models.[43]

In particular we specify on a new protocol for packing glassy polymer with intermediate control of free volume distribution for four polyether-ether-ketones, an amorphous poly(oxa-*p*-phenylene-3,3-phthalido-*p*-phenylenxoxa-*p*-phenyl-enexoxi-*p*-phenylene) (PEEK-WC), the dimethyl PEEK-WC (DMPEEK), the tetramethyl PEEK-WC (TMPEEK) and the di-isopropyl dimethyl PEEK-WC (DIDMPEEK).

Modified polyether-ether-ketones[135–138] are of considerable interest due to their excellent mechanical toughness, thermo-oxidative stability, solvent resistance and high transition temperature. In the last decade considerable effort has been spent to introduce chemical modifications in this class of polymers in order to obtain better physical properties and to build up membranes for electro-dialysis, gas dehumidification and gas separation. Relatively few atomistic simulations have been performed on this class of polymers.[37,139] The monomer structure, experimental density and the glass transition temperature of the four poly(ether ether ketone)s is reported in Figure 1.5.

1.3.1 Force Field and Choice of Ensemble

A force field provides a mathematical description of the potential energy as a function of the atomistic configuration of the polymer. The amorphous packing of the PEEK-WC polymer have been constructed using the force field COMPASS (condensed-phase optimized molecular potentials for atomistic

Monomer Structure	ρ (g cm^{-3})	T_g (K)
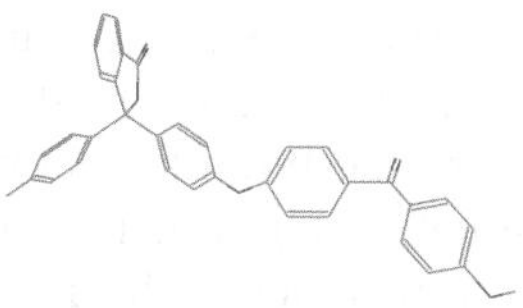 PEEK-WC	1.25	501
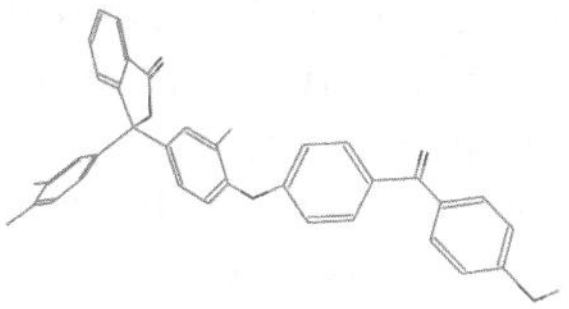 DMPEEK	1.247	512
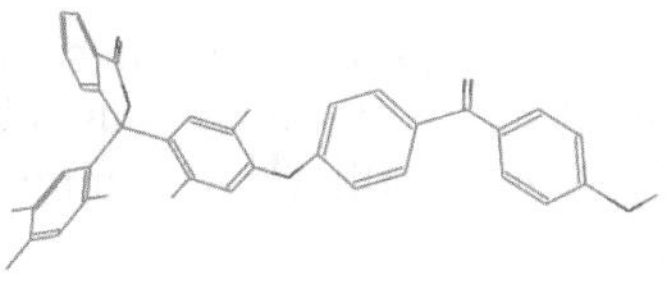 TMPEEK	1.195	519
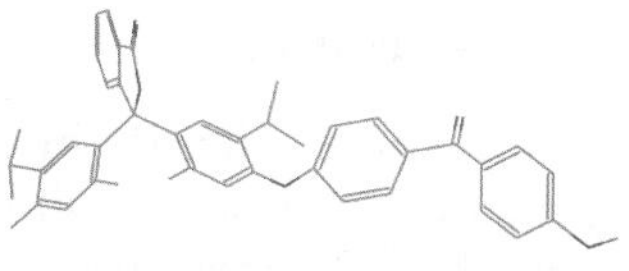 DIDMPEEK	1.140	514

Figure 1.5 Monomer structures and experimental data of selected PEEKs. Reprinted with permission from Tocci, E., Pullumbi, P. Molecular simulation of realistic membrane models of alkylated PEEK membranes. *Mol. Simul.*, **32**, 145–154 (2006). Copyright © 2006 Taylor & Francis.

simulation studies) force field[140] which is particularly adapted to the simulation of polymeric systems.

1.3.2 Generation of Amorphous Cell Packings

In order to minimize chain end effects, each simulation box contains only one single polymeric chain rather than several confined to the same volume, which would lead to increased density of chain ends. The use of single-chain polymers representing bulk amorphous systems is common and has been proven to be quite accurate in replicating the behavior of experimental polymeric systems.[27,139,142] Due to the limited lateral dimensions of the packing models of few nanometers makes it impossible to simulate complete membranes or other polymer-based samples.

For this reason the bulk models are typically cubic volume elements of some nanometers side length and they represent a part cut out of the interior of a polymer membrane. The cubic basic volume element first are filled with segments of a growing chain under periodic boundary conditions following a combination of the Theodorou–Suter[106] chain-generation approach and the Meirovitch's scanning method[143] reproducing the natural distribution of conformation angles.

Careful consideration of the construction and equilibration procedures is necessary to assure a physically realistic cell. In the case of polymers without aromatic moieties the size of the volume element can be chosen so as to reproduce the experimentally observed or theoretically predicted macroscopic density of the relevant polymer. For partly aromatic polymers, however, the chain packing stage has to be performed at very low densities to avoid ring-catenation and 'spearing' effects.

The incorporation of small 'spacer' molecules such as methane, carbon dioxide *etc.* inserted during the initial phase of amorphous cell and removed in subsequent equilibration steps after cell building[42] can minimize the effects of ring–ring catenation.

1.3.3 Realistic Amorphous Cell Selection

The spacer-free packing models, with a reduced density in comparison to the experimental value, have been subjected to extensive equilibration procedures of NVT-MD (constant particle number, temperature and volume) and annealing simulations combined with force field parameter (torsion, non-bonded and coulomb interactions) scaling steps.

The goodness of each model has been analyzed, before the complete equilibration procedure has been performed at an intermediate stage when the density of the packing boxes was at a value corresponding at the 90% of the experimental value. The variation of surface to accessible volume ratio as well as its gradient with the probe radius has been used to analyze the accessible volume of each cell.

The calculations have been repeated systematically by varying the probe radius in the interval between 1.2 and 2.1 Å, with a 0.1 Å as step.

Several cycles of NPT-MD (constant particle number, temperature and pressure) runs at pressures of thousands of bars the density of the systems has been increased. Besides, simulated annealing runs with temperatures up to 1000 K and NVT dynamics at 303 K were used to further relax the polymer structure. The successive step, besides the check on the quality of boxes at an intermediate stage, has been the reaching of the experimental density by increasing the pressure with several cycles of NPT-MD (constant particle number, temperature and pressure) runs at pressures of thousands of bars.

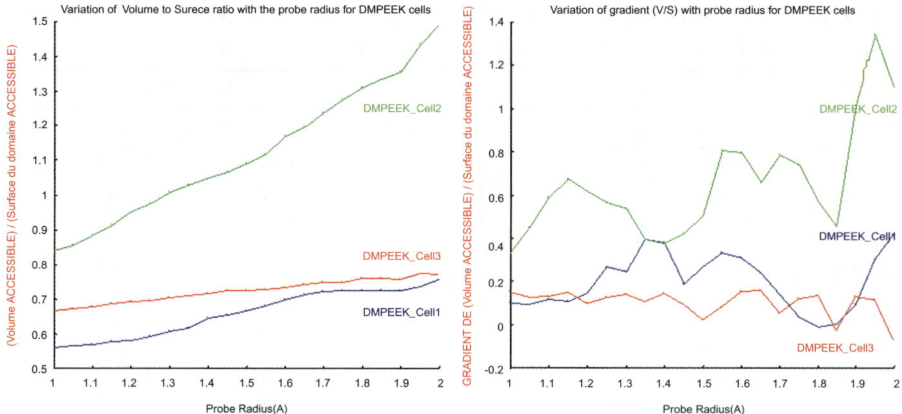

Figure 1.6 Variation of volume-to-surface ratio and its gradient with probe radius. Reprinted with permission from Tocci, E., Pullumbi, P. Molecular simulation of realistic membrane models of alkylated PEEK membranes. *Mol. Simul.*, **32**, 145–154 (2006). Copyright © 2006 Taylor & Francis.

Moreover, simulated annealing runs with temperatures up to 1000 K and NVT dynamics at 303 K have been used to further relax the polymeric structures.

Equilibration and density adjustment of the polymer system have been achieved through a final 300 ps MD run (Figure 1.6).

1.3.4 Estimation of Gas Transport Properties through Amorphous Cells

The estimation of the diffusion coefficient, D, of the solubility coefficient, S (and consequently of permeability, Pe) of small permeant molecules in a polymeric membrane is strongly dependent on the quality of the amorphous cell used in the calculation. This is particularly true in the case of stiff chain polymers containing aromatic moieties.

The TST approach of Gusev–Suter methodology was used to calculate diffusion and solubility coefficients for N_2, O_2, and CO_2 gas molecules in the generated membrane models.

The comparison of modelled values of modified PEEK-WC with the experimental ones indicates that the new approach of packing produces amorphous cells show less scatter in predicted S and D values with respect to other previous published papers dealing with these membranes.[37,139] The simulated values are more self-consistent showing less scatter than in recently reported papers dealing with modeling of alkylated PEEK-WC membranes.[37,139] A relatively good agreement is obtained for O_2 and N_2 molecules for diffusion coefficient as well as the solubility.

A comparison of predicted to experimental diffusion and solubility selectivities reinforces the conclusions obtained from the analysis of Table 1.1.

Table 1.1 Comparison of predicted to experimental diffusion selectivities and solubility selectivities for (O_2, N_2) in selected PEEKs.

Polymer	Model	*Diffusion selectivity*		*Solubility selectivity*	
		Exp (O_2/N_2)	*Sim (O_2/N_2)*	*Exp (O_2/N_2)*	*Sim (O_2/N_2)*
PEEK-WC	cell_1	4.05	4.31	1.28	1.69
	cell_2	4.05	5.04	1.28	1.69
	cell_3	4.05	4.03	1.28	1.76
DMPEEK	cell_1	6.64	6.53	1.37	1.69
	cell_2	6.64	6.03	1.37	1.66
	cell_3	6.64	6.94	1.37	1.65
TMPEEK	cell_1	6.41	6.22	1.29	1.77
	cell_2	6.41	6.43	1.29	1.83
	cell_3	6.41	5.75	1.29	1.88
DIDMPEEK	cell_1	3.48	3.92	1.41	1.73
	cell_2	3.48	3.88	1.41	1.88
	cell_3	3.48	3.89	1.41	1.70

The predicted diffusion selectivity nicely reproduces the experimentally observed trend for O_2 and N_2 diffusion. The predicted trend for solubility selectivity for O_2 and N_2 is in perfect agreement with the experimental observation with a slightly higher estimation of relative O_2 solubility.

1.4   Concluding Remarks

The optimal design of membrane units for gas separation processes is by its nature very complex due to different phenomena taking place at different scales and a real multidisciplinary approach is needed to deal with the different facets that are frontiers of active research in many interconnected fields like materials discovery, membrane fabrication process, optimal fluid flow/distribution inside the membrane module and membrane unit process simulation just to mention some of these. It is to notice that each of these fields of research is by itself complex and implies phenomena that comport a variety of time and length scales. In this contribution we have illustrated through the PEEK-WC polymer membrane simulations the use and limits of detailed atomistic description of gas-membrane system for predicting gas transport properties through polymer membrane models. The need to develop a multi-scale simulation methodology becomes evident even for this case where for estimating the diffusivity of gases through membrane amorphous cell models is difficult to capture using molecular dynamics simulations. The use of TST for estimating gas transport properties of small gas molecules through these glassy polymer membranes has been proven useful for the case studied but it is limited to relatively simple gas permeates that could be treated as single Lennard-Jones sphere without an explicit consideration of the effect of the electrostatic interactions. Recently, improved hierarchical modeling approaches taking into account not only the atomistic description of the gas molecules including the partial charges on each atom but also combine the analysis of accessible volume within model glassy polymer configurations, followed by the identification of transition paths for elementary diffusive jumps in the multidimensional configuration space allowing the polymer matrix to locally relax when the permeate molecule is placed in the transition state and estimate rate constants for these jumps based on TST.[20,113,114]

The field of atomistic modeling of gas transport properties in membranes has experienced a rapid progress due not only to the well known Moore's law on computational power growth but also to the development of clever and efficient algorithms dealing with the simulation of different time-length phenomena that are important for the realistic description of gas permeation through membranes. It is worth noting that even with the recent developments the methodology used for the generation of the realistic models of the models of the membrane and described in this contribution is computationally quite intensive and can be hardly applied for the systematic search of novel polymers with desired properties. Several attempts[144–146] on generalizing simulation protocols based on coarse-graining methodologies have been reported in the literature but to our knowledge the difficult part is to develop methods that automatically would define the coarse grained groups and define their interaction parameters from their composition.

The gas permeation through the polymer membrane as a function of the structural features of the polymer chain and its repeat unit is only one of the aspects of the research of better membranes for a given gas separation. The properties of the membranes are defined also by the process of its formation. This process controls the morphology of the membrane. Most commercial membranes are prepared by immersion precipitation in which the polymer solution is cast on a substrate that could be a flat or hollow fiber and then immersed into a coagulation bath containing the non-solvent (preferably water). The kinetics of diffusion of the water into the polymer solution and of the solvent into the coagulation bath defines the membrane formed by a phase separation process followed the solidification of the polymer rich phase. The performance of the membrane strongly depends on the morphology obtained during the phase separation and solidification. Mesoscale methods have been recently[147] used to simulate this process. The simulation results look very promising in terms of providing a deeper understanding of the immersion precipitation process and serve for the guidance of experiments for designing better membrane formation.

The design of optimal membrane modules for a given gas separation depends also on the fluid distribution through the numerical modeling of the hydrodynamics, mass transfer for single module configurations and especially the pressure distribution in the case of pressure-driven processes. The hydrodynamics allows the increase of the shear stress near the wall, thus allowing the enhancement of membrane separation processes. The commercial available numerical methods for treating fluid distribution and transport phenomena at the level of the module geometry are the finite element and the computational fluid dynamics (CFD) methods.[148–153] CFD methods have been used to optimise membrane separation processes.[148–151] General models of hollow-fiber and spiral-wound membrane modules reported[148,149] were developed from rigorous mass, momentum and energy balances. The CFD analysis of transport phenomena basically assumes the continuity of matter and deals with problems on the length scale significantly higher than molecular dimensions. For the cases where the hypothesis of continuity of the fluid matter cannot be made the adapted approach is based on the lattice-Boltzmann model.[152,153] The fundamental equations of the model stem from the kinetic theory of gases and constitute the basic law employed in the stochastic fluid mechanics. Many engineering problems are concerned with fluid flow and heat transfer in microscale systems dealing with length scales of sub-micron order. For these cases the MD simulation method helps to get microscopic insight to the region near the interface and transfer it to the upper level of the lattice-Boltzmann simulation. The influence of long-range intermolecular forces between the porous membrane surface and fluid in the immediate contact with the pores and the membrane on the can be evaluated using microscopic information and contribute towards understanding of the complex physics of fluid flow.

The rational research for better gas separation polymer membrane systems necessitates a comprehensive understanding of phenomena taking place at different time and length scales. Multi-scale simulation is emerging as the adequate technique for the description of such complex systems. During the last decades several computational methods have been successfully developed and used to complement experimental studies of different aspects of membrane science and engineering. In this connection, many traditional simulation techniques (Monte Carlo, molecular dynamics, transition state theory, lattice-Boltzmann, computational fluid dynamics, finite element) have been employed by several groups for obtaining responses for various time and length scales from atomistic, to mesoscale (coarse grains, particle beads) and to macroscale domains (continuum, computational fluid dynamics and finite element). During the last years hierarchical computational approaches for coupling methods used to describe different spatial or temporal domains have been developed. Their use needs long preparation protocols and the simulations are computationally very demanding, thus not so applicable to the prediction of properties of complex systems like membrane modules. The development of methodologies that would allow the automatic definition of each calculation domain and the coupling protocols between each adjacent domain. Despite the progress made over the past years, there are a number of challenges in computer modeling and simulation. In general, these challenges represent the work in two directions. First, there is a need to develop new and improved simulation techniques at individual time and length scales. Second, it is important to integrate the developed methods at wider range of time and length scales, spanning from quantum mechanical domain (a few atoms) to the molecular level. Developing such a multi-scale methodology is very challenging but it will open bright perspectives for numerical simulation, not only in polymer membrane systems but in many other fields concerned by a large span of spatial and temporal domains. New computational tools need be developed in the future to make multi-scale modeling a useful approach for industrial design and predicting properties of materials used in complex systems like the membranes for gas separation.

Acknowledgements

The authors are indebted to the European Commission for financial support through the PERMOD project (Project Number G5RD-CT-2000-00200) and MULTIMATDESIGN project (Project Number NMP3-CT-2005-013644) and to all partners of both projects for their collaboration. Interactions with Dr Dieter Hofmann and Dr Matthias Heuchel of GKSS, Teltow, Germany, are deeply appreciated.

References

1. K. E. Gubbins and J. D. Moore, *Ind. Eng. Chem. Res.*, 2010, **49**, 3026.
2. D. N. Theodorou, in *Challenges in Molecular Simulations: Bridging the Time-Scale and Length-Scale Gap (SIMU)* 19-40 (The European Science Foundation programme, 2002).
3. J. C. Charpentier, *Comput. Chem. Eng.*, 2009, **33**, 936.
4. K. Binder, W. Paul, S. Santos and U. W. Suter, in *Simulation Methods for Polymers*, ed. M. Kotelyanskii and D. N. Theodorou, Marcel Decker, 2004, p. 491.
5. R. A. van Santen and P. Sautet, in *Computational Methods in Catalysis and Materials Science: An Introduction for Scientists and Engineers,* ed. R. A. van Santen and P. Sautet, Wiley-VCH Verlag, Weinheim, 2009, p. 441.
6. C. R. A. Catlow, in *Computer Modelling in Inorganic Crystallography*, ed. C. R. A. Catlow, Academic Press, 1997.
7. M. Tafipolsky, S. Amirjalayer and R. Schmid, *J. Phys. Chem. C*, 2010, **114**, 14402.
8. R. Krishna and J. van Baten, *Langmuir*, 2010, **26**, 8450.
9. P. M. Mathias, H. Cheng, S. J. Cook, H. C. Klotz and V. S. Parekh, *Fluid Phase Equilib.*, 1996, **116**, 225.
10. K. Rajan, C. Suh, A. Rajagopalan and X. Li, in *Materials Research Society Symposium Proceedings*, 2002, **223**.
11. A. Z. Panagiotopoulos, N. Quirke, M. Stapleton and D. J. Tildesley, *Mol. Phys.*, 1988, **63**, 527.
12. P. I. Ravikovitch, A. Vishnyakov and A. V. Neimark, *Phys. Rev. E: Stat., Nonlin. Soft Matter Phys.*, 2001, **64**, 1.
13. M. B. Sweatman and N. Quirke, *Mol. Sim.*, 2001, **27**, 295.
14. B. Liu and B. Smit, *J. Phys. Chem. C*, 2010, **114**, 8515.
15. I. C. Roman, R. W. Ubersax and G. K. Fleming, *Chim. Ind. (Milan, Italy)*, 2001, **83**.

16. S. A. Stern and W. J. Koros, *Chim. Nouv.*, 2000, VOL, 3201.

17. L. B. Robeson, *J. Membr. Sci.*, 2008, **320**, 390.

18. Yu. P. Yampolskii, *Russ. Chem. Rev.*, 2007, **76**, 59.

19. D. N. Theodorou, *Lect. Notes Phys.*, 2003, VOL, 67.

20. M. L. Greenfield and D. N. Theodorou, *Macromolecules*, 2001, **34**, 8541.

21. F. Mueller-Plathe, *Soft Mater.*, 2003, **1**, 1.

22. A. Uhlherr, V. G. Mavrantzas, M. Doxastakis and D. N. Theodorou, *Macromolecules*, 2001, **34**, 8554.

23. A. Uhlherr, S. J. Leak, N. E. Adams, P. E. Nyberg, M. Doxastakis, V. G. Mavrantzas and D. N. Theodorou, *Comput. Phys. Commun.*, 2002, **144**, 1.

24. P. V. K. Pant and R. H. Boyd, *Macromolecules*, 1992, **114**, 494.

25. P. V. K. Pant and R. H. Boyd, *Macromolecules*, 1993, **26**, 679.

26. F. Müller-Plathe, *Acta Polymerica*, 1994, **45**, 259.

27. F. Muller-Plathe, W. F. Van Gunsteren and U. W. Suter, *Adv. Polym. Sci.*, 1994, **116**, 207.

28. P. V. K. Pant and D. N. Theodorou, *Polym. Prepr.*, 1994, **35**, 165.

29. D. Hofmann, J. Ulbritch, D. Fritsch and D. Paul, *Polymer*, 1996, **37**, 4773.

30. H. Takeuchi and K. Okazaki, *Mol. Simul.*, 1996, **16**, 59.

31. J. R. Fried and D. K. Goyal, *J. Polym. Sci. Part B: Polym. Phys.*, 1998, **36**, 519.

32. R. K. Bharadwaj and R. H. Boyd, *Polymer*, 1999, **40**, 4229.

33. D. Hofmann, L. Fritz, J. Ulbrich, C. Schepers and M. Bohning, *Macromol. Theor. Simul.*, 2000, **9**, 293.

34. M. Lopez-Gonzalez, E. Saiz, J. Guzman and E. Riande, *J. Chem. Phys.*, 2001, **115**, 6728.

35. F. Muller-Plathe, *DECHEMA Monograph.*, 2001, **137**, 163.

36. R. Shanks and D. Pavel, *Mol. Simul.*, 2002, **28**, 939.

37. E. Tocci, E. Bellacchio, N. Russo and E. Drioli, *J. Membr. Sci.*, 2002, **206**, 389.

38. M. Heuchel, D. Hofmann and P. Pullumbi, *Macromolecules*, 2004, **37**, 201.

39. N. C. Karayiannis, V. G. Mavrantzas and D. N. Theodorou, *Macromolecules*, 2004, **37**, 2978.

40. S. Neyertz and D. Brown, *Macromolecules*, 2004, **37**, 10109.

41. V. E. Raptis, I. G. Economou, D. N. Theodorou, J. Petrou and J. H. Petropoulos, *Macromolecules*, 2004, **37**, 1102.

42. D. Hofmann, M. Entrialgo-Castano, A. Lerbret, M. Heuchel and Yu. Yampolskii, *Macromolecules*, 2003, **36**, 8528.

43. E. Tocci and P. Pullumbi, *Mol. Simul.*, 2006, **32**, 145.

44. B. D. Freeman, *Macromolecules*, 1999, **32**, 375.

45. L. M. Robeson, W. F. Burgoyne, M. Langsam, A. C. Savoca and C. F. Tien, *Polymer*, 1994, **35**, 4970.

46. S. A. Stern, *J. Membr. Sci.*, 1994, **94**, 1.

47. C. M. Zimmerman, A. Singh and W. J. Koros, *J. Membr. Sci.*, 1997, **137**, 145.

48. W. J. Koros and G. K. Fleming, *J. Membr. Sci.*, 1993, **83**, 1.
49. L. M. Robeson, *J. Membr. Sci.*, 1991, **92**, 165.
50. W. J. Koros, M. R. Coleman and D. R. B. Walker, *Annu. Rev. Mater.*, 1992, **22**, 47.
51. D. R. Paul and M. R. Pixton, *Macromol. Symp.*, 1997, **118**, 401.
52. J. K. Park and D. R. Paul, *J. Membr. Sci.*, 1997, **125**, 23.
53. D. Hofmann, M. Heuchel, Yu. Yampolskii, V. Khotimskii and V. Shantarovich, *Macromolecules,* 2002, **35**, 2129.
54. B. R. Wilks, Thesis, Georgia Institute of Technology, Atlanta, 2002.
55. J. C. Jansen, M. Macchione, E. Tocci, L. De Lorenzo, Yu. P. Yampolskii, O. Sanfirova, V. P. Shantarovich, M. Heuchel, D. Hofmann and E. Drioli, *Macromolecules*, 2009, **42**, 7589.
56. V. N. Burganos, in *Comprehensive Membrane Science and Engineering*, ed. E. Drioli and L. Giorno, Elsevier, Oxford, 2010, p. 29.
57. M. Fermeglia and S. Pricl, *Org. Coat.*, 2007, **58**, 187.
58. J. Grotendorst, N. Attig, S. Bluegel and D. Marx, in *Multiscale Simulation Methods in Molecular Sciences*, NIC Series 42, 2009.
59. P. J. Hoogerbrugge and J. M. V. A. Koelman, *Europhys. Lett.*, 1992, **19**, 155.
60. J. M. V. A. Koelman and P. J. Hoogerbrugge, *Europhys. Lett.*, 1993, **21**, 363.
61. R. D. Groot and P. B. Warren, *J. Chem. Phys.*, 1997, **107**, 4423.
62. J. G. E. M. Fraaije, B. A. C. van Vlimmeren, N. M. Maurits, M. Postma, O. A. Evers, C. Hoffman, P. Altevogt and G. Goldbeck-Wood, *J. Chem. Phys.*, 1997, **106**, 4260.
63. P. Altevogt, O. A. Evers, J. G. E. M. Fraaije, N. M. Maurits and B. A. C. van Vlimmeren, *J. Mol. Struct. (THEOCHEM)*, 1999, **463**, 139.
64. Y. Kong, C. W. Manke, W. G. Madden and A. G. Schlijper, *Int. J. Thermophys.*, 1994, **15**, 1093.
65. A. G. Schlijper, P. J. Hoogerbrugge and C. W. Manke, *J. Rheol.*, 1995, **39**, 567.
66. P. Espanol and P. B. Warren, *Europhys. Lett.*, 1995, **30**, 191.
67. R. D. Groot and T. J. Madden, *J. Chem. Phys.*, 1998, **108**, 8713.
68. A. F. Jakobsen, *J. Chem. Phys.*, 2005, **122**, 124901.
69. R. D. Groot, *J. Chem. Phys.*, 2003, **118**, 11265.
70. K. P. Travis, M. Bankhead, K. Good and S. L. Owens, *J. Chem. Phys.*, 2007, **127**, 14109.
71. P. M. Chaikin and T. C. Lubensky, *Principles of Condensed Matter Physics*, Cambridge University Press, Cambridge, 1995.
72. B. A. C. van Vlimmeren, N. M. Maurits, A. V. Zvelindovsky, G. J. A. Sevink and J. G. E. M. Fraaije, *Macromolecules*, 1999, **32**, 646.
73. G. J. A. Sevink, A. V. Zvelindovsky, B. A. C. van Vlimmeren, N. M. Maurits and J. G. E. M. Fraaije, *J. Chem. Phys.*, 1999, **110**, 2250.
74. N. M. Maurits, A. V. Zvelindovsky, G. J. A. Sevink, B. A. C. van Vlimmeren and J. G. E. M. Fraaije, *J. Chem. Phys.*, 1998, **108**, 9150.
75. A. V. Zvelindovsky, G. J. A. Sevink, B. A. C. van Vlimmeren, N. M. Maurits and J. G. E. M. Fraaije, *Phys. Rev.*, 1998, **57**, 4879.

76. A. Knoll, A. Horvat, K. S. Lyakhova, G. Krausch, G. J. A. Sevink, A. V. Zvelindovsky and R. Magerle, *Phys. Rev. Lett.*, 2002, **89**, 35501.
77. S. S. Jawalkar, K. V. S. N. Raju, S. B. Halligudi, M. Sairam and T. M. Aminabhavi, *J. Phys. Chem. B*, 2007, **111**, 2431.
78. T. Schmatko, H. Hervet and L. Leger, *Phys. Rev. Lett.*, 2005, **94**, 275501.
79. G. Csanyi, T. Albaret, M. C. Payne and A. D. Vita, *Phys. Rev. Lett.*, 2004, **93**, 175503.
80. G. Wagner, E. Flekkoy, J. Feder and T. Jossang, *Comput. Phys. Commun.*, 2002, **147**, 670.
81. J. Koplik and J. R. Banavar, *Annu. Rev. Fluid Mech.*, 1995, **27**, 257.
82. A. Akthakul, W. F. MacDonald and A. M. Mayes, *J. Membr. Sci.*, 2002, **208**, 147.
83. A. Akthakul, PhD thesis, MIT, 2003.
84. B. Zhou and A. C. Powel, *J. Membr. Sci.*, 2006, **268**, 150.
85. M. Cossi, V. Barone, R. Cammi and J. Tomasi, *Chem. Phys. Lett.*, 1996, **255**, 327.
86. C. J. Cramer and D. G. Truhlar, *Chem. Rev.*, 1999, **99**, 2161.
87. A. Klamt, *Cosmo-RS: From Quantum Chemistry to Fluid Phase Thermodynamics and Drug Design*, Elsevier BV., 2005.
88. F. Maseras and K. Morokuma, *J. Comp. Chem.*, 1995, **16**, 1170.
89. P. H. König, M. Hoffmann, Th. Frauenheim and Q. Cui, *J. Phys. Chem. B*, 2005, **109**, 9082.
90. N. Reuter, A. Dejaegere, B. Maigret and M. Karplus, *J. Phys. Chem. A*, 2000, **104**, 1720.
91. H. M. Senn and W. Thiel, *Curr. Opin. Chem. Biol.*, 2007, **11**, 182.
92. T. Schlick, in *Molecular Modeling and Simulation – An Interdisciplinary Guide*, Springer, New York, 2002.
93. S. Trohalaki, A. Kloczkowski, J. E. Mark, R. J. Roe and D. Rigby, *Comp. Theor. Polym. Sci.*, 1992, **2**, 147.
94. J. R. Fried, *J. Inorg. Organomet. Polym. Mater.*, 2006, **16**, 407.
95. J. R. Fried, in *Molecular Simulation of Gas and Vapor Transport in Highly Permeable Polymers in Materials Science of Membranes for Gas and Vapor Separation*, ed. Y. Yampolskii, I. Pinnau and B. D. Freeman, John Wiley & Sons, 2006, 95 and references therein cited.
96. S. Neyertz, *Macromol. Theor. Simul.*, 2007, **16**, 513.
97. O. Holck, M. Heuchel, M. Bohning and D. Hofmann, *J. Polym. Sci. Part B: Polym. Phys.*, 2008, **46**, 59.
98. M. Heuchel, D. Fritsch, P. M. Budd, N. B. McKeown and D. Hofmann, *J. Membr. Sci.*, 2008, **318**, 84.
99. Lu, Chunhai; Ni, Shijun; Chen, Wenkai; Liao, Junsheng and Zhang, Chengjiang, *Comput. Mater. Sci.*, 2010, **49**, 65.
100. E. J. Maginn and J. R. Elliott, *Ind. Eng. Chem. Res.*, 2010, **49**, 3059.
101. J. M. Haile, in *Molecular Dynamics Simulations, Elementary Methods*, Wiley-Interscience, New York: 1992.
102. M. Karplus and J. A. McCammon, *Nat. Struct. Biol.*, 2002, **9**, 646.
103. N. F. A. Van der Vegt, *Macromolecules*, 2000, **33**, 3153.

104. R. M. Sok, H. J. C. Berendsen and W. F. van Gunsteren, *J. Chem. Phys.*, 1992, **96**, 4699.
105. F. Muller-Plathe, *J. Chem. Phys.*, 1991, **9**, 3192.
106. D. N. Theodorou and U. W. Suter, *Macromolecules*, 1985, **18**, 1467.
107. M. P. Allen and D.J. Tildesley, *Computer Simulation of Liquids*, Clarendon Press, Oxford, UK, 1987.
108. D. Frenkel and B. Smit, *Understanding Molecular Simulations*, Academic Press, New York, NY, 1996.
109. D. N. Theodorou, *Ind. Eng. Chem Res.*, 2010, **49**, 3047.
110. F. Jagodic, B. Borstnik and A. Azman, *Makromol. Chem.*, 1973, **173**, 221.
111. A. A. Gusev, S. Arizzi, U. W. Suter and D. J. Moll, *J. Chem. Phys.*, 1993, **99**, 2221.
112. A. A. Gusev, S. Arizzi and U. W. Suter, *J. Chem. Phys.*, 1993, **99**, 2228.
113. M. L. Greenfield and D. N. Theodorou, *Macromolecules*, 1998, **31**, 7068.
114. P. S. Rallabandi, A. P. Thompson and D. M. Ford, *Macromolecules*, 2000, **33**, 3142.
115. E. S. Goll and P. C. Jurs, *J. Chem. Inf. Comput. Sci.*, 1999, **39**, 1081.
116. J. Huuskonen, *J. Chem. Inf. Comput. Sci.*, 2000, **40**, 773.
117. J. M. Sutter and P. C. Jurs, *J. Chem. Inf. Comput. Sci.*, 1996, **36**, 100.
118. B. E. Mitchell and P. C. Jurs, *J. Chem. Inf. Comput. Sci.*, 1998, **38**, 489.
119. J. Tetteh, T. Suzuki, E. Metcalfe and S. Howells, *J. Chem. Inf. Comput. Sci.*, 1999, **39**, 491.
120. S. S. Liu, H. L. Liu, Z. N. Xia, C. Z. Cao and Z. L. Li, *J. Chem. Inf. Comput. Sci.*, 1999, **39**, 951.
121. Qianfeng Li, Xingguo Chen and Zhide Hu, *Chemom. Intell. Lab. Syst.*, 2004, **72**, 93.
122. J. C. Eslick, Q. Yeb, J. Park, E. M. Topp, P. Spencer and K. V. Camarda, *Comput. Chem. Eng.*, 2009, **33**, 954.
123. V. Kholodovych, J. R. Smith, D. Knight, S. Abramson and J. Kohn, *Polymer*, 2004, **45**, 7367.
124. F. Luan, X. Y. Zhang, H. X. Zhang, R. S. Zhang, M. C. Liu, Z. D. Hu and B. T. Fan, *Comput. Mater. Sci.*, 2006, **36**, 386.
125. P. Pullumbi, E. Tocci, M. Heuchel, S. Pelzer and D. Hofmann,. International Congress on Membrane Processes (ICOM 2008). Honolulu, Hawai. 12-18, July 2008.
126. A. A. D'Archivio, A. Incani, P. Mazzeo and F. Ruggieri, *Anal. Chim. Acta*, 2009, **650**, 175.
127. J. S. Tokarski, A. J. Hopfinger, J. David Hobbs, D. M. Ford and J. L. M. Faulon, *Comput. Theor. Polym. Sci.*, 1997, **7**, 199.
128. D. W. van Krevelen, *Properties of Polymers, Their Estimation and Correlation with Chemical Structure*, Elsevier, Amsterdam, 1976.
129. J. Bicerano, *Prediction of Polymers Properties*, Marcel Dekker Inc, New York, 2nd edn, 1996.
130. X. Yu, Z. Xie, B. Yi, X. Wang and F. Liu, *Eur. Polym. J.*, 2007, **43**, 818.
131. A. G. Mercader, P. R. Duchowicz, M. A. Sanservino, F. M. Fernandez and E. A. Castro, *J. Fluorine Chem.*, 2007, **128**, 484.

132. P. Pullumbi and I. Milosavljevic, *Organohalogen Compd.*, 1999, **553**.
133. A. R. Katritzky, S. H. Slavov, D. A. Dobchev and M. Karelson, *Comput. Chem. Eng.*, 2007, **31**, 1123.
134. A. Afantitis, G. Melagraki, H. Sarimveis, P. A. Koutentis, J. Markopoulos and O. Igglessi-Markopoulou, *Polymer*, 2006, **47**, 3240.
135. K. Liu, H. C. Zhang and T. L. Chen, Chinese Patent 85,101,721.
136. H. C. Zhang, T. L. Chen and Y. G. Yuan, Chinese Patent 85,108,751.
137. Z. Wang, T. Chen and J. Xu, *Macromolecules*, 2000, **33**, 5672.
138. J. C. Jansen and E. Drioli, *Polym. Sci. Ser. A*, 2009, **51**, 1355.
139. E. Tocci, D. Hofmann, D. Paul, N. Russo and E. Drioli, *Polymer*, 2001, **42**, 521.
140. D. Rigby, H. Sun and B. E. Eichinger, *Polym. Int.*, 1997, **44**, 311.
141. H. Sun, S. J. Mumby, J. R. Maple and A. T. Hagler, *J. Am. Chem. Soc.*, 1994, **116**, 2978.
142. R. H. Gee, L. E. Fried and R. C. Cook, *Macromolecules*, 2001, **34**, 3050.
143. H. J. Meriovitch, *J. Chem. Phys.*, 1983, **79**, 502.
144. D. Reith, M. Puetz and F. Mueller-Plathe, *J. Comput. Chem.*, 2003, **24**, 1624.
145. N. Zacharopoulos, N. Vergadou and D. N. Theodorou, *J. Chem. Phys.*, 2005, **122**, 244111.
146. T. Spyriouni, C. Tzoumanekas, D. N. Theodorou, F. Müller-Plathe and G. Milano, *Macromolecules*, 2007, **40**, 3876.
147. B. Zhou and A. C. Powell, *J. Membr. Sci.*, 2006, **268**, 150.
148. J. Marriott, E. Sorensen and I. D. L. Bogle, *Comput. Chem. Eng.*, 2001, **25**, 693.
149. J. Marriott and E. Sorensen, *Chem. Eng. Sci.*, 2003, **58**, 4975.
150. V. Nassehi, *Chem. Eng. Sci.*, 1998, **53**, 1253.
151. D. B. Das, V. Nassehi and R. J. Wakeman, *Adv. Environ. Res.*, 2002, **7**, 35.
152. S. Succi, *The Lattice Boltzmann Equation for Fluid Dynamics and Beyond*, Clarendon Press, Oxford, 2001.
153. M. C. Sukop and D. T. Thorne, *Lattice Boltzmann Modelling: An Introduction for Geoscientists and Engineers*, SpringerVerlag, Berlin, 2006.

CHAPTER 2

Simulation of Polymeric Membrane Systems for CO₂ Capture

ERIC FAVRE

Laboratoire Réactions et Génie des Procédés (UPR CNRS 3349), Nancy Université, 1 rue Grandville, 54001 Nancy, France

2.1 Introduction

2.1.1 Global Warming and Carbon Capture

Global warming is now clearly established to result, to a large extent, from the increase of carbon dioxide concentration in the atmosphere due to fossil fuel usage.[1] According to different scenarios, CO_2 concentration in the atmosphere is expected to double by around 2050, if no special actions are taken. Among the different options that can prevent carbon dioxide build-up (such as processes with improved energy efficiency, an increased use of renewable energy sources or the development of non-CO_2 emitting energy sources), carbon capture and storage is considered as a key issue. Nevertheless, a formidable challenge has to be addressed in order to achieve this target: recover (as a priority from large emission sources), transport and store CO_2 in geological formations on long time scales, with a minimal cost and environmental impact. In that context, the identification of the most efficient capture process, which would enable to typically recover 80–90% of the CO_2 from a given source, has attracted considerable attention for several years, due to the fact that this step

Membrane Engineering for the Treatment of Gases, Volume 1:
Gas-separation Problems with Membranes
Edited by Enrico Drioli and Giuseppe Barbieri
© Royal Society of Chemistry 2011
Published by the Royal Society of Chemistry, www.rsc.org

Table 2.1 Carbon capture strategies, currently best available capture technologies, and tentative breakthrough membrane separation processes

Carbon capture strategy	*Gas mixture*	*Currently best available separation process*	*Possible breakthrough membrane process*
Oxycombustion	O_2/N_2	Cryogeny	Ion transfer membranes (ITMs)
Precombustion	CO_2/H_2	Gas–liquid absorption in physical solvent	Membrane reactor
Postcombustion	CO_2/N_2	Gas–liquid absorption in chemical solvent (MEA)	1. *Hybrid process:* Membrane contactors (intensified gas–liquid absorption) 2. *Membrane separation process:* (a) Gas separation membranes (b) Fixed Site Carrier Membranes (c) Liquid membranes 3. *Ancillary operations:* (a) Flue gas drying (b) Compressor purge treatment (c) Minor species removal

concentrates 60–80% of the overall carbon capture and storage cost, both in terms of capital (CAPEX) and operating (OPEX) contributions. A very large number of capture processes has been already explored and, basically, three main strategies,[1] summarized in Table 2.1, can be proposed.

First, *oxycombustion*, uses a concentrated (typically 95%) oxygen stream, in place of air, for the combustion of the fuel (gas, oil, coal, *etc.*). Consequently, a CO_2/H_2O mixture is produced as flue gas and a relatively pure CO_2 stream (90–95%) can be obtained after water condensation, before the compression, injection and storage steps.

Second, *precombustion*, produces a syngas mixture (CO/H_2) by conversion of the fuel through partial oxidation, followed by a water gas shift. The separation of a CO_2/H_2 mixture under high pressure and high temperature conditions is the key issue in that case.

Third, for the last option, namely the *post-combustion* approach, the flue gas is directly treated after the combustion step. A CO_2 capture process is designed in order to concentrate and purify CO_2 up to the target specifications for transportation and storage. This strategy can be applied in principle to any existing plant where a CO_2 containing stream is produced (a so-called retrofit property, which cannot be achieved by oxycombustion or precombustion). Furthermore, post-combustion can be proposed not only for power plants like the two previous options, but also for any industrial plant (steel, cement, petrochemical plants, *etc.*). These two strong advantages have, however, to be balanced by the fact that a diluted, low pressure, wet CO_2/N_2 mixture has to be treated. As a consequence, post-combustion CO_2 capture can be anticipated as an outstanding process design challenge.[2]

2.1.2 Membrane Processes and Carbon Capture

Generally speaking, membrane processes are not considered today as adequate separation processes in any of the three carbon capture strategies listed above.[1] Equilibrium-based processes, such as cryogeny or gas–liquid absorption are, in fact, proposed for the first generation technologies in oxycombustion, precombustion or post-combustion (Table 2.1). Membrane processes, which are based on a non-equilibrium separation principle, do not, in fact, belong to the so-called equilibrium staged process family in the separations spectrum. Various membrane processes, mentioned in the last column in Table 2.1, are, however, currently investigated for each situation and could be possible breakthrough technologies for the second generation.

For post-combustion situations only, different membrane processes can be proposed, corresponding to different roles and functions.

First, *membrane contactors* can first be used in order to achieve intensified gas–liquid absorption.[3] In that case, no separation function is expected from the membrane and the energy requirement of the process will remain essentially unchanged compared to a packed tower. The main interest of using porous, non-selective membranes in that case is due to the possibility to decrease the size (and weight) of the installations and consequently the capital costs (CAPEX) of the capture process.[4]

Second, *other types* of membrane separation processes can also be proposed for the CO_2/N_2 separation step, in order to play the same role as the absorption process. In that case, it is absolutely necessary to carefully estimate the operating costs (OPEX) and capital costs (CAPEX) of the operation, so that an overall CO_2 capture cost (in euros or \$ per ton of recovered CO_2) can be precisely estimated. This data will obviously be of utmost importance in order to evaluate the chances of the membrane process to possibly compete with the standard capture process. A CO_2/N_2 separation efficiency is clearly needed in that case. Depending on the membrane material, this characteristic can be obtained based on a physical mechanism (solution–diffusion in a dense polymer, for instance, which corresponds to a classical membrane gas separation operation[5]), or a chemical reaction in a polymeric matrix based on fixed sites (so-called fixed site carrier membranes[6]) or with membranes making use of a mobile selective carrier (liquid membranes[7]).

Third, finally, membrane processes can also be proposed for ancillary operations such as flue gas drying,[8] compressor purge treatment, or polishing operations (minor species removal, for instance). These different aspects have, however, not been investigated in detail in the open literature up to now.

Coming back to the overview of the carbon capture landscape summarized in Table 2.1 and the simplified classification of separation processes among first generation and breakthrough processes, it is interesting to notice that a somehow similar situation took place decades ago for other applications. For large scale, challenging, separation processes, such as air separation or water desalination, for instance, equilibrium-based separation processes (namely cryogeny for air separation and multi-stage evaporation for desalination) were first proposed and exclusively applied on an industrial scale. In these two cases, however, membrane processes (namely gas permeation and reverse osmosis) finally found a place as a second generation process and are nowadays classically selected. It might be that a similar evolution occurs for post-combustion carbon capture, but a rigorous evaluation of the possibilities and limitations of membrane processes is first absolutely necessary. Experiments (ideally on a pilot or industrial scale using real flue gases), associated with engineering design and simulation studies, are the classical complementary tools which are proposed for a realistic and possibly optimal solution to be determined.

This chapter will be dedicated to a state of the art analysis of simulation of polymeric membrane separation processes for post-combustion CO_2 capture.

2.2 Membrane Module Simulation Framework

2.2.1 Identifying Capture Step Boundary Conditions

Similarly to any separation problem, the definition of the system boundary conditions is an essential prerequisite before simulations are performed. In the case of post-combustion carbon capture, the boundary conditions will correspond to the flue gas composition at the inlet, and the target purity specifications for pipeline transport at the outlet (Figure 2.1).

While the first ones are rather easy to obtain and define, the purity specifications, which are dictated mostly by the transport constraints, are more

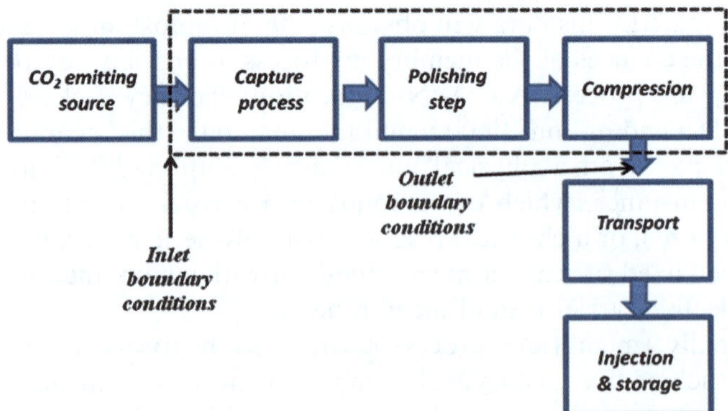

Figure 2.1 Schematic representation of the different steps included in the carbon capture and storage chain, showing the inlet and outlet boundary conditions of the capture step.

Table 2.2 Simulation framework inlet boundary conditions: characteristics and flue gases compositions of the main large emitting sources for post-combustion capture

	Power plant, gas	*Power plant, coal*	*Blast furnace (steel production)*	*Cement production*
Flue gas flow rate $(N\ m^3\ h^{-1})$	$\sim 10^6$	$\sim 10^6$	$\sim 5 \times 10^5$	$\sim 10^5$
Pressure (Bar)	~ 1	~ 1	~ 3	~ 0.8
CO_2 (%)	~ 3	~ 14	19–30	14–25
O_2 (%)	~ 14	~ 6	—	3–10
N_2 (%)	75	71–75	9–40	65
H_2O (%)	~ 7	~ 7	15	5–11
Other compounds	—	SO_x, NO_x	Ar, NO, H_2S, COS, HCN	Ar, SO_x, NO_x

Table 2.3 Simulation framework outlet boundary conditions: characteristic CO_2 transportation constraints in pipelines according to different sources (HC = hydrocarbons)

	IPCC (2005)[1]	*Barry (1985)[9]*	*Johnson (1985)[10]*	*de Visser (2006)[11]*
CO_2	>95%	>95%	97–99%	>95.5%
O_2	<10 ppm	—	—	0.1–4 ppm
N_2	<4%	<4%	<3%	<4%
H_2O	<600 ppm	<600 ppm	<600 ppm	<500 ppm
Other compounds	Total HC<5%	Total HC<5% H_2S <20 ppm	Total HC <3%	NO_x<100 ppm SO_2<100 ppm CH_4 2–4 % Ar<4%

controversial. These two sets of data are summarized in Table 2.2 for inlet conditions (typical flue gases compositions and characteristics) and Table 2.3 for outlet conditions (examples of target purity data for CO_2 transport in pipes[9–11]).

Several comments can be made at this stage. First, the pressure, temperature and carbon dioxide partial pressure conditions vary to a large extent depending on the emitting source. The latter variable will logically play a key role for membrane capture processes, because it governs the 'intrinsic' driving force of the feed mixture. Nevertheless, a very large majority of carbon capture studies is exclusively devoted to coal power plants, with 14% CO_2 content taken as the reference flue gas composition.[1] Even though coal power plants clearly represent the largest potential for carbon capture and storage deployment in terms of tonnage, it should be kept in mind that other CO_2 sources offer very different pressure and composition conditions, which can completely modify the application interest of membrane processes. Unfortunately, apart from a few exceptions,[12] most studies reported with membrane processes and post-combustion carbon capture consider coal power plant as a target application.

Second, flue gases contain different species and are usually saturated in humidity. Apart from few exceptions, the capture processes will first be experimentally tested or simulated based on conditions which postulate a dry, binary (CO_2/N_2) flue gas. This simplifying assumption clearly enables preliminary calculations to be performed, but the complexity of the real situation should be finally taken into account.[13]

Third, in terms of outlet boundary conditions, a high enough CO_2 purity (95% or more) is most often considered as necessary (Table 2.3). It should be stressed, however, that a large number of sources for CO_2 transport concern the enhanced oil recovery sector, which is a specific, demanding application. For CO_2 storage in an aquifer, for instance, the CO_2 target purity results from a trade-off between the compression and pipeline costs (which increase when CO_2 purity decreases) and the complications induced when inert gases or water exceed a critical concentration. Since no precise quality standards for CO_2 transport exist, the target CO_2 purity is usually explored in a certain range (roughly 80–99%). Similarly, the controversial guidelines on so-called minor compounds (especially O_2 and H_2O) can completely change the situation in terms of capture process or polishing step selection.[14]

A general sketch of the inlet–outlet boundary conditions which will be defined for the carbon dioxide capture process, according to the previous analysis, is presented on Figure 2.1. The membrane process simulation framework will be presented in the next paragraph, based on a binary feed mixture in a first step, as explained above.

2.2.2 Membrane Module Design: a Simplified Framework

The general problem of establishing key equations that describe the separations performances of a gas separation module can lead to complex formulations if an exhaustive and rigorous approach is taken[15,16] (*i.e.* mass, heat and energy transfer, multi-component fluxes including coupling effects, *etc.*). Fortunately, a simple development can be easily derived for a binary gas mixture, provided that a series of simplifying assumptions is proposed. This strategy enables an analytical solution to be derived, where the roles of the different key variables in the separation performances can be explicitly described.

The starting hypotheses are the following:

- Steady state conditions
- Isothermal conditions (same temperature everywhere in the system)
- Isobaric conditions in each compartment (no pressure drop on the upstream and downstream side)
- Perfect gas law
- Constant permeability of the species in the membrane (CO_2 and N_2)
- No flux coupling (the driving force of each compound is its partial pressure difference)
- Perfect mixing conditions on both the upstream and downstream sides.

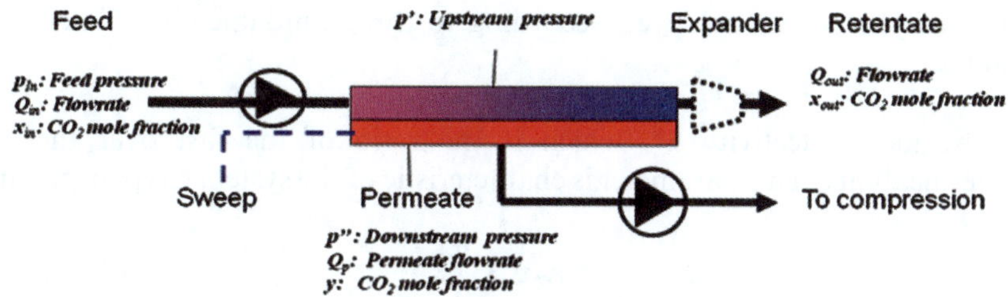

Figure 2.2   Schematic representation and list of variables for a single stage membrane module for post-combustion carbon capture.

The permeate composition, y, can be determined depending on various parameters such as the feed composition (x_{in}), the upstream (p') and downstream (p'') pressure, the permeability to the fast (P_i) and slow (P_j) permeant. The mixture composition is usually expressed in mole fractions (which equal the volume fractions for a perfect gas mixture) with the fast compound as a reference. Figure 2.2 summarizes the main process variables which have to be taken into account for a single membrane module design study.

The different hypotheses allow a series of expressions to be proposed. The steady state assumption is first used to express the permeate composition as:

$$y = \frac{Q_{p,i}}{Q_p} = \frac{Q_{p,i}}{Q_{p,j} + Q_{p,i}} \tag{2.1}$$

where Q stands for specific or global flowrate. Similarly, steady state conditions lead to the following mass balance equation for the fast compound:

$$Q_{in} \cdot x_{in} = Q_{out} \cdot x_{out} + Q_p \cdot y \tag{2.2}$$

and the global mass balance is written as:

$$Q_{in} = Q_{out} + Q_p \tag{2.3}$$

For each compound, a classical permeability mass transfer expression, which makes use of the compound permeability in the membrane material, P, is written:

$$Q_{p,i} = \frac{A \cdot Pe_i}{z} \cdot [P' \cdot x_{out} - P'' \cdot y] \tag{2.4}$$

$$Q_{p,j} = \frac{A \cdot Pe_j}{z} \cdot [P' \cdot (1 - x_{out}) - P'' \cdot (1 - y)] \tag{2.5}$$

z is the membrane thickness and A the membrane surface area.

The analysis is best achieved according to three fundamental adimensional variables:

- The ideal selectivity α^*, which is the ratio of the fast over the slow permeability. This parameter is characteristic of the system gas pair/polymer.

$$\alpha^* = \frac{Pe_i}{Pe_j} \tag{2.6}$$

- The stage cut, θ, corresponds to the ratio of the total permeate to the flow rate of the feed. It is linked to the productivity of the module, and can be seen as a key design variable:

$$\theta = \frac{Q_p}{Q_{\text{in}}} \tag{2.7}$$

- The driving force is exclusively expressed through the pressure ratio and not through the absolute pressure or pressure difference. It is a major operating variable governing, among others, the energy requirement of the module (*i.e.* to a large extent, the operating costs):

$$\psi = \frac{P''}{P'} \tag{2.8}$$

From the permeate composition and mass transfer expressions, it is possible to obtain the following expression:

$$\frac{y}{1-y} = \alpha^* \cdot \left(\frac{x_{\text{out}} - \psi \cdot y}{1 - x_{\text{out}} - \psi \cdot (1-y)} \right) \tag{2.9}$$

If the three previous expressions (permeate composition, mass transfer and mass balance) are combined, a quadratic equation is obtained. This result is interesting in that it can be used to determine the permeate composition (y) as a function of the feed composition (x_{in}), the ideal selectivity (α^*), the stage cut (θ) and the pressure ratio (Φ):

$$ay^2 + by + c = 0 \tag{2.10}$$

with:

$$a = \left(\frac{\theta}{1-\theta} + \psi \right) \cdot (\alpha^* - 1) \tag{2.11}$$

$$b = (1 - \alpha^*) \left(\frac{\theta}{1-\theta} + \psi + \frac{x_{\text{in}}}{1-\theta} \right) - \frac{1}{1-\theta}$$

$$c = \alpha^* \cdot \left(\frac{x_{\text{in}}}{1-\theta} \right)$$

The positive root of the second order polynomial ranging from 0 to 1 is the only root which is physically consistent and it is taken as the solution to the problem.

The conditions that will maximize the permeate mole fraction of the fast compound (y), *i.e.* the maximal separation performances can further be identified. A vanishing pressure ratio (*i.e.* a perfect vacuum at the downstream side, $p'' = 0$) and stage cut ($\theta \sim 0$, which corresponds to an infinitely small permeate flow rate and thus a vanishing productivity) are required in order to get the highest y value. The second condition implies a constant composition in the feed stream ($x_{in} \sim x_{out} = x$). Somehow, this mode of operation corresponds to a distillation column operated under infinite reflux conditions: a maximal selectivity can be obtained at the expense of a vanishingly small productivity. The final expression that applies for $p'' \sim 0$ and $\theta \sim 0$ is:

$$y = \frac{\alpha^* \cdot x}{1 + (\alpha^* - 1) \cdot x} \tag{2.12}$$

A series of curves generated by the previous expression is shown on Figure 2.3.

Logically, no separation is obtained when the ideal separation factor (α^*) is 1. For a given feed composition x, an increasing separation effect is found when the fast to slow permeability ratio α^* increases. Interestingly, the same mathematical expression and equivalent master curves are obtained for vapor liquid equilibrium of a perfect binary mixture. In this case, the saturated vapor pressure ratio plays the same role as the ideal selectivity. This simple expression is of great help for shortcut calculations since, as already mentioned, it provides the highest mole fraction that can be obtained in a single stage membrane unit

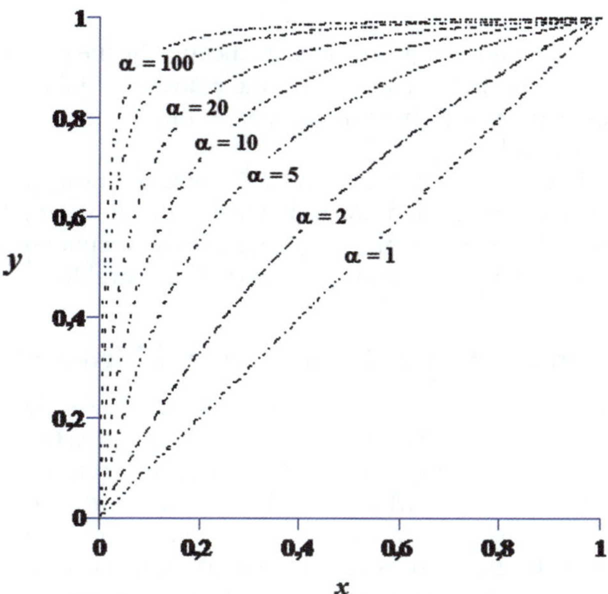

Figure 2.3　Examples of permeate (y) *versus* feed mixture (x) composition curves, based on the membrane ideal selectivity (α^*) eqn (2.12).

on the permeate side. Since the attainment of a target purity is often imposed (classically $0.8 < y < 0.99$), this expression can be used in order to verify whether the specification can be obtained with a given polymer material and a known permeability ratio. Furthermore, it can also be used for the determination of the minimal target material selectivity (α^*) if x and y are fixed. Nevertheless, it is important to notice that the hypotheses underlying this calculation are far away from the real carbon capture framework, for which a significant capture ratio has to be achieved and a vanishingly small downstream pressure is unachievable.

Another asymptotic solution can be derived for perfect mixing conditions if an infinitely small stage cut is assumed ($\theta \sim 0$). In this case, the mole fraction of the outlet retentate is again identical to the feed mole fraction ($x_{\text{out}} \sim x_{\text{in}} = x$). This expression is of interest as it permits the analysis of the respective influence of the ideal selectivity and the pressure ratio on the separation performances of the module.

$$y = \frac{2}{\psi} \cdot \left(x + \psi + \frac{1}{\alpha^* - 1} - \sqrt{\left(x + \psi + \frac{1}{\alpha^* - 1} \right)^2 - \frac{4 \cdot \alpha^* \cdot \psi \cdot x}{(\alpha^* - 1)}} \right) \tag{2.13}$$

For a binary system, eqn (2.13) has two limiting cases. When $\alpha^* \gg 1/\Phi$, the selectivity is no longer important, and, under the condition that the value of y can never be greater than unity, the permeate composition simply becomes:

$$y \approx \frac{x}{\psi} \tag{2.14}$$

In this case, the separation performances are dictated by the process conditions, and the pressure ratio that is available for the industrial application becomes a key issue. This situation is typical for separations where a very high ideal selectivity is achievable.

Conversely, if $\alpha^* \ll 1/\Phi$, the influence of the pressure ratio can be neglected and the permeate composition becomes limited by the polymer selectivity. This situation occurs when the membrane is operated with vacuum on the permeate side (*i.e.* $\Phi \to 0$) or when only a low selectivity is achievable.

2.2.3   Membrane Module Design: Classical Methodology

Historically, the basis of gas permeation module design was first proposed by Weller and Steiner[17] in 1950. Nowadays, modern computation techniques provide numerical solutions to the problems thanks to dedicated routines.[18] Orthogonal collocation methods[19] or perturbation methods[20] are reported to be particularly attractive when a minimum resolution time and computational efforts are required. Several of these routines have been implemented in commercial process simulation software, where advantage can be taken of thermodynamics or unit operation design packages in order to simulate hybrid or multi-stage operations with gas separation membranes.[21] Nevertheless, much effort has been devoted to the search of asymptotic analytical solutions to the problem before computing facilities where available. Approximate analytical solutions are still occasionally proposed and they can be of interest, for instance when a broad and systematic sensitivity analysis is desired.

A general methodology was proposed above in order to predict the performances of a membrane separation module based on a series of simplifying assumptions. The main interest of the approach was to offer the possibility of obtaining a simple, easy-to-handle analytical expression. Nevertheless, the solution obtained through this simplified approach should be taken as a rough estimate, and the hypothesis of perfect mixing conditions must be reconsidered if one expects a more realistic answer with regard to the performance of an industrial module. Similarly to the framework proposed in chemical reactors, heat exchangers or mass transfer operations, a completely different hypothesis, namely that of a perfect plug flow, is first proposed in order to estimate the incidence of hydrodynamics conditions on the separation performances. This modification necessarily leads to a system of ordinary differential equations and, strictly speaking, a generic analytical solution can no longer be derived. Basically, different configurations can be proposed. Nevertheless, the cross plug flow model, which remains the cornerstone when it comes to models in this field, is detailed in the following so as to offer a generic framework for module design. As depicted in Figure 2.4, the cross plug flow model assumes a plug flow on the upstream, feed side of the membrane and free flow on the downstream, permeate side.

Apart from the flow conditions, the hypotheses proposed for the derivation of the solution in the case with perfect mixing conditions are maintained in the under-mentioned analysis.

A mass balance over a differential surface area, dS, leads to the following differential equation:

$$Q^* \cdot \frac{dx}{dS} = -\left(x - \psi \cdot y + x \cdot \frac{dQ^*}{dS} \right) \qquad (2.15)$$

where x is the upstream fast compound mole fraction, and S is the dimensionless membrane area:

$$S = \frac{A \cdot Pe_i \cdot P'}{z \cdot Q_{\text{in}}} \qquad (2.16)$$

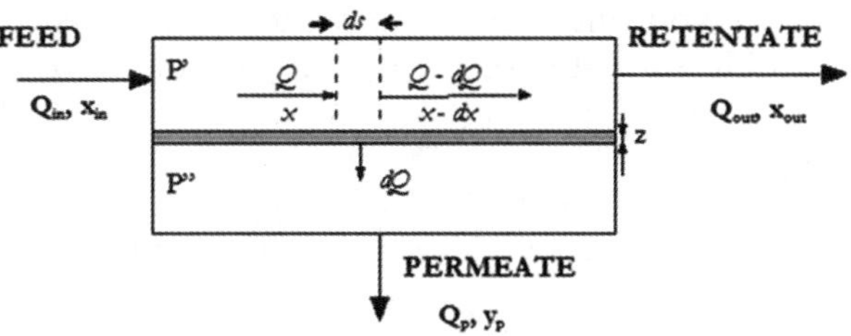

Figure 2.4   Schematic representation and list of variables for a single stage membrane module working under cross-plug flow conditions.

Q^* corresponds to the local stage cut $Q^* = Q/Q_{in}$. We can thus write the flux relationships for each component according to:

$$-d(Qx) = \frac{Pei \cdot P'}{z}(x - y\psi)dS \tag{2.17}$$

$$-d[Q(1 - x)] = \frac{Pej \cdot P'}{z} \cdot [(1 - x) - (1 - y)\psi]dS \tag{2.18}$$

These two equations can be combined, giving rise to the following expression:

$$\frac{dQ^*}{dS} = -\left(x - \psi \cdot y + \frac{1}{\alpha^*}\left(1 - x - \psi \cdot (1 - y)\right)\right) \tag{2.19}$$

Finally, from the definition of the equations of local permeate composition and the mass transfer:

$$\frac{y}{1 - y} = \alpha^* \cdot \left(\frac{x - \psi \cdot y}{1 - x - \psi \cdot (1 - y)}\right) \tag{2.20}$$

The set of equations (2.15), (2.19) and (2.20) can be solved numerically and they represent the basis of the cross plug flow model. Similar sets of equations have, with the corresponding boundary conditions, been applied by various authors in order to predict separation performances for co-current or counter-current plug flow modes. A correct to excellent agreement has been obtained when predictions are compared to experimental laboratory-scale data.[22] For multi-component permeation situations, several modeling studies have been reported. Generally, and in more practical terms, it is interesting to notice that experimental data from on-site pilot modules are often reported as being intermediate between the perfectly mixed and the cross plug flow predictions. These two situations have been detailed above.

2.3   Simulation Studies for Post-combustion CO_2 Capture by a Membrane Gas Separation Module

Based on the basic modeling methodology which has been detailed in the previous section, different issues of the post-combustion CO_2 capture can be tackled. Basically, the main questions that have to be addressed can be summarized as follows.

First, what is the minimal selectivity of the membrane material which enables the purity and capture ratio to be achieved? This first target clearly corresponds to a *material challenge*.

Second, what will be the energy requirement of the process when the previous conditions are fulfilled? The *energy challenge*, one of the most stringent selection criteria of carbon capture processes has to be stressed here.

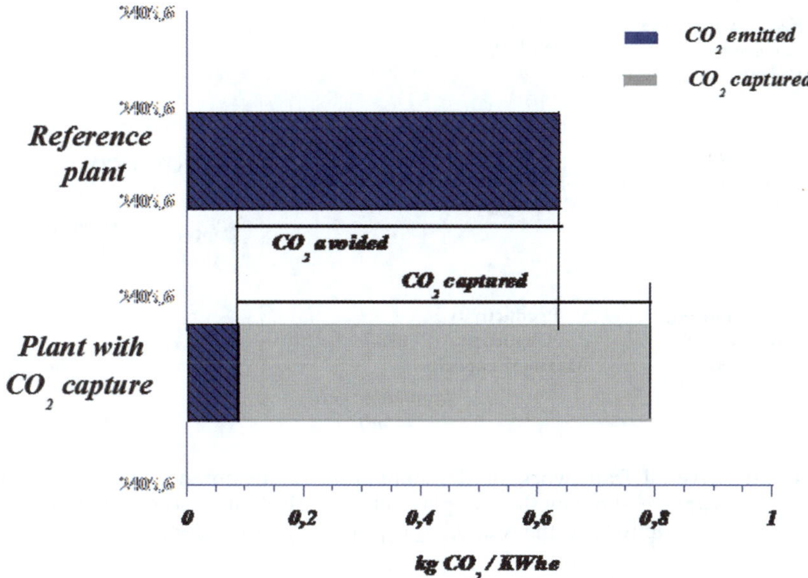

Figure 2.5 Schematic representation of the CO$_2$ avoided and CO$_2$ captured concept. The difference between the reference plant emissions and the plant with capture emissions corresponds to the secondary emissions due to the capture process.

This characteristic will govern, to a large extent, the operating expenses (OPEX) of the process. Standard post-combustion capture processes, such as gas liquid absorption in a chemical solvent (MEA), typically require 3.5 GJ (thermal basis) per ton of recovered CO$_2$, and 2 GJ ton^{-1} is often considered as a target.[23] The importance of energy requirement considerations in the carbon capture process selection results from the secondary CO$_2$ emissions which are generated by the capture process. Figure 2.5 summarizes the interplay between CO$_2$ emissions by the reference plant and secondary, additional CO$_2$ emissions due to the capture step.

Third, what is the corresponding membrane surface area? This variable, together with compressors and/or vacuum pumps, will play a key role in the capital expenses (CAPEX) of the process. A *design challenge* takes place here, since an interplay between the operating conditions (pressure difference in the module) and the corresponding surface area has to be addressed here, in order to minimize the overall capture cost. The membrane productivity (*i.e.* permeance), which includes the intrinsic permeability and effective active layer thickness obviously plays a key role here. This point will be discussed afterwards.

The three major challenges listed above, and the associated design methodology leading to the overall capture cost are shown in Figure 2.6. In a first step, these three issues will be examined based on simulations for a single

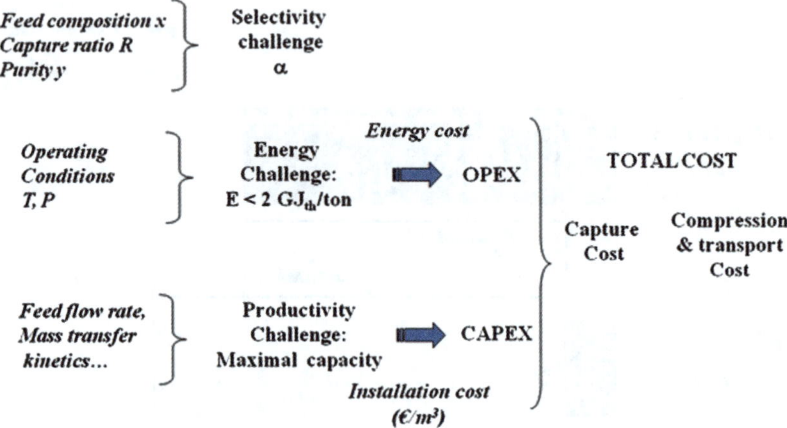

Figure 2.6 Overall framework of the simulation of membrane processes for CO_2 capture, showing the interplay between selectivity, energy requirement and productivity in the total CO_2 capture cost estimation.

membrane module. Generally speaking, the CO_2 permeate purity (y) and the CO_2 capture ratio (R) are the imposed variables. R is expressed as:

$$R = \theta \cdot \frac{y}{x_{in}} \tag{2.21}$$

It should be noted that, similarly to other studies in the field, CO_2 is systematically considered as the fast permeating species for polymeric membrane simulations. In fact, apart from a very specific observation, no polymeric material has been reported with N_2 permeating faster than CO_2. The decisive advantages that a nitrogen selective material would offer for CO_2 post-combustion capture have been discussed on a theoretical basis.[24]

2.3.1 Addressing the Separation Problem: Selectivity Challenge

The relative permeability of the compounds present in the feed mixture is a key material characteristic and it is called the ideal separation factor, or membrane selectivity (α^*). It is recalled that, similarly to a convention used in distillation for the relative volatility, it is usually expressed so as to be larger than 1:

$$\alpha^* = \frac{Pe_i}{Pe_j} \geq 1 \tag{2.22}$$

The identification of the minimal effective selectivity which enables the target separation performances to be attained is often one of the first problems to be solved in the design of membrane processes. For the post-combustion carbon capture challenge, the minimal selectivity can be determined for the set of process performances to be attained: a capture ratio, R, of 0.8 (or more) and a permeate purity, y, of 0.8 (or more). Based on the assumptions detailed in the previous section, the pressure ratio Φ, the adimensional membrane area, S, can be identified when the inlet feed content, x_{in}, the permeate purity, y, and the membrane selectivity, α, are defined. A series of master curves showing the evolution of these different variables, for different feed inlet compositions, when a target permeate purity of 0.9 is imposed, are shown on Figure 2.7(a–d).

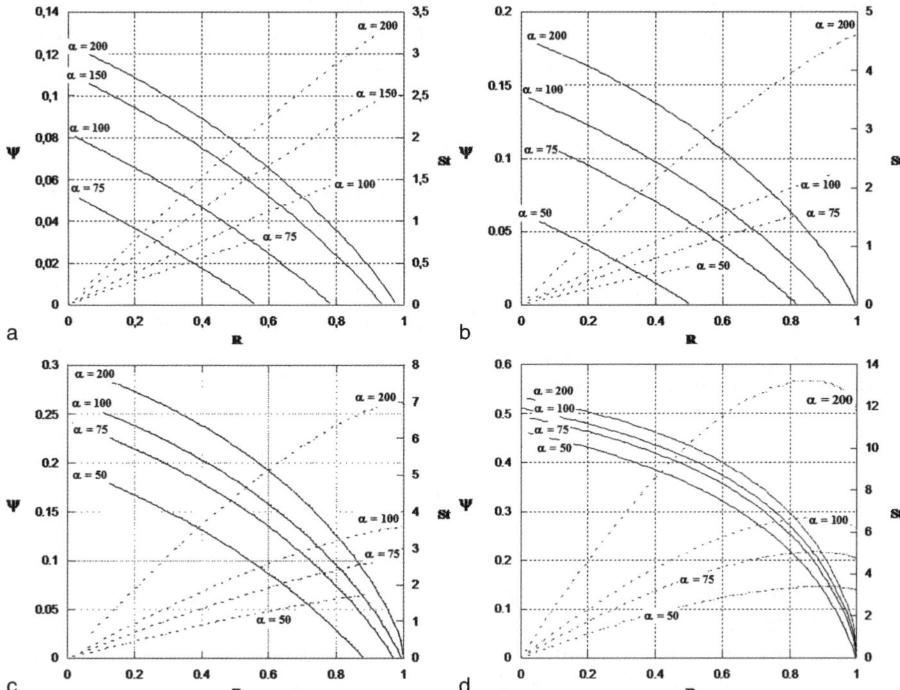

Figure 2.7   (a) General chart showing the influence of the capture ratio, R, on the pressure ratio, Ψ (continuous lines), and adimensional membrane surface St (dotted lines) for different CO_2/N_2 selectivity α. Simulation conditions: cross-plug flow model with negligible pressure drop, binary feed mixture (CO_2/N_2), inlet CO_2 content 15%, permeate CO_2 purity 90%. (b) General chart showing the influence of the capture ratio, R, on the pressure ratio, Ψ (continuous lines), and adimensional membrane surface St (dotted lines) for different CO_2/N_2 selectivity α. Simulation conditions: cross-plug flow model with negligible pressure drop, binary feed mixture (CO_2/N_2), inlet CO_2 content 20%, permeate CO_2 purity 90%. (c) General chart showing the influence of the capture ratio, R, on the pressure ratio, Ψ (continuous lines), and adimensional membrane surface St (dotted lines) for different CO_2/N_2 selectivity α. Simulation conditions: cross-plug flow model with negligible pressure drop, binary feed mixture (CO_2/N_2), inlet CO_2 content 30%, permeate CO_2 purity 90%. (d) General chart showing the influence of the capture ratio, R, on the pressure ratio, Ψ (continuous lines), and adimensional membrane surface St (dotted lines) for different CO_2/N_2 selectivity α. Simulation conditions: cross-plug flow model with negligible pressure drop, binary feed mixture (CO_2/N_2), inlet CO_2 content 50%, permeate CO_2 purity 90%.

Taking into consideration the coal power plant reference case with an inlet CO_2 content of 15% (Figure 2.7(a)), it can be concluded that a membrane selectivity of 150 or more is required if a capture ratio of 0.8 or more is aimed. A comparison of this key data with Figure 7(b–d) shows the very large parametric sensitivity of membrane processes for this application: a membrane selectivity of 100 is indeed large enough in order to achieve the same performances for a 20% CO_2 inlet concentration and this data drops below 75 for more concentrated flue gases. These simulation results will be compared to the performances of existing membrane materials later.

2.3.2 Tackling the Energy Requirement Issue

The estimation of the energy requirement has to be achieved in a second step. For a single stage process, the evaluation is straightforward. Since the pressure ratio only plays a role in the analysis, feed compression or vacuum pumping at the permeate side can, in principle, be indifferently applied. For instance, for a feed compression with atmospheric pressure at the permeate side ($p'' = 1$) and no energy recovery system at the retentate (such as an expander), the energy requirement can be estimated as:

$$E = Q_{in} \cdot \frac{\gamma \cdot RT}{\eta \cdot (\gamma - 1)} \cdot \left[\left(\frac{1}{\psi} \right)^{\frac{\gamma-1}{\gamma}} - 1 \right] \qquad (2.23)$$

Here, γ is the adiabatic expansion factor of the gas mixture (*e.g.* $\gamma = 1.4$ J mol^{-1} K^{-1} for nitrogen, for instance), η the isentropic efficiency and T is the inlet temperature (K).

For a permeate vacuum pumping strategy with atmospheric pressure at the retentate side ($p' = 1$), the energy requirement becomes:

$$E' = Q_P \cdot \frac{\gamma \cdot RT}{\eta' \cdot (\gamma - 1)} \cdot \left[\left(\frac{1}{\psi} \right)^{\frac{\gamma-1}{\gamma}} - 1 \right] = \theta \cdot E \qquad (2.24)$$

Taking into account the fact that, for post-combustion carbon capture, the target purity, y ($0.8 < y < 0.99$) is close to the capture ratio, R ($0.8 < R < 0.9$), a small stage cut value, close to the inlet CO_2 feed content is obtained ($\theta \sim x_{in}$). As a consequence, Q_P being much smaller than Q_{in}, E' will be much smaller than E and a vacuum pumping should be systematically applied, if (and only if) a minimal energy requirement only is aimed. Figure 2.8 shows the influence of membrane selectivity on the energy requirement for a single stage membrane module when a capture ratio (R) of 0.8 and a permeate purity (y) of 0.8 is imposed.

Two different inlet CO_2 compositions have been compared in order to show again the very strong parametric sensitivity. The equivalent energy requirement, on a thermal basis, of 2 GJ ton^{-1} of recovered CO_2 has been added in order to visualize the possibility of membrane processes to possibly attain this level of performance. It can be seen that, even with highly selective membrane

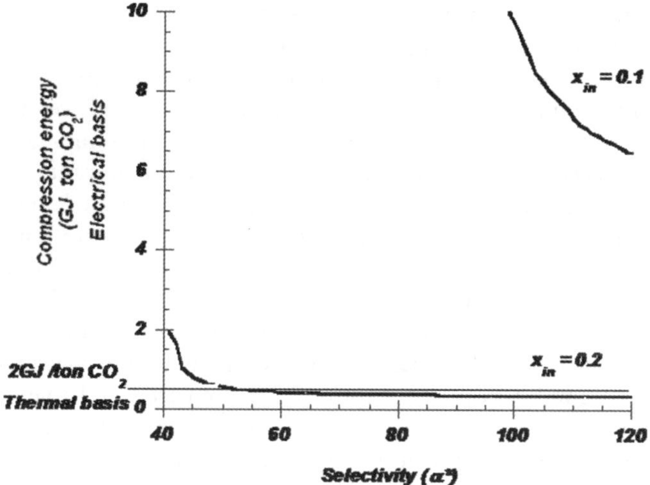

Figure 2.8 Influence of membrane selectivity on energy requirement for 10% and 20% CO_2 content in feed mixture. Operating conditions: vacuum pumping. Separation targets: capture ratio (R) 80%, CO_2 purity (y) 0.8.

materials and a vacuum-based driving force, it is impossible to reach the target energy requirement for a 10% feed inlet CO_2 content. A completely different situation is obtained for a 20% CO_2 content in the flue gas, since a very low energy requirement (0.5 GJ ton^{-1} or less) can be attained as soon as a selectivity of 50 or more is achieved by the membrane material. Interestingly, more selective membrane materials do not provide a significant decrease of the energy requirement. This observation is of importance: in fact, more selective polymeric materials for gas separation applications usually show decreased permeabilities, a situation leading to so-called trade-off curves.[25] Consequently, the search for very selective membranes could simultaneously generate unacceptable membrane surface areas, without a clear advantage in terms of energy requirement. This behavior corresponds to the simple preliminary analysis presented before: a very low energy requirement being acceptable, a limited pressure ratio can be tolerated, typically 0.1 to 0.5. In that case, the ideal selectivity can easily be much larger than the inverse of the pressure ratio ($\alpha^* \gg 1/\Phi$), and selectivity will no longer play a key role. Consequently, some authors[26] recommend to select polymeric materials with a reasonable CO_2/N_2 selectivity (typically around 50), so that the previous condition $\alpha^* \gg 1/\Phi$ is fulfilled, and later push the material productivity performances (*i.e.* permeance) in order to minimize the capital expenses.

2.3.3 The Energy Requirement/Membrane Area Trade-off

The vacuum pumping strategy has been shown to generate the smallest energy requirement. Nevertheless, this mode of operation simultaneously generates a much smaller driving force than the feed compression; a much larger surface area is required, thus giving rise to a considerably higher capital cost.[27]

Various arguments in favor of feed compression can be proposed: the difficulty of achieving vacuum at an industrial scale, the lower energy efficiency of vacuum pumps as compared to compressors, and the large foot prints of vacuum pumps compared to compressors. As a consequence, the selection of the most efficient driving force strategy, which combines a low energy requirement together with a reasonable surface of membrane has attracted attention; other options such as the possibility to make use of an expander on the retentate side in order to recover part of the compression energy, or the application of a condensable gas sweep on the permeate side in order to provide a driving force without the need to apply vacuum (two options shown by dotted lines in Figure 2.2) have been investigated through simulations by some authors.[28–35] A list of the studies devoted to single membrane module simulations for post-combustion CO_2 capture is summarized in Table 2.4.

A summary of the main conclusions of simulations for a single stage membrane process for post-combustion CO_2 capture can be proposed as follows:

- Vacuum is clearly the best choice in terms of overall energy requirement. Nevertheless, a low downstream pressure, usually considered to be unrealistic for industrial applications (typically 2 mbar), is needed in order to achieve a reasonable CO_2 purity and recovery.
- When energy requirement and membrane surface area are simultaneously taken into account, a combination of moderate feed compression (typically 2–3 bar) together with a moderate vacuum (typically 0.2 bar) usually offers the best compromise.
- The development of membranes with improved permeance is more important than the development of more selective materials. For polymeric materials with a physical separation process, the best performances reported up to now can be defined as a CO_2/N_2 selectivity of 50–70 and a permeance of 1000 GPU (gas permeation unit).

Table 2.4 Summary of simulation studies for CO_2 post-combustion capture by a single stage membrane process

Module type	Operating conditions	Author	Reference
Single stage with expander	Compression (54 Bar)	Van der Sluis *et al.* (1992)	29
Single stage no expander	Compression (4-16 Bar), or vacuum (0.1 Bar) or sweep	Hägg *et al.* (2005)	30
Single stage no expander	Compression or vacuum	Bounaceur *et al.* (2005)	31
Single stage	Compression (20 Bar)	Ho *et al.* (2006)	32
Single stage no expander	Vacuum (0.03 Bar)	Zhao *et al.* (2008)	33
Single stage	Compression (1.5 Bar) or vacuum (5 mBar)	Kotowicz *et al.* (2010)	34
Single stage no expander	Compression or vacuum	Brunetti *et al.* (2009)	35

- Recycle streams offer the possibility to attain improved performances, but an exhaustive and rigorous analysis of the numerous options which are possible remains to be achieved.

2.3.4   Towards Multi-stage Processes

The limitations of single stage membrane processes have been discussed in the previous section. A classical solution which is proposed in order to overcome these drawbacks consists to design multistage processes. It should be noted first that, generally speaking, multi-staged membrane processes for gas separation applications are relatively rare.[36] A recompression step has in fact to be inserted between two membrane stages, and the capital cost increases considerably. Consequently, multi-stage membrane systems are usually limited to two stages[37] (except from the very specific case of uranium enrichment by gaseous diffusion). From the simulation point of view, this strategy opens a very large number of design possibilities which have been only partially explored up to now. Options such as driving force selection (compression, vacuum or sweep) or recycling mode (permeate, retentate, stage number) lead to a complex framework[38] which are summarized in Figure 2.9.

The general problem of membrane cascades has been investigated by several authors, where mixing of streams of similar or different compositions have been compared. Even though no generic optimal solution can be proposed, some guidelines, only partly explored for the carbon capture problem, can be proposed:

- Membrane cascades without recycling.
- Retentate recycling is attractive for membranes which show to too low selectivity. This option can lead to the highest efficiency, especially when the composition of the retentate corresponds to the composition of the respective feed stream. Otherwise, mixing losses decrease the overall efficiency of the process.
- Permeate recycling, which can be attractive in terms of increased recovery, usually generates an increase in membrane area and energy requirement.
- Different membranes can be selected for the different stages, offering a supplementary degree of freedom. For a two-stage cascade, the membrane in the second stage should exhibit a selectivity similar or higher than the

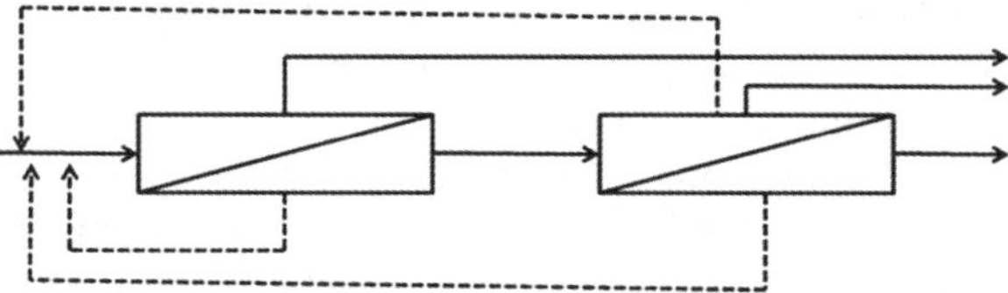

Figure 2.9   Example of the different association possibilities for a two stage membrane separation process. Compressors, vacuum pumps and expanders are not shown. Dotted lines correspond to options such as permeate recycle from the first and second stage and retentate recycle from the second stage.

Table 2.5 Summary of simulation studies for CO_2 post-combustion capture by multi-stage membrane processes

Module type	Operating conditions	Author	Reference
Two stage with recycle	Compression (21.4 Bar)	Herzog *et al.* (1991)	28
Two stage with expander	Compression (54 Bar)	Van der Sluis *et al.* (1992)	29
Two stage with recycle	Compression (1.5 Bar) and vacuum (80 mBar)	Ho *et al.* (2008)	39
Multistage with or without recycle	Compression (10 bar), vacuum (0.03 Bar)	Zhao *et al.* (2009)	33
Two stage with recycle	Compression (3 Bar) and vacuum (0.2 Bar)	Merkel *et al.* (2009)	26
Two stage with or without sweep	Compression (2–5 Bar) or vacuum (25–125 mBar)	Hussain *et al.* (2010)	41

selectivity of the first stage. This strategy leads to a bulk removal in the first stage based on a reasonable surface area, and a enriching step in the second stage up to the target purity. Unfortunately, for carbon capture, mixed membrane combinations have not been addressed so far.

Table 2.5 shows a summary of the limited number of studies which have addressed the design of multi-stage membrane processes for post-combustion CO_2 capture. It can be noticed that the number of stages remains effectively limited to two, while different options such as compression only[29], a combination of compression and vacuum[26,39–41] have been explored.

In the last section, the challenges some challenges in relationship with the simulation of membrane processes for post-combustion CO_2 capture will be discussed.

2.4   Scientific and Technological Challenges

2.4.1   Improved Materials: Selectivity and Productivity

Selectivity obviously plays a key role for CO_2 capture by membrane processes. In the simulation studies discussed in the previous sections, it has been concluded that a CO_2/N_2 selectivity of 150 or more is required for a single stage membrane process to fit the purity and recovery target when a 15% CO_2 flue gas is treated. It is interesting to compare, at this stage, this value with real selectivity data of different membrane materials.

A first category of materials, which corresponds to the most widely used ones for membrane gas separations corresponds to dense polymeric membranes, without any chemical reactivity property. In that case, taking into account the definition of permeability and the well established solution–diffusion framework, membrane selectivity can be expressed as:

$$\alpha^* = \frac{Pe_i}{Pe_j} = \frac{S_i \cdot D_i}{S_j \cdot D_j} = \alpha_S^* \cdot \alpha_D^* \tag{2.25}$$

It can be seen that a high selectivity can be achieved from a high solubility ratio α^*_S or a high diffusivity ratio α^*_D, or both. For the CO_2/N_2 mixture, theoretical computations[42] show that a moderate diffusion selectivity only can be expected, while a quite large effect can result from the solubility effect. Combining these two phenomena enable a CO_2/N_2 selectivity up to 70 to be attained. Different types of materials, mostly based on block copolymers with polyethyleneglycol type moieties have been reported with selectivity performances which are close to this theoretical value.[26,43,44] Few results have been also reported on the selectivity of inorganic membrane materials;[45–47] theoretical computations based on molecular modeling predict an impressive CO_2/N_2 selectivity, up to 1000, to be possibly achievable for zeolite-based membranes.[48] Nevertheless, while pilot scale studies with polymeric membrane materials are already performed,[26,49] no commercially inorganic membrane is available today for post-combustion carbon capture.

In addition to the selectivity, the question of the productivity of a membrane separation system should logically also be addressed. For a given membrane area and driving force, the productivity is usually referred to the so-called permeance of the membrane. For a dense polymeric membrane, the permeance corresponds to the ratio of the gas permeability (expressed, for instance in Barrer) to the dense layer thickness. A derived and arbitrary unit it the so-called gas permeation unit which corresponds to a polymer showing a permeability of 1 Barrer, with an active layer of 1 µm. It is logically interesting to examine the intrinsic selectivity and productivity performances of various polymers for the CO_2/N_2 gas pair. This type analysis has been detailedly investigated by Robeson for gas separation applications.[25,50] The systematic antagonism between productivity (*i.e.* membrane permeability) and selectivity (expressed through the ratio of the gas permeabilities, called ideal selectivity, α^*, and defined above) is evidenced through this type of chart.

More interestingly, the application to different gas pairs shows that an empirical limit seems to be characteristic for each gas mixture. This observation has been further explored by Freeman[51] and a fundamental analysis of the factors playing a role in the position of this limit has been proposed. Figure 2.10 shows a comparison of the trade-off curves with different experimental data for polymeric membrane materials towards the CO_2/N_2 gas pair.

For these physically based separation membranes, the most promising materials show a CO_2/N_2 selectivity in the range of 50 to 70, together with a CO_2 permeability of several hundreds of Barrer.

As detailed in the simulation section above, this level of selectivity requires a two-stage membrane process for the CO_2 capture ratio and purity to be attained. For a higher selectivity to be attainable, chemically reactive materials are necessary. Numerous studies have addressed this issue[7,52–55] and the different performances which can be obtained by the various types of membrane materials are summarized in Table 2.6.

It is important to stress, however, at this stage that, while selectivity performances for classical, physically based gas separation membranes are obtained with dry CO_2/N_2 mixtures, fixed site wet mixtures are absolutely

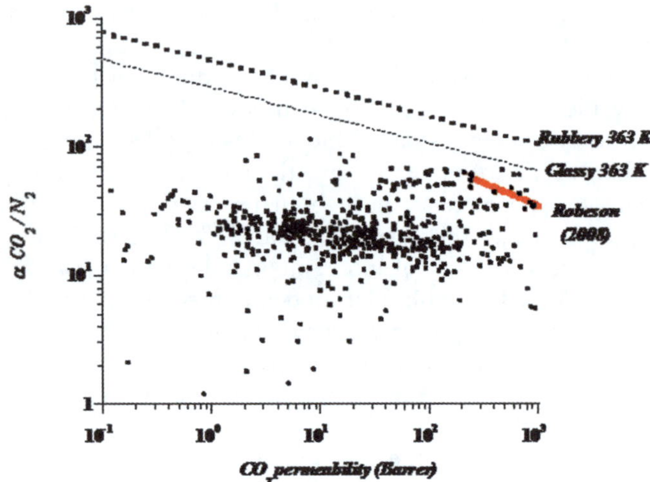

Figure 2.10 CO_2/N_2 selectivity *versus* CO_2 permeability for a series of polymeric membrane materials and corresponding trade-off curves based on theoretical (dotted lines) or empirical (bold line) approaches.

Table 2.6   Summary of the selectivity and permeability performances of different membrane materials for CO_2/N_2 separation

Membrane type	Material and/or carrier	CO_2/N_2 selectivity	CO_2 permeability (Barrer) or permeance (GPU)
Gas separation membrane (dense polymers)	PEO-PBT	70	120 Barrer
	PEG/Pebax©	47	151 Barrer
	PEG-DME/ Pebax©	43	600 Barrer
	PEGDA/PEGMEA	41	570 Barrer
	Polaris™	50	1000 GPU
Fixed Site Carrier Membrane (FSCM)	PAAM-PVA/PS	80	24 GPU
	PVAm/PVA	145	212 GPU
	PEI/PVA	230	1 GPU
	PDMA/PS	53	30 GPU
	PDMAMA	80	5 GPU
Liquid Membrane (LM)	PVAm-PVA/PS	90	22 GPU
	PVAm/PVA	90	15 GPU
	Amines/PVA	500	250 GPU
	Carbonic anhydrase	250	80 GPU
	Amines/PVA	493	693 Barrer

required for carrier membranes and liquid membranes to work. This fundamental difference necessarily induces pros and cons for the two categories of materials which have not been examined in detail for carbon dioxide capture.

2.4.2　Beyond Model Mixtures

The role of water for chemically reactive membranes logically addresses the question of the influence of humidity on membrane separation performances. It has been stated before that, while flue gases are systematically wet, CO_2/N_2 separation studies by membrane materials most often report on dry mixtures. Apart from chemically reactive and liquid membranes, the analysis of wet CO_2/N_2 mixtures by dense polymers has been indeed seldom reported.[8,56] It might be anticipated, however, that the presence of water will have a significant impact in terms of separation performances. First, water can play a significant role in terms of plasticization for glassy membrane materials.[57] Even though most membrane materials for CO_2 capture are of rubbery type, major modifications could occur compared to the dry mixture results.

Second, water removal is generally accompanied by strong energy effects, leading to thermal polarization phenomena. This addresses a challenge for simulation, namely coupled heat and mass transfer, which has not been developed in detail for CO_2 capture up to now.

Third, more specifically, water is a fast permeating species and impressive H_2O/CO_2 selectivity data can be observed for polymeric materials. Some examples of experimental data for different polymers are shown on Figure 2.11. Interestingly, more CO_2 permeable polymers seem to show smaller H_2O/CO_2 selectivity. The implications of the various situations, in terms of separation performances, covered by the large spectrum of H_2O, CO_2 and N_2 permeabilities remains to be investigated.

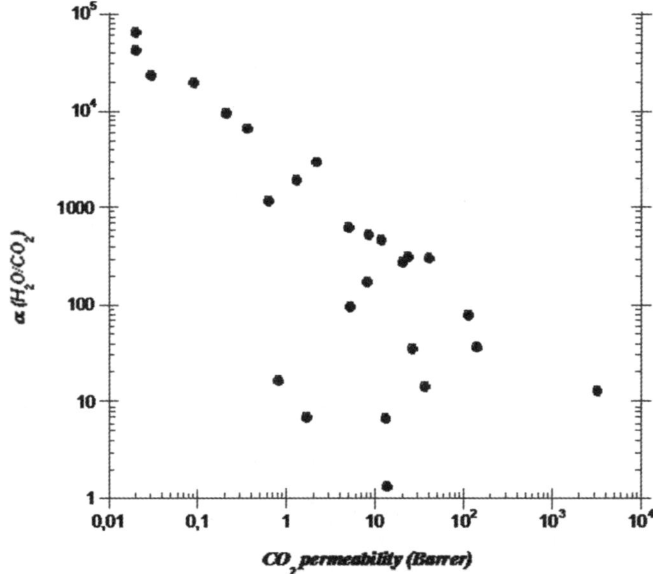

Figure 2.11　H_2O/CO_2 selectivity *versus* CO_2 permeability for a series of polymeric membranes (experimental data obtained on pure compounds at 298 K).

As stated in the introductory part of this chapter, other minor species such as O_2, Ar, SO_x and NO_x should also be taken into account in order to evaluate the ability of membrane processes to reach the composition targets listed in Table 2.3. One difficulty comes from the fact that, apart from oxygen polymer permeability for these compounds are poorly documented.[58]

2.4.3　Alternative Approaches and Prospects

Membrane processes only have been considered in the preceding sections. It might be that a better use of membranes for CO_2 post-combustion capture corresponds to a combination with another technology. Such a strategy leads to hybrid processes, which have been shown to give in some cases the best performances. For natural gas treatment for instance, a combination of membrane processes for bulk removal, followed by absorption (amine washing) is occasionally applied.[59] Surprisingly, apart from very specific case of membrane contactors for intensified CO_2 absorption in chemical solvents, very few studies have addressed the problem of hybrid systems such as membrane separation together with cryogeny, adsorption or absorption for carbon capture. This type of technology clearly opens a very large spectrum of simulation and optimization problems.

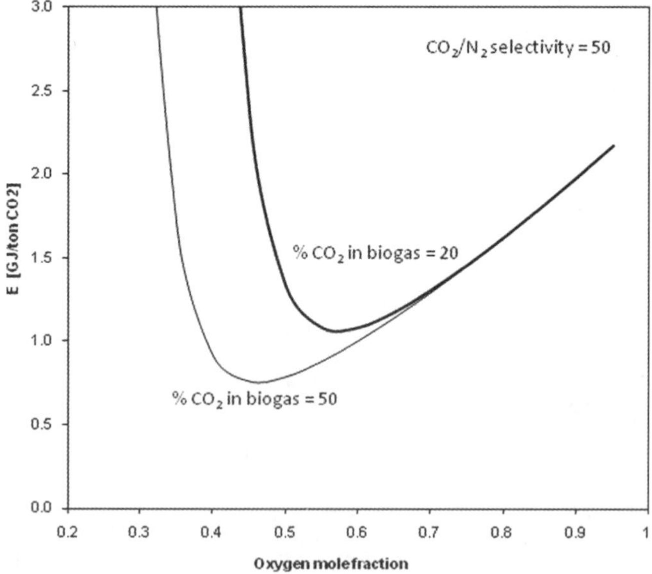

Figure 2.12　Evolution of the overall energy requirement of a hybrid process based on oxygen enriched air combustion and post combustion CO_2 capture by a membrane unit applied to biogas with a CO_2/N_2 selectivity of 50. A minimal energy requirement is obtained for an oxygen content in air around 0.45 for a CO_2 content of biogas of 50% and 0.55 for a CO_2 content of biogas of 20%.

Finally, another hybrid process possibility consists to combine an oxygen enriched air combustion step with a membrane post-combustion capture process. The key concept behind this approach comes from the strong parametric sensitivity of membrane processes towards the CO_2 content in the feed (*i.e.* flue gas), discussed in the previous sections. Basically, it has been shown a 15% CO_2 content is not favorable for single stage membrane CO_2 capture. A high membrane selectivity is required and a high energy requirement results. A completely different situation occurs however, as soon as the CO_2 content exceeds 20%. In that case, the capture ratio and purity targets can be attained in a single stage, without an unacceptable energy requirement. Some preliminary simulation studies of this hybrid concept have been recently reported.[60] An oxygen content in air around 60% seems to induce require the lowest overall energy requirement. The application to the same concept to biogas has been also investigated through a simulation study and promising performances in terms of energy requirement have been obtained for a CO_2/N_2 selectivity of 50 only.[61] An example of simulation results for CO_2 capture from a biogas power unit is shown on Figure 2.12.

2.5 Concluding Remarks

The objective of this review was to provide a state of the art of simulations studies for CO_2 post-combustion capture by membrane processes. Similarly to other engineering challenges, simulation studies can play a dual role.

First, they can identify the most relevant and efficient schemes which fits the separation targets. Optimization techniques and effective data (*i.e.* membrane permeances for the different species) are absolutely necessary in order to achieve that target. As discussed above, the uncertainties on purity targets, the lack of data obtained under real flue gas treatment conditions have impeded up to now the identification of the optimal place and role of membrane processes in a post-combustion framework. Additionally, the inherent complexity of multi-component membrane separations where coupled heat and mass transfer phenomena should be taken into account, the multi-faceted problems associated to multi-staged or hybrid processes clearly complicate the picture.

The second major role of simulation corresponds to the identification of target materials and cost performances. The ultimate objective is here to produce cost curves which enable the key variables such as membrane selectivity, permeance or specific production cost to be represented in master curves, and the overall technology cost to be estimated as a function of the variables. This level of knowledge has been achieved for some applications such as nitrogen or oxygen production. The priorities for membrane development can thus be clearly identified, as well as the chances to compete with established technologies. This step has clearly not been achieved so far for post-combustion carbon dioxide capture by polymeric membrane processes. A significant effort in terms of simulations in link with technico-economical analyses is required in order to achieve that purpose. The best place and role of membrane processes will be clearly defined only when this complex and demanding task is performed.

2.6 List of Symbols

Symbol	Definition	Units
A	Membrane area	m^2
D	Diffusion coefficient	$m^2\,s^{-1}$
E	Energy requirement	$J\,mol^{-1}$
p'	Membrane module upstream side pressure	bar
p''	Membrane module downstream side pressure	bar
p	Pressure	bar
Pe	Permeability coefficient	Barrer or, in SI, $mol\,m^{-1}\,s^{-1}\,Pa^{-1}$
Q	Flow rate	$m^3\,s^{-1}$
R	Recovery ratio	dimensionless
R	Perfect gas constant	$8.314\,J\,mol^{-1}\,K^{-1}$
S	Non-dimensional membrane surface area	dimensionless
T	Temperature	K
V	Volume	m^3
x_{in}	Mole fraction in the feed mixture	dimensionless
x_{out}	Mole fraction in the retentate	dimensionless
y	Mole fraction in the permeate side	dimensionless
z	Membrane thickness	m
α^*	Ideal separation factor	dimensionless
γ	Adiabatic gas expansion coefficient	dimensionless
Φ	Pressure ratio	dimensionless
θ	Stage cut	dimensionless

Acknowledgements

The author sincerely thanks his colleagues Roda Bounaceur and Denis Roizard for their help in computer simulations and for stimulating discussions. The support of CNRS through the Energie program (grant COCASE) is also gratefully acknowledged.

References

1. B. Metz and Intergovernmental Panel on Climate Change. IPCC special report on carbon dioxide capture and storage. Cambridge University Press (for the Intergovernmental Panel on Climate Change), Cambridge, 2005.
2. R. Steeneveldt, CO_2 capture and storage: Closing the knowing-doing gap, *Chem. Eng. Res. Des.*, 2006, **84**, 739–763.
3. A. Gabelman, Hollow fiber membrane contactors, *J. Membr. Sci.*, 1999, **159**, 61–106.
4. P. Feron, CO_2 separation with polyolefin membrane contactors and dedicated absorption liquids: Performances and prospects, *Sep. Purif. Technol.*, 2002, **27**, 231–242.

5. S. Stern, Polymers for gas separations: The next decade, *J. Membr. Sci.*, 1994, **94**, 1–65.

6. M. Sandru, High molecular fixed-site-carrier PVAm membrane for CO$_2$ capture, *Desalination*, 2009, **240**, 298–300.

7. J. Zou, CO$_2$-selective polymeric membranes containing amines in cross-linked poly(vinyl alcohol), *J. Membr. Sci.*, 2006, **286**, 310–321.

8. S. Reijerkerk and S. Highly hydrophilic, rubbery membranes for CO$_2$ capture and dehydration of flue gas, *Int. J. Greenhouse Gas Contr.*, 2011, **1**, 26–36.

9. D. W. Barry, Design of Cortez CO$_2$ system detailed, *Oil Gas J.*, 1985, **83**, 96–104.

10. J. Johnson, Gas processing needs for EOR, *Hydrocarbon Process*, 1985, **64**(10), 62–66.

11. E. de Visser *et al.*, Dynamis CO$_2$ quality recommendations, *Int. J. Greenhouse Gas Contr.*, 2008, **2**, 478–484.

12. J. Lie, Optimization of a membrane process for CO$_2$ capture in the steel-making industry, *Int. J. Greenhouse Gas Contr.*, 2007, **1**, 309–317.

13. H. Herzog, What future for carbon capture and sequestration?, *Environ. Sci. Technol.*, 2001, **35**, 148A–153A.

14. A. Aspelund, Gas conditioning-The interface between CO$_2$ capture and transport, *Int. J. Greenhouse Gas Contr.*, 2007, **1**, 343–354.

15. D. Coker, Modeling multicomponent gas separation using hollow-fiber membrane contactors, *AIChE J.*, 1998, **44**, 1289–1300.

16. D. Coker, Nonisothermal model for gas separation hollow-fiber membranes, *AIChE J.*, 1999, **45**, 1451–1468.

17. S. Weller, Steiner, Separation of gases by fractional permeation through membranes, *J. Appl. Phys.*, 1950, **21**, 279–283.

18. R. Rautenbach, *Membrane Processes*, Wiley, Chichester, 1989.

19. S. Kaldis, Simulation of binary gas separation in hollow fiber asymmetric membranes by orthogonal collocation, *J. Membr. Sci.*, 1998, **142**, 43–59.

20. D. Chang, Perturbation solution of hollow-fiber membrane module for pure gas permeation, *J. Membr. Sci.*, 1998, **143**, 53–64.

21. R. Rautenbach, Auslegung von Membranprozessen mit dem Simulationsprogramm Aspen Plus-Teil 1: Optimierung eines Benzindampfrückgewinnungsprozesses. *Chem.-Ing.-Tech.*, 1996.

22. A. Kovvali, Models and analyses of membrane gas permeators, *J. Membr. Sci.*, 1992, **73**, 1–23.

23. E. Favre, Carbon dioxide recovery from post-combustion processes: Can gas permeation membranes compete with absorption?, *J. Membr. Sci.*, 2007, **294**, 50–59.

24. E. Favre, CO$_2$/N$_2$ reverse selective gas separation membranes: Technological opportunities and scientific challenges, *Ind. Eng. Chem. Res.*, 2009, **48**, 3700–3701.

25. L. Robeson, Correlation of separation factor versus permeability for polymeric membranes, *J. Membr. Sci.*, 1991, **62**, 165–185.

26. T. Merkel, Power plant post-combustion carbon dioxide capture: An opportunity for membranes, *J. Membr. Sci.*, 2010, **359**, 126–139.
27. W. Koros, Membrane-based gas separation, *J. Membr. Sci*, 1993, **83**, 1–80.
28. H. Herzog, Feasibility, modeling and economics of sequestering power plant CO_2 emissions in the deep ocean, *Enviro. Prog.*, 1991, **10**, 64–74.
29. J. Van Der Sluijs, Feasibility of polymer membranes for carbon dioxide recovery from flue gases, *Energy Conversion Manage*, 1992, **33**, 429–436.
30. M. Hägg, CO_2 capture from natural gas fired power plants by using membrane technology, *Ind. Eng. Chem. Res.*, 2005, **44**, 7668–7675.
31. R. Bounaceur, R. Membrane processes for post-combustion carbon dioxide capture: A parametric study, *Energy*, 2006, **31**, 2220–2234.
32. M. Ho, Economies of CO_2 and mixed gas geosequestration of flue gas using gas separation membranes, *Ind. Eng. Chem. Res.*, 2006, **45**, 2546–2552.
33. L. Zhao, A parametric study of CO_2/N_2 gas separation membrane processes for post-combustion capture, *J. Membr. Sci.*, 2008, **325**, 284–294.
34. J. Kotowicz, The influence of membrane CO_2 separation on the efficiency of a coal-fired power plant, *Energy*, 2010, **35**, 841–850.
35. A. Brunetti, Membrane technologies for CO_2 separation, *J. Membr. Sci.*, 2010, **359**, 115–125.
36. R. Baker, *Membrane Technology and Applications*, J. Wiley, Chichester, 2004.
37. S. Matson, Separation of gases with synthetic membranes, *Chem. Eng. Sci.*, 1983, **38**, 503–524.
38. L. Zanderighi, Evaluation of the performance of multistage membrane separation cascades, *Sep. Sci. Technol.*, 1996, **31**, 1291–1308.
39. M. Ho, Reducing the cost of CO_2 capture from flue gases using membrane technology, *Ind. Eng. Chem. Res.*, 2008, **47**, 1562–1568.
40. L. Zhao, Multi-stage gas separation membrane processes used in post-combustion capture: Energetic and economic analyses, *J. Membr. Sci.*, 2010, **359**, 160–172.
41. A. Hussain, A feasibility study of CO_2 capture from flue gas by a facilitated transport membrane, *J. Membr. Sci.*, 2010, **359**, 140–148.
42. H. Lin, Materials selection guidelines for membranes that remove CO_2 from gas mixtures, *J. Mol. Struct.*, 2005, **739**, 57–74.
43. W. Yave, CO_2-Philic polymer membrane with extremely high separation performance, *Macromolecules*, 2010, **43**, 326–333.
44. H. Lin, Effect of copolymer composition, temperature, and carbon dioxide fugacity on pure- and mixed-gas permeability in poly(ethylene glycol)-based materials: Free volume interpretation, *J. Membr. Sci.*, 2007, **291**, 131–139.
45. A. Ebner, State-of-the-art adsorption and membrane separation processes for carbon dioxide production from carbon dioxide emitting industries, *Sep. Sci. Technol.*, 2009, **44**, 1273–1421.

46. M. Pera-Titus, Nanocomposite MFI-alumina membranes: High-flux hollow fibers for CO$_2$ capture from internal combustion vehicles, *Ind. Eng. Chem. Res.*, 2009, **48**, 9215–9223.

47. G. Xomeritakis, Microporous sol-gel derived aminosilicate membrane for enhanced carbon dioxide separation, *Sep. Purif. Technol.*, 2005, **42**, 249–257.

48. R. Krishna, In silico screening of zeolite membranes for CO$_2$ capture, *J. Membr. Sci.*, 2010, **360**, 323–333.

49. M. Bram, Testing of nanostructured gas separation membranes in the flue gas of a post-combustion power plant, *Int. J. Greenhouse Gas Contr.*, 2011, **1**, 37–48.

50. L. Robeson, The upper bound revisited, *J. Membr. Sci.*, 2008, **320**, 390–400.

51. B. Freeman, Basis of permeability/selectivity tradeoff relations in polymeric gas separation membranes, *Macromolecules*, 1999, **32**, 375–380.

52. L. Deng, PVA/PVAm blend FSC membrane for CO$_2$-capture, *Desalination*, 2006, **199**, 523–524.

53. D. Grainger, Techno-economic evaluation of a PVAm CO$_2$-selective membrane in an IGCC power plant with CO$_2$ capture, *Fuel*, 2008, **87**, 14–24.

54. J. Huang, Carbon dioxide capture using a CO$_2$-selective facilitated transport membrane, *Ind. Eng. Chem. Res.*, 2008, **47**, 1261–1267.

55. M. Trachtenberg, Carbon dioxide transport by proteic and facilitated transport membranes, *Life Support Biosphere Sci.: Int. J. Earth Space*, 1999, **6**, 293–302.

56. J. Potreck, Mixed water vapor/gas transport through the rubbery polymer PEBAX$^{®}$ 1074, *J. Membr. Sci.*, 2009, **338**, 11–16.

57. C. Pan, Permeation of water vapor through cellulose triacetate membranes in hollow fiber form, *J. Appl. Polym. Sci.*, 1978, **22**, 2307–2323.

58. C. Scholes, Effects of minor components in carbon dioxide capture using polymeric gas separation membranes, *Sep. Purif. Rev.*, 2009, **38**, 1–44.

59. R. McKee, CO$_2$ removal: membrane plus amine, *Hydrocarbon Process*, 1991, **70**(4), 63–65.

60. E. Favre, A hybrid process combining oxygen enriched air combustion and membrane separation for post-combustion carbon dioxide capture, *Sep. Purif. Technol.*, 2009, **68**, 30–36.

61. E. Favre, Biogas, membranes and carbon dioxide capture, *J. Membr. Sci.*, 2009, **328**, 11–14.

Physical Aging of Membranes for Gas Separations

B.W. ROWE,[a,b] B.D. FREEMAN[a] AND D.R. PAUL[*a]

[a] Department of Chemical Engineering, Texas Materials Institute and Center for Energy and Environmental Resources, The University of Texas at Austin, Austin, Texas 78712, USA; [b] Polymers Division, National Institute of Standards and Technology, Gaithersburg, Maryland 20899, USA

3.1 Introduction

The polymers that comprise most high-performance gas separation membranes are amorphous, glassy materials.[1] These materials are selected based on their superior permeability and selectivity characteristics and their ability to form rigid structures that enable the development of self-supporting architectures, such as asymmetric hollow fibers.[2–4] The performance of gas separation membranes change with time due to the inherent non-equilibrium nature of glassy polymeric materials. As non-equilibrium materials, glassy polymers undergo physical aging, a spontaneous, but typically slow, evolution toward an equilibrium state.[5–7] However, the changes in material properties caused by physical aging are thermo-reversible, *i.e.* the changes can be reversed by heating the sample above the glass transition of the material. This important feature of physical aging separates the phenomenon from other factors that may influence behavior over time such as chemical aging, degradation, or contamination. Extensive studies on the influence of physical aging on glassy polymer behavior, *e.g.* mechanical properties, volume relaxation, and thermodynamic behavior, have been reported in the literature.[6–11] This chapter does not attempt to

Membrane Engineering for the Treatment of Gases, Volume 1:
Gas-separation Problems with Membranes
Edited by Enrico Drioli and Giuseppe Barbieri

Published by the Royal Society of Chemistry, www.rsc.org

summarize the entire field of physical aging, but will focus on the influence of aging on transport behavior of gas separation membranes and related material properties. Aging in glassy polymers causes permeability to decrease over time, requiring additional membrane area to treat a given process stream than originally needed based on initial performance. It is essential to understand the influence of physical aging on the performance of gas separation membranes when designing a membrane system to ensure satisfactory performance over long service periods.

Over the last several decades, research and development of selective ultra-thin membranes has gained tremendous momentum in an attempt to compete with alternative separation technologies, such as pressure swing adsorption and cryogenic distillation.[12–14] Because membrane permeance is inversely proportional to thickness, the development of ultra-thin active selective layers is required to maximize membrane productivity. The membrane support structure must also be optimized to minimize additional mass transfer resistance.[15] Commercially, the active separating layer of gas separation membranes can be on the order of 100 nm, with ongoing efforts to reduce this layer thickness to further increase membrane permeance. Ultra-thin selective layers are most commonly produced by coating a dilute polymer solution onto a microporous support to form composite membranes or by solution spinning techniques to produce asymmetric hollow fiber membranes.[4,16] Creating these thin film structures with limited defects is often the dominant challenge in membrane manufacturing, due to the extremely fragile nature of these ultra-thin films. A nearly defect-free film is imperative for high selectivity since even a small area fraction, on the order of 10^{-6}, can destroy the native gas selectivity of a material.[17]

The influence of film thickness and confinement on polymer behavior is an active research topic in polymer physics due to the importance of these factors in the development of a diverse range of technologies, including nanocomposite materials, microelectronics, coatings, and optics.[18–23] Although there is some debate regarding the influences of sample preparation and experimental conditions on the observed confinement effects, results indicate that the glass transition temperature of ultra-thin films can deviate substantially from bulk values.[24,25] These differences have been attributed to interfacial interactions between the polymer and substrate, or free surface.[26,27] Free standing films or films on non-attractive substrates exhibit decreasing glassy transition temperature (T_g) with film thickness, while strongly attractive substrates have been shown to increase the T_g of thin films. Furthermore, interfacial interactions have also been shown to influence aging behavior. For example, slower aging was observed for a 20 nm thick poly(methyl methacrylate) (PMMA) film on a silicon substrate as compared to 500 nm thick films; the reduced aging rate was attributed to restricted polymer chain mobility induced by attractive interfacial interactions.[28]

Thick film, *i.e.* bulk, samples are typically used when screening the transport properties of polymeric membrane materials, due to ease of handling and creation of defect-free films. Except for ultra-high free volume polymers such as

Table 3.1   Influence of film thickness and aging time on the transport properties of Matrimid®[31]

Matrimid®		*Bulk* *(Literature values)*	*550 nm*		*18 nm*	
			1 hr	*1000 hr*	*1 hr*	*1000 hr*
Permeability (Barrer)	O_2	2.12	3.35	1.95	1.68	0.57
	N_2	0.32	0.55	0.28	0.23	0.07
	CH_4	0.28	0.47	0.21	0.18	0.05
Selectivity	O_2/N_2	6.6	6.1	7.0	7.3	8.1
	N_2/CH_4	1.14	1.17	1.33	1.28	1.40

poly[1-(trimethyl-silyl)propyne] (PTMSP), which ages rapidly even in the bulk state, the transport properties of these thick glassy polymer films do not change dramatically with time due to physical aging.[29,30] Therefore, the reported permeabilities for a given polymer are typically within an acceptable range to allow comparison with other materials. However, when the glassy material is prepared in an ultra-thin film structure for industrial application, the polymer properties can differ significantly from bulk behavior. Table 3.1 illustrates the influence of film thickness and aging time on the gas transport properties of Matrimid a commercially available polyimide of interest for gas separation applications.[31]

Depending on film thickness, the initial permeability coefficients, measured at 1 h of aging, can be above or below the reported bulk values. After aging for approximately 1000 h, both thin films have lower permeability than the bulk. In the case of the 18 nm film, O_2 permeability decreased to ~25% of the bulk value after 1000 h of aging. These aging induced changes increase with penetrant size, as the larger molecules are more strongly affected by the loss of free volume that occurs during aging. This behavior causes pure gas selectivity to increase with aging, as seen in Table 3.1. Clearly, large deviations from bulk properties exist in ultra-thin films, and predicting the behavior of ultra-thin films from bulk measurements is only a rough approximation of the true behavior. These differences also highlight possible errors in the current methods used to estimate the selective layer thickness of asymmetric membranes using bulk permeability values. Furthermore, because these deviations from bulk behavior are not well understood, accurately predicting ultra-thin film behavior is particularly challenging. The goals of this chapter are to describe the recent discoveries related to accelerated aging in thin glassy films, highlight the advances in experimental techniques that have enabled these studies, and to discuss the current state of understanding of this phenomenon and the questions and opportunities that remain.

3.2   Aging Behavior in Thin and Ultra-thin Films

The influence of thickness on aging behavior in gas separation membrane materials was first reported in the early 1990s; however, these aging effects may have also been partially responsible for other time-dependent transport properties described in earlier reports.[32,33] For instance, although not originally considered as a controlling factor, physical aging may have contributed to the flux decline in asymmetric cellulose acetate membranes described by Baayens and Rosen.[34] Early reports on the influence of thickness on aging behavior compared bulk behavior to thin, complex membrane structures with ill-defined thermal histories and lacked accurate methods to determine film thickness. Rezac *et al.* reported the influence of aging time on the gas transport properties of thin and thick films of 6FDA-IPDA polyimide and tetramethyl-hexafluoropolycarbonate (TMHFPC).[32] While a 71 µm thick film of the 6FDA-IPDA polyimide maintained 98% of the initial nitrogen flux after 2 months of aging, the nitrogen flux through a thin film was reduced to 20% of the initial value after just 20 days of aging. The thickness of the thin film, which was formed on a porous ceramic support, was estimated to be ~300 nm based on the initial permeance. This thickness was reportedly in fair agreement with values estimated from scanning electron microscopy (SEM) measurements and material balance calculations, *i.e.* thickness calculated based on the mass deposited and the polymer density. While the methods of thickness determination were approximate, the difference in the aging behavior of the thin film as compared to the bulk state was a clear indication that physical aging is influenced by film thickness. This work also showed the initial selectivity of the thin films started below the bulk value, but increased beyond the bulk value with aging time. Other studies showed similar influences of film thickness on aging behavior by comparing the time-dependent transport properties of thin-skinned asymmetric membranes and bulk films.[33,35]

Several experimental advances have enabled more systematic studies on the influence of film thickness on physical aging. Accurate film thickness measurements are essential in developing a fundamental understanding of these effects. Film thickness measurements of ceramic composite membranes using SEM were improved over previous techniques by comparing the back-scattered electron image, which only shows the higher atomic mass ceramic support layer, to the secondary electron image, which shows the support and film layers.[36] More recently, further improvement in thickness measurements were realized by using variable angle spectroscopic ellipsometry.[37] Ellipsometry measures the changes in elliptically polarized light that is reflected through a sample and can provide detailed information about thin films and surfaces. Although this technique has been applied in other fields, such as microelectronics, for some time, it has not been widely used to characterize gas separation membranes until recently. The application of this technique to the membrane field was a significant development in the ability to characterize thin films. In addition to very accurate thickness measurements, ellipsometry can provide optical properties, such as refractive index, that can be related to physical properties as another method for tracking physical aging.[38,39] The use of ellipsometry to study physical aging will be discussed in more detail later in the chapter. These improvements in thin film characterization also aided in the development of models for describing the aging response of thin glassy films.[40]

Because the state of a glassy polymer and, therefore, its aging behavior, depend on the material's previous history in addition to the immediate experimental conditions, it is important to consider how materials are treated before testing in physical aging studies.[41,42] Unless otherwise specified, the results presented in this chapter were collected using films created by spin-coating that were subsequently heated above their glass transition temperature in a free standing state to erase any influence of previous history and then rapidly quenched to the aging temperature to define a reproducible starting time for the aging experiments. While these procedures may exaggerate the physical aging response by beginning with the material in a high free volume state, the consistency in the initial conditions is critical. Some discussion on the influences of different thermal histories and experimental procedures on physical aging in thin films will be given in Section 3.4. The chemical structures and bulk material properties of the polymers investigated in this chapter are given in Table 3.2.

Table 3.2   Bulk material properties

Polymer	T_g	P_{O_2}	P_{N_2}	P_{CH_4}	f_b
Polysulfone	186°C	1.4	0.24	0.29	0.144
Matrimid®	317°C	2.12	0.32	0.28	0.170
PPO	210°C	14.6	3.5	4.1	0.183
6FDA-DAM	372°C	55	17	15.1	0.19
PTMSP	>280°C	9000	6600	15000	0.29

Permeability values are given in Barrers: 1 Barrer $= 1 \times 10^{-10}$ [cm^3(STP)·cm/(cm^2·s·cmHg)]
PPO = poly(2,6-dimethyl-1,4-phenylene oxide).

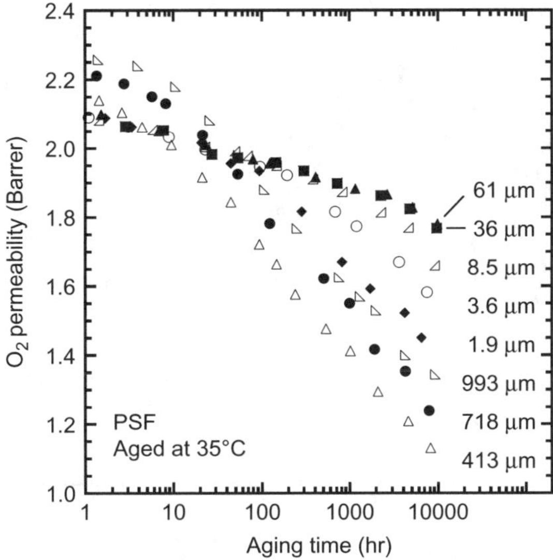

Figure 3.1 Influence of aging time on oxygen permeability coefficients in PSF films with thicknesses ranging from 61 μm (*i.e.* bulk) to 413 nm. Reproduced with permission of Elsevier.[43]

These polymers were chosen based on their importance in the gas separation field and because they are deep in the glassy state, *i.e.* far below T_g, at the temperatures of interest, which is approximately ambient for most applications.

The foundation for understanding how film thickness influences physical aging behavior begins with a systematic study on films across a broad thickness range with identical thermal histories. Figure 3.1 shows the influence of physical aging at 35 °C on the O_2 permeability coefficient of polysulfone (PSF) films with thicknesses (*l*) ranging from 61 μm to 413 nm.[43] While the relatively thick films, *i.e.* $l > 10$ μm, show only a modest decrease in permeability with aging time, the thin films, *i.e.* $l < 10$ μm, are strongly affected by physical aging.

As film thickness is reduced below 10 μm, the rate of O_2 permeability loss is greatly accelerated as compared to the thick film behavior. Additionally, the initial permeability of the thin films is higher than the bulk value; this result is presumed to be a result of the higher free volume state of the thin films after the rapid quench from above T_g. Surprisingly, while changes in the glass transition typically are not observed in films greater than ~100 nm in thickness, physical aging can depend on dimensions that are on the order of microns. Similar to the results in Table 3.1, the pure gas selectivity through these films increased with physical aging as the material became more size selective due to the loss of free volume.

In addition to the results presented for PSF, analogous thickness-dependent aging behavior has been reported for other polymers.[36,43,44] Figure 3.2 compares the influence of aging time on the relative oxygen permeability of

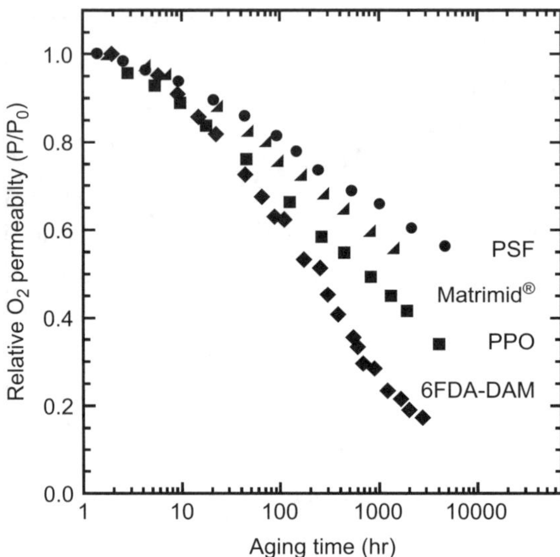

Figure 3.2 Influence of aging time on relative oxygen permeability (based on permeability after 1 h of aging) in thin films of various polymers about 400 nm thick.

~400 nm thick films made from PSF, Matrimid, poly(2,6-dimethyl-1,4-phenylene oxide) (PPO), and 6FDA-DAM.

Again, the chemical structures and bulk properties, including fractional free volume, of these polymers are given in Table 3.2. Each of these materials shows a strong decrease in permeability with aging time when prepared as a thin film, and the magnitude of the aging-induced changes is highly dependent on the polymer structure.

In the case of PSF, the relative oxygen permeability decreased to 60% of the initial value after ~2000 h of aging, while the 6FDA-DAM polyimide retained less than 20% of its initial permeability after a similar aging period. The physical aging rate, as determined by the rate of relative permeability decline, of these polymers follows the same trend as the bulk fractional free volume, *i.e.* 6FDA-DAM > PPO > Matrimid > PSF. This trend could be expected, as the rate of physical aging is intimately related to the excess free volume in a material.[40] This comparison highlights the current need to study each material of interest to understand their behavior in thin film form. If connections between the bulk polymer properties and thin film behavior can be defined, better predictions of the influence of physical aging effects in thin films could be developed.

The extreme difficulty in creating and handling delicate thin films for gas permeation measurement early on limited the study of single-layer free standing films to a minimum thickness of 300–400 nm. Microscopic pinhole defects that destroy the selective of thin polymer membranes form with increasing frequency as film thickness is decreased.[45] To overcome this challenge, a

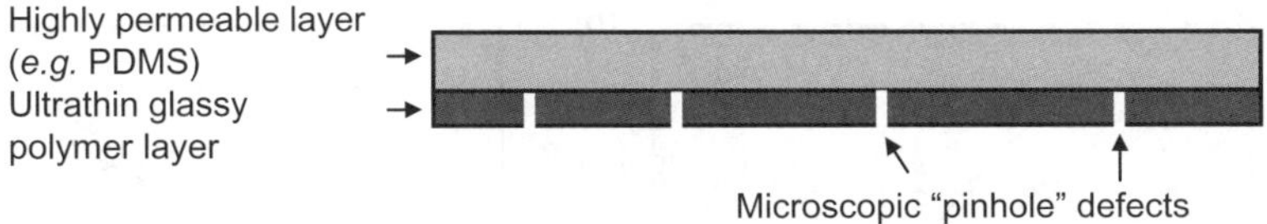

Figure 3.3   Diagram of film structure used by Rowe *et al.*, which enabled the study of gas permeability properties and physical aging behavior of ultra-thin glassy films. Reproduced with permission of Elsevier.[31]

coating technique, similar to that which initially facilitated the application of polymeric membranes to industrial application, was developed.[17,46] The thin glassy films were prepared using the spin coating methods described previously. Before removing the films from the silicon substrate, a thin layer of highly permeable poly(dimethylsiloxane) (PDMS) was coated directly on top of the thin glassy layer.

Figure 3.3 presents a schematic of the film structure utilized in this procedure. This rubbery layer effectively blocks convective flow through any pinhole defects but adds very little mass transfer resistance due to its high permeability.[47] Furthermore, the additional mass transfer resistance, which is constant over time for the rubbery layer, can be accounted for using a series resistance model:[15]

$$\frac{l_{composite}}{P_{composite}} = \frac{l_{PDMS}}{P_{PDMS}} + \frac{l_{Glassy}}{P_{Glassy}} \tag{3.1}$$

where l_{PDMS} and l_{Glassy} are the thicknesses of the PDMS and glassy polymer layers, respectively, and P_{PDMS} and P_{Glassy} are the permeability coefficients of the PDMS and glassy polymer, respectively. The total thickness of the composite film is $l_{composite} = l_{PDMS} + l_{Glassy}$, and the permeability of the composite film is $P_{composite}$. By measuring the thickness of each layer and knowing the permeability of PDMS, eqn (3.1) can be used to calculate the permeability of the glassy layer from measurements on the composite film.

The development of this coating technique has enabled the study of ultra-thin films with thicknesses that are of interest in gas separation membrane applications.[31] Figure 3.4(a) presents the influence of aging time on the oxygen permeability coefficients of Matrimid films down to 18 nm in thickness prepared with the coating technique. Note that the quoted film thickness is the thickness of the glassy layer.

The films were aged at 35 °C, which is 280 °C below the T_g of bulk Matrimid, in a dry environment between measurements. Despite being far below T_g, dramatic aging effects on gas permeability are evident in the ultra-thin films. The oxygen permeability decreased rapidly with aging time, to $\sim$50% of the initial value after 1000 h, in all films as the material evolved towards an equilibrium state. Interestingly, the initial permeability coefficient measured at 1 h of aging time decreases with film thickness in the ultra-thin films. This initially lower gas permeability is attributed to the rapid physical aging

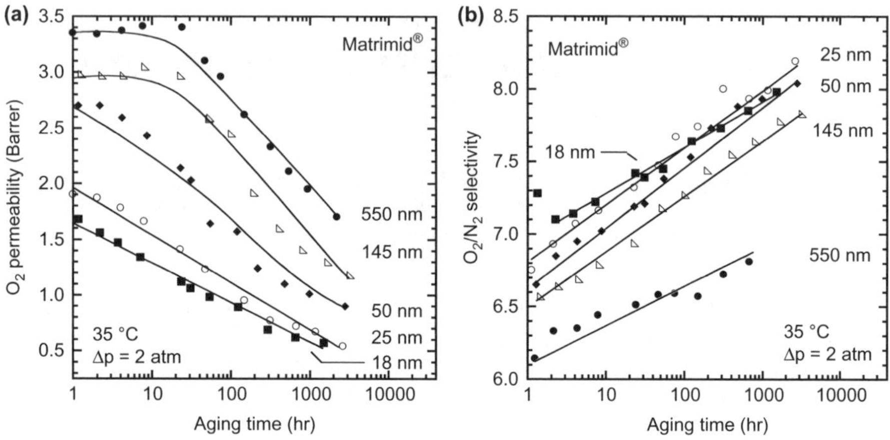

Figure 3.4 Influence of aging time on (a) oxygen permeability coefficients, and (b) O_2/N_2 pure gas selectivity in Matrimid® films with thicknesses ranging from 550 nm to 18 nm. Reproduced with permission of Elsevier.[31]

that occured during the first hour after the quench from above T_g; this time is required to prepare the sample for permeation testing using the current experimental techniques hence no data can be collected before 1 h. The notion that the initially measured permeabilities are different for the ultra-thin films is supported by recent modeling work, which will be described in Section 3.5.[31]

Figure 3.4(b) shows the O_2/N_2 pure gas selectivity as a function of aging time for the ultra-thin Matrimid films.[31] All films exhibit selectivities that are near or above the bulk selectivity value, 6.6, indicating the films are essentially defect free. The O_2/N_2 selectivity increases with physical aging, as expected; the reduction in free volume caused by physical aging makes the material more size selective by reducing the diffusion of the larger N_2 molecules more than that of smaller O_2 molecules. The increase in selectivity is fairly substantial; for instance, the O_2/N_2 selectivity of the 25 nm thick Matrimid film increases from an initial value of 6.75 to above 8.0 after 1000 h of aging. Additionally, selectivity increases with decreasing film thickness, consistent with the observation of decreasing permeability in thinner films described previously.

The ability to study ultra-thin films provides important guidelines for understanding the behavior of membranes used in gas separation applications. These films, with thicknesses similar to estimates of the skin layer in asymmetric hollow fibers, demonstrate that deviations from bulk behavior continue as film thickness is reduced. Of course, some questions remain regarding the behavior of more complex membrane structures and applications, *e.g.* how does the underlying structure of an asymmetric membrane affect aging, and how does exposure to high pressure multi-component feed streams with highly sorbing penetrants impact long-term performance? Due to the immense complexity of these issues, careful development of current and future experimental techniques, in addition to improved theoretical models, is essential to fully understand how these factors influence thin film behavior.

3.3 Additional Experimental Methods used to Study Physical Aging

In addition to providing accurate measurements of film thickness, ellipsometry can also be used to determine the optical properties of thin films, *e.g.* refractive index, thus providing additional means to track the influence of aging on film properties.[38,39,48,49] The Lorentz–Lorenz equation provides a fundamental connection between refractive index and density:[50]

$$\frac{\bar{n}^2 - 1}{\bar{n}^2 + 2} = \frac{\rho N_{av}\alpha}{3M_0\varepsilon_0} \tag{3.2}$$

where ρ is the polymer density, N_{av} is Avogadro's number, α is the average polarizability of the polymer repeat unit, M_0 is the polymer repeat unit molecular weight, and ε_0 is the permittivity of free space. Only refractive index and density are anticipated to change with physical aging, reducing the Lorentz–Lorenz equation to:

$$L = \frac{\bar{n}^2 - 1}{\bar{n}^2 + 2} = \rho C \tag{3.3}$$

where L is defined as the Lorentz–Lorenz parameter. Because the specific refraction, C, is constant during physical aging, this relationship can be used to compare the relative density of materials during physical aging from measured refractive index values, even if the specific refraction is not known. Specific refraction parameters can be calculated from bulk values of density and refractive index, allowing density to be calculated as a function of aging time. Furthermore, density can be related to fractional free volume, FFV, as follows:

$$\text{FFV} = \frac{v - v_o}{v} = 1 - \rho v_o \tag{3.4}$$

where v is specific volume, (equivalent to $1/\rho$), and v_o is the occupied volume, which can be estimated by the Bondi method (*i.e.* $v_o = 1.3v_w$, where v_w is the van der Waals volume estimated using the group contribution method).[51] This method to determine free volume can be used to track free volume changes over time and aid in the development of theoretical physical aging models.

Figure 3.5 presents the relative density, calculated using eqn (3.3), of 400 nm thick films made from four different polymers as a function of aging time at 35 °C.[38,49] The results from these ellipsometry studies directly show that these thin films densify during the physical aging process. The rate of densification depends strongly on the polymer type. Again, the rate of aging follows the same

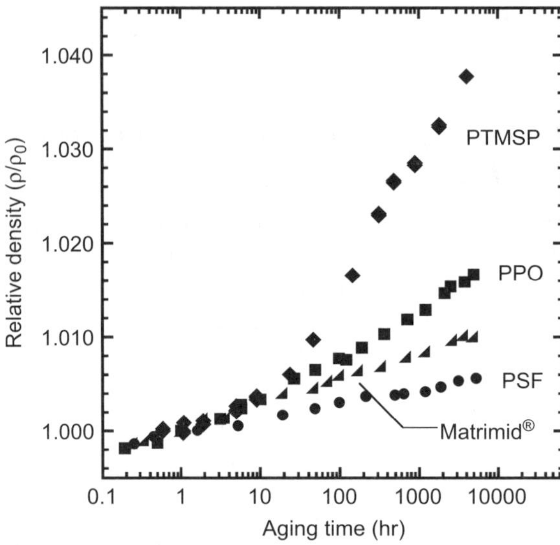

Figure 3.5   Influence of aging time on relative density (calculated using the Lorentz–
Lorenz equation and based on density after 1 h of aging) in thin films of
various polymers about 400 nm thick.

order as the fractional free volume, listed in Table 3.2, *i.e.* poly(1-trimethylsilyl-1-propyne) (PTMSP) > PPO > Matrimid > PSF. The ultra-high free volume polymer, PTMSP, shows significantly faster physical aging than the other glassy materials. The density of PTMSP increases by nearly 4% from its initial value after several thousand hours of aging, while the other materials show density increases of 0.5–1.5%. These films were all approximately 400 nm thick and prepared by spin coating onto silicon wafer supports. Three of the polymers, PSF, PPO, and Matrimid where heated above their T_g prior to the aging studies; however, PTMSP decomposes before reaching its glass transition.[52] Therefore, the PTMSP film did not begin the aging experiment from a freshly quenched state and was heated to 180 for 90 min to remove any residual solvent prior to the aging study; both of these factors would presumably retard the aging response of a material in comparison to the freshly quenched state. Considering these factors, PTMSP might be expected to age even faster than represented by these results if the experiments were conducted using a PTMSP sample with higher initial free volume.

In addition to characterizing the influence of aging on density, results from aging studies using ellipsometry have also been shown to correlate well with gas permeability measurements.[40,44] Direct comparison of volumetric aging rates determined by ellipsometry to rates of permeability decline shows a strong correlation to the properties probed using these two different techniques.

Another experimental technique that has been useful in studying physical aging of glassy polymers is positron annihilation lifetime spectroscopy

(PALS).[9,10,53,54] PALS provides unique information about the free volume properties of amorphous materials by measuring the lifetime of injected positrons.[55,56] The lifetime of ortho-positronium (*o*-Ps), τ_3, can be related to the size of free volume elements in the sample by assuming a spherical cavity shape and using the relationship given in eqn (3.5):

$$\tau_3 = \frac{1}{2}\left[1 - \frac{r}{r + \Delta r} + \frac{1}{2\pi}\sin\left(\frac{2\pi r}{r + \Delta r}\right)\right]^{-1} \tag{3.5}$$

where r is the cavity radius and Δr is the e$^-$ layer thickness (1.656 Å).[57,58] Furthermore, the relative number of positrons that annihilate as *o*-Ps, the *o*-Ps intensity (I_3), is related to the concentration of free volume elements in a material.

While PALS has been used in numerous studies to investigate physical aging of polymeric membrane materials, the standard technique is limited to studying bulk samples because of the high energy of the source positrons.[56] By combining positron annihilation lifetime spectroscopy with a slow mono-energetic positron beam, thin films of interest for gas separation applications can be studied. Additionally, the energy, and therefore, implantation depth, of the positron beam can be controlled, allowing the free volume properties of a film to be studied as a function of distance from the free surface.[59,60] The ability to probe the free volume property profile as a function of aging time provides vital information for developing and validating mechanistic physical aging models. In addition to aging studies on dense thin films, this variable energy PALS technique has been used to study asymmetric thin film structures.[61–63] The ability to study samples in a wet condition, during what would typically be a high vacuum experiment, was recently achieved by using a plasma-enhanced chemical vapor deposition technique to seal the sample prior to study.[63] These unique experimental techniques provide important information for understanding the influences of film thickness and interfacial interactions on membrane behavior by probing material properties that are difficult or impossible to measure by other means.

Figure 3.6 shows the average *o*-Ps lifetime, τ_3, as a function of mean implantation depth for thin PSF films ($l = 450$ nm) after aging for various times on silicon wafer supports.[64] Each lifetime profile represents a different PSF film that was aged at 35 °C for the time indicated in the figure. The *o*-Ps lifetime decreases with physical aging at all implantation depths. The decrease in τ_3 corresponds to a decrease in the average free volume element size with aging, which is expected as the material approaches a more dense, equilibrium state. Interestingly, the average free volume element size, as related to τ_3, is reduced near the film surface as compared to the center of the film. This intriguing result is thought to be caused by the initially rapid physical aging that occurs near the film surface, thus reducing τ_3 near the surface as compared to the remainder of the film. While the τ_3 profile is not symmetric about the center of the film, the *o*-Ps lifetime is also lower near the silicon substrate as compared to the film center. This lack of symmetry in the apparent τ_3 profile may be related to

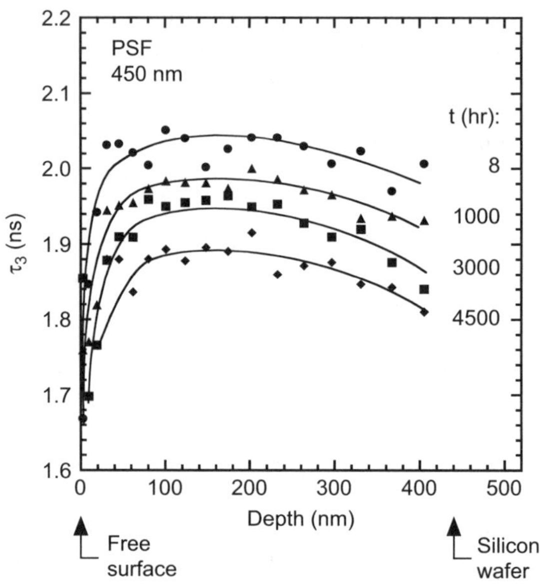

Figure 3.6 Influence of positron implantation depth and physical aging on *o*-Ps lifetime in thin PSF films. Lines are provided to guide the eye. Reproduced with permission of Elsevier.[64]

the change in implantation depth probability with beam energy, *i.e.* the depth–sensitivity of the positron beam decreases as the implantation depth increases.[65] It should be noted that while these films were studied on a silicon support, they were prepared and aged in the free standing state to allow comparison with results from gas permeability studies. Furthermore, no influence of the silicon support on aging behavior was detected using this technique by studying films that had been aged with and without the silicon support.[64]

The variable energy PALS measurements also gave information on the *o*-Ps intensity, which can be related to the concentration of free volume elements in a sample, as a function of aging time and implantation depth; however, no influence of aging on the *o*-Ps intensity profile was detected.[64] The constant value of I_3 with aging time suggests that the loss of free volume in these thin films with aging was caused solely by shrinking of the free volume elements. This result indicates that the accelerated aging in these thin films, as compared to the bulk material, is caused by a thickness-dependent lattice contraction type mechanism that allows the material near the surface to age more rapidly than the bulk, thus resulting in the lower free volume element sizes near the film surface. Although diffusion based models can often describe thin film aging behavior, this type of mechanism would suggest that physical aging causes a reduction in the number of elements, and no significant change in the con-centration of free volume elements was measured using this technique.

The depth profile of free volume properties measured using variable energy PALS provides an opportunity to investigate the aging behavior as a function of depth from the free surface. The average *o*-Ps lifetime across a specific section of the film, $\bar{\tau}_3$ can be calculated by:

$$\bar{\tau}_3 = \frac{\sum(\tau_{3i} \times \Delta l_i)}{l} \tag{3.6}$$

where τ_{3i} is the *o*-Ps lifetime across distance Δl_i and l is the total thickness of the film section of interest. Figure 3.7 illustrates the influence of physical aging on the average *o*-Ps lifetime across three different sections of the PSF film, *i.e.* 50 nm from the free surface, 125 nm from the free surface, and the entire 450 nm film.

The sections of interest are highlighted in the inset of Figure 3.7. Initially, at 8 h of aging, the average *o*-Ps lifetime decreases with decreasing penetration depth. This result is attributed to the rapid aging that occurs in the near-surface region of the film during the first 8 h of aging, and is consistent with observations from the gas permeability studies, *i.e.* the initial gas permeability decreased with decreasing film thickness. Due to the time required to measure the positron lifetime spectra, the first measurements were conducted after 8 h of aging to minimize any changes that may occur over the experimental time scale due to physical aging. The average *o*-Ps lifetime decreased similarly in each section of the film from 8 to 3000 h of aging.

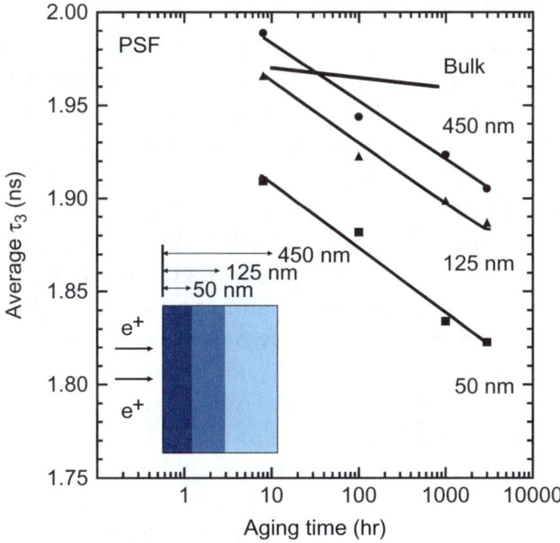

Figure 3.7	Influence of physical aging on the *o*-Ps lifetime averaged across different penetration depths of a 450 nm thick PSF film. Bulk aging behavior included for comparison. Inset schematic shows the film regions of interest. Lines are provided to guide the eye. Reproduced with permission of Elsevier.[64]

As seen from these results, variable energy PALS can provide useful information in determining the influences of film thickness and interfacial interactions on physical aging behavior. Furthermore, free volume properties, calculated from the variable energy PALS results, were used to estimate the gas transport behavior of the different sections of the PSF film.[64] Good agreement between the PALS-estimated permeabilities and experimentally measured values highlights the connection between changes in free volume and transport properties caused by physical aging.

Numerous other experimental techniques have been used to investigate the properties and behavior of thin polymeric films, including broadband dielectric spectroscopy, fluorescence spectroscopy, X-ray and neutron reflectivity, and more.[24,66–68] While most of these techniques have not been utilized to investigate the physical aging behavior of gas separation membrane materials, they offer unique perspectives into thin film behavior. Fluorescence spectroscopy has been used to study physical aging of thin polymeric films of PMMA and polystyrene.[28] Furthermore, clever experimental design and sample preparation techniques also enabled the study of polymer properties at specific locations relative to film interfaces using fluorescence spectroscopy.[21] Continued development of experimental techniques and application to gas separation membrane materials will certainly add to the understanding of ultra-thin film behavior.

3.4   Influence of Previous History and Experimental Conditions on Aging

The techniques used to create gas separation membranes, *e.g.* choice of casting solvent, drying treatment, *etc.*, can strongly affect their performance. For melt-processed films, or films heated above the glass transition during preparation, the rate at which the material is cooled from above T_g also has a significant influence on the final physical properties.[69–71] For example, rapidly quenched glasses exhibit accelerated aging when compared to slow-cooled samples, due to their initially higher free volume state.[72] After the polymer vitrifies, *i.e.* enters the glassy state, various treatments can alter its thermodynamic state, such as, mechanical stresses, thermal annealing, and exposure to highly sorbing penetrants.[41,73–75] For instance, in bulk glassy polymers, exposure to high pressure CO_2 causes the glassy polymer to swell, and after removal of the CO_2, the polymer structure does not immediately return to its initial state, effectively reducing its density.[75,76] While the effects of various conditioning treatments on polymer properties have been demonstrated in bulk materials, it is not well understood how thin films would respond to similar treatments.

Because glassy polymer properties depend strongly on previous history and experimental conditions, studies on these materials require samples with well-defined histories and consistent methodologies for meaningful comparisons. In the aging studies described previously in this chapter, a controlled quench from above the glass transition was used to define the experimental starting time; however, actual application of these materials in gas separation membranes can involve very complex histories. The ultra-thin glassy films used in gas separation membranes are often products of elaborate phase inversion processes; additionally, the membranes may be subjected to further treatment steps and long periods of storage before use in the field. Moreover, while the earlier studies focused on films that were aged in a controlled, dry environment and exposed to only permanent gases, *i.e.* O_2, N_2, He, *etc.*, commercial gas separation membranes are exposed to a wide variety of process gases and contaminants that can strongly affect their performance. Because of these complexities, it is important to understand how aging behavior depends on a material's previous history and experimental conditions to accurately predict the long-term performance of practical gas separation membranes.

In an effort to understand how the initial state of a thin film affects its aging behavior, PSF samples were annealed for times ranging from 2 to 160 h at 170 °C, ~15 °C below its T_g, prior to gas permeability and aging measurements.[77] Figure 3.8 shows the O_2 permeability coefficients as a function of aging time for the 125 nm thick PSF films that were annealed at 170 °C for the specified times after quenching from above T_g.

The base case represents a sample of identical thickness that began aging at 35 °C immediately following the quench from above T_g. As in the previous studies, all films were aged at 35 °C in a dry environment between measurements. The initial O_2 permeability decreased with longer annealing times at 170 °C, as expected, due to the influence of aging the material at the elevated temperature. While the aging responses were similar between all PSF films, the

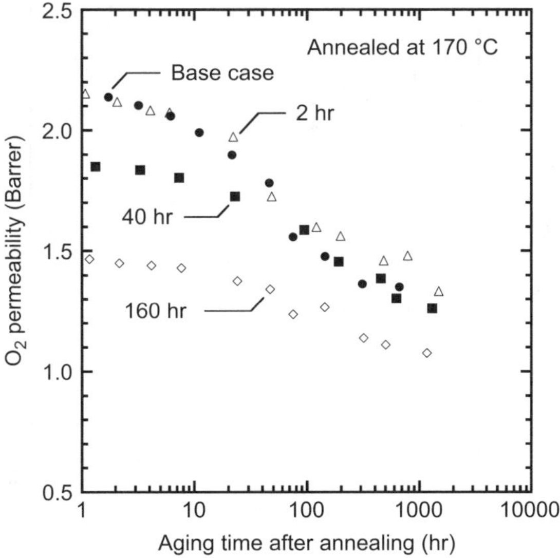

Figure 3.8 Influence of aging time at 35 °C on O_2 permeability coefficients in thin PSF films annealed for specified time at 170 °C after a quench from above T_g. Reproduced with permission of Elsevier.[77]

magnitude of aging induced permeability changes decreased with annealing time at 170 °C. The slower apparent aging rate in the annealed samples is consistent with the loss in driving force toward equilibrium, *i.e.* the annealed films had less excess free volume than the base case material. These observed trends in the aging behavior of annealed samples were well described by recent modeling efforts that account for the material's altered initial free volume state while using the same material properties as the base case.[77] Analogous behavior was seen for other probe gases, including N_2 and CH_4.

In addition to the conditioning effects that exposure to high pressure CO_2 has on glassy polymer properties, exposure to low pressures of highly sorbing gas can also affect membrane performance and aging behavior.[78,79] Figure 3.9 displays the influence of intermittent exposure to low pressure CO_2 (2 bar) on the aging behavior of thin 6FDA-DAM films, as tracked by methane permeability.[78]

The periods of exposure to low pressure CO_2 were spaced every 10–20 h or longer, and exposure times were about 30 min each. Surprisingly, with exposure pressure to only 2 bar of CO_2 over times that were short in comparison with the aging time scales, significant effects on the aging behavior are apparent. Both 6FDA-DAM films show decreasing CH_4 permeability with aging time; however, the influence of aging is notably reduced in the case of the film that had been intermittently exposed to CO_2. The impact of this exposure protocol on permeability changes with aging also depends on the size of the probe molecule.

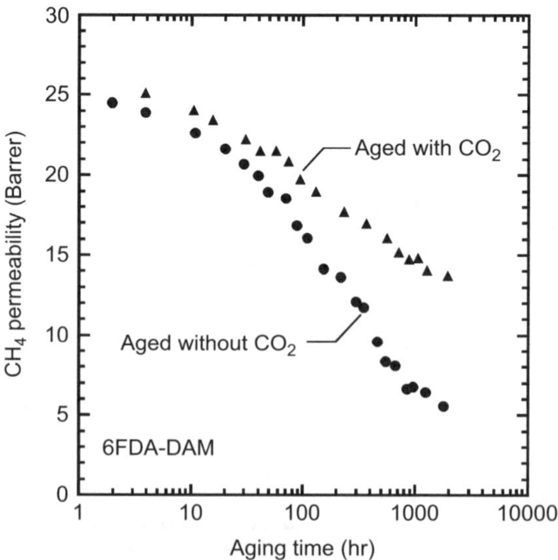

Figure 3.9   Influence of exposure to CO_2 during physical aging on the aging response in 6FDA-DAM as tracked by methane permeability. Throughout the aging process, the film was periodically exposed to CO_2 for short periods ($\sim$10 min) at 2 atm pressure. Reproduced with permission of Elsevier.[78]

While CO_2 exposure strongly affected the aging behavior of 6FDA-DAM as tracked by N_2 and CH_4 permeability, it did not significantly influence the He or O_2 permeability responses. Because the exposure to CO_2 slowed the permeability decline of larger molecules more than that of smaller molecules, the pure gas selectivity of these films actually decreased with aging time.[78] The strong influence that CO_2 exposure has on the aging behavior of thin films highlights the importance of understanding its role in the behavior of gas separation membranes.

While the properties of bulk membrane materials are typically only affected by exposure to CO_2 at relatively high pressures (tens of bars), thin film behavior can be affected at much lower pressures with exaggerated or even qualitatively different responses than those seen in thick films. This difference in response to CO_2 exposure is another example of the deviations from bulk behavior that manifest in thin glassy films. Because of the complex, and sometimes overlapping, effects of physical aging, plasticization, and penetrant-induced structural relaxation on thin film behavior, carefully designed experiments are required to identify the mechanisms responsible for the observed changes. Some evidence in the literature suggests thin films are more easily plasticized than bulk materials, and current research efforts show that this behavior is also dependent on sample history.[80–82] As might be expected, thin films display more pronounced hysteresis than thick films exposed to the same experimental cycle.[82] The impact that CO_2 has on the behavior of thin gas separation membranes indicates that these systems will also be strongly affected by other highly sorbing species that may exist in practical processing streams.

3.5 Modeling Physical Aging Behavior

While several models have been proposed to describe accelerated aging in thin polymer films, including the diffusion of free volume and thickness-dependent lattice contraction, efforts to more fully understand the mechanisms responsible for this behavior are ongoing. Physical aging models are typically characterized by a relaxation time, which represents chain mobility, and a measure of the current deviation from the predicted equilibrium, which represents the driving force. Models commonly used to describe physical aging, or isothermal relaxation, have foundations in work by Tool, Kovacs and others.[83–85] These models typically follow the form:

$$\text{aging rate} = \frac{\mathrm{d}v}{\mathrm{d}t} = \frac{-(v - v_\infty)}{\tau(v, T_g - T)} \tag{3.7}$$

where v and v_∞ are the polymer specific volumes at time t and at equilibrium, respectively, T is the system temperature, and τ is a characteristic relaxation time. A diffusion-based model developed by Curro *et al.*, based on Fick's second law, estimates free volume mobility using the well-known Doolittle relationship and implicitly includes the dependence of thickness on aging.[86] A related model with an alternative method for calculating vacancy diffusion, termed the empirically derived vacancy diffusion model, has been developed more recently.[87] Although these models have been used to describe a variety of aging data in a phenomenological manner, the mechanisms that cause accelerated aging remain unclear.[40,88]

A straightforward model to describe aging in ultra-thin films was recently proposed based on the self-retarding aging model developed by Struik, which described the variation in many physical properties (*e.g.* specific volume, impact strength, and creep compliance) of bulk glassy polymers with aging.[6] According to this model, the change in free volume with aging time can be described by:

$$\frac{d\Delta f}{dt} = \frac{-\Delta f}{\tau} = \frac{-\Delta f}{\tau_\infty \exp(-\gamma \Delta f)} \tag{3.8}$$

where Δf is the excess fractional free volume (*i.e.* $f(t) - f_e$, where f_e is the equilibrium free volume), τ is the relaxation time at time t, τ_∞ is the relaxation time at equilibrium, and γ is a material constant characterizing the sensitivity of relaxation time to excess fractional free volume. Fractional free volume, FFV or f, can be calculated using eqn (3.4). While this model was validated using data from bulk polymers, where thickness effects on physical aging are not observed, τ_∞ can be allowed to depend on film thickness as a means to capture thickness-dependent aging behavior.

The fractional free volume, f, calculated as a function of aging time using the Struik model can then be used in the following correlation to give gas permeability:

$$P = Ae^{-B/f} \tag{3.9}$$

where A and B are constants based on bulk membrane permeability measurements.[89] The combination of this correlation between permeability and free volume and the thickness-dependent Struik model has been referred to as the modified Struik model.[31]

Figure 3.10(a) compares the experimentally measured permeability coefficients to predictions based on the modified Struik model for PSF films from 465 to 20 nm thick.[31] The model accurately captures the aging response when τ_∞ is allowed to change with films thickness; the γ value used was 350. The aging behavior of the ultra-thin PSF films is similar to that of the Matrimid[®] material shown in Figure 3.4(a), which is also well represented by the modified Struik model.[31] The τ_∞ values used in the model are shown in Figure 3.10(b) as a function of film thickness. The strong decrease of τ_∞ with film thickness below about 100 nm suggests that these ultra-thin films age more rapidly than thick films due to an enhanced rate of relaxation. Interestingly, the τ_∞ values show a similar sensitivity to film thickness as the glass transition temperature measured using ellipsometry as reported by Kim *et al.*, suggesting that the mechanisms that cause an apparent reduction in T_g are related to the accelerated aging of ultra-thin films.[90]

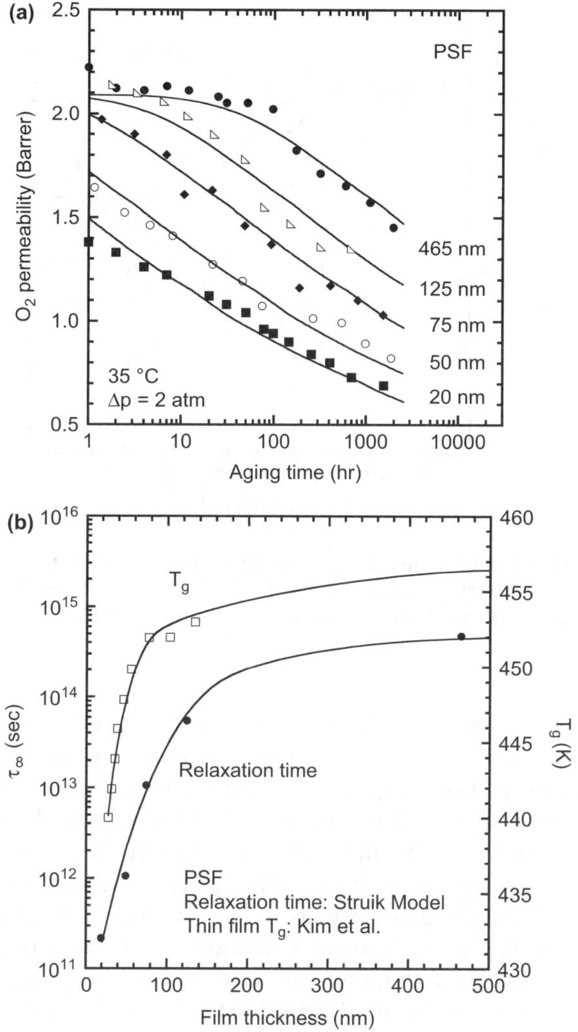

Figure 3.10   Influence of film thickness on physical aging and relaxation rates. (a) Effect of aging time on oxygen permeability coefficients in PSF films ranging from 465 nm to 20 nm in thickness. Lines were generated from the modified Struik model. (b) Dependence of τ_∞ and T_g, from Kim *et al.*, on PSF film thickness. Lines drawn to guide the eye in Figure 3.8(b). Reproduced with permission of Elsevier.[31]

The modified Struik model, as applied to these ultra-thin PSF films, used the same initial conditions for each film, *i.e.* at a short, finite time following the quench from above T_g, permeability was independent of film thickness. Figure 3.11 presents the model predicted O$_2$ permeability coefficients for the PSF films, including aging times $\ll 1$ h.

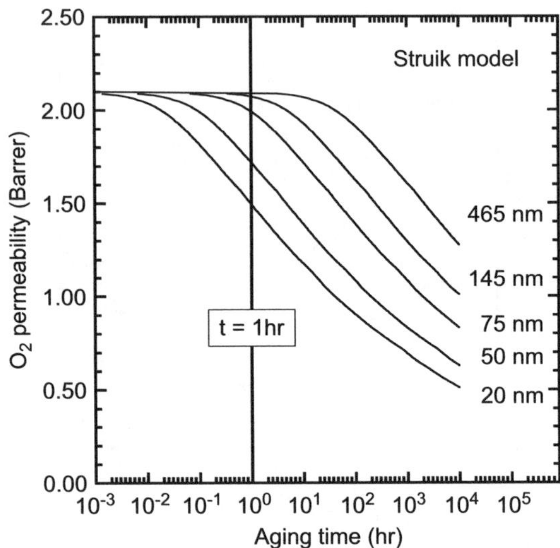

Figure 3.11   Influence of aging time and film thickness on the predicted oxygen permeability behavior of the PSF films studied based on the modified Struik model. Reproduced with permission of Elsevier.[31]

The vertical line drawn at 1 h of aging time highlights why the initial gas transport measurements show decreasing permeability as film thickness is reduced. The applied model is also consistent with the results of the variable energy PALS study discussed previously.[64] It was suggested that the smaller free volume elements near the film surface resulted from enhanced mobility in the near-surface region over the bulk. While, standard aging models, which do not account for thickness-dependent behavior, might suggest that the smaller free volume elements near the surface would age more slowly (due to decreased driving force) as compared to larger vacancies in the film center, the enhanced mobility of the near-surface region suggested by the model, counterbalances the reduced driving force, resulting in more rapid aging.

3.6   Concluding Remarks

Understanding the influences of film thickness and confinement on the physical aging behavior of glassy materials utilized in gas separation membranes is essential for predicting their long-term performance. Since physical aging was first recognized to depend on film thickness, multiple experimental advances have enabled more accurate characterization of these materials and their behavior. Techniques related to creating the defect-free films needed for gas transport measurements, precisely measuring film thickness, and handling the delicate structures have all contributed to the improved understanding of thin membrane behavior.

The results presented in this chapter suggest that the accelerated aging of ultra-thin films of gas separation membrane materials is caused by an enhanced mobility near the film surface that allows the polymer to relax towards an equilibrium state more rapidly than bulk samples. This deviation from thick film behavior is also strongly dependent on the polymer structure, complicating efforts to predict thin film behavior from bulk properties. Although some modeling efforts capture thickness dependent properties, estimating thin film behavior from bulk measurements only provides a first-order approximation.

The majority of systematic aging studies have focused on tracking material properties immediately following a quench from above T_g to define a consistent time zero and equivalent thermal history for different samples. The application of this technique is complicated for composite or asymmetric materials that may collapse or undergo macroscopic changes at elevated temperatures. Additionally, some high-performance materials have glass transition temperatures that are above the onset of decomposition, *e.g.* PTMSP, polymers of intrinsic microporosity, and some thermally rearranged polymers.[91–93] These complex structures and materials require alternative methods to reset or at least standardize their thermal history and create a meaningful starting point for aging studies. Physical aging studies on PTMSP have used a swelling non-solvent, *e.g.* methanol, to minimize aging effects prior to measurements and in efforts to create similar initial conditions between different samples.[94] These challenges also provide opportunities to better understand the influence of processing conditions and previous history on subsequent aging behavior in membrane materials.

Significant changes in physical properties occur over shorter times as the time scale for structural relaxation decreases with decreasing membrane thickness. This behavior is highlighted by the gas permeability results that show decreasing permeability coefficients with decreasing thickness in the ultra-thin films after just 1 h of aging.[31] Rapid, or *in situ*, experimental measurements are needed to accurately characterize the state of ultra-thin membranes immediately following a quench from above T_g, and to test the assumption that this method provides a consistent initial state for films across a wide thickness range.

Some of the initial investigations that reported thickness-dependent aging behavior were conducted using asymmetric hollow fiber membranes with complex histories and structures. With goals to better understand the role of film thickness on aging, research efforts shifted focus to dense isotropic membranes with well defined histories and well characterized thickness. While these studies provide valuable information regarding the influence of film thickness and the free surface on physical aging, the behavior of these isotropic membranes may still be different than that of the complex structures used in commercially produced membranes. For example, the rapid quench from above T_g in the free standing film studies accentuates the membrane's initial free volume state, and, subsequently, accelerates its aging behavior. Unraveling the influences of the film formation conditions and the connection to the underlying asymmetric structure on physical aging is perhaps the most significant challenge in understanding the long term performance of gas separation membranes. Additionally, the exposure to condensable gases and vapors that may be present in process feed streams can strongly influence the aging behavior of ultra-thin membrane materials, even at low pressures. If the influences that these factors have on aging are better understood with new advances in experimental techniques, more accurate long-term performance predictions can be made, and perhaps ways to slow or arrest physical aging can be identified.

References

1. R. W. Baker, *Ind. Eng. Chem. Res.*, 2002, **41**, 1393.
2. L. M. Robeson, *J. Membr. Sci.*, 1991, **62**, 165.
3. L. M. Robeson, *J. Membr. Sci.*, 2008, **320**, 390.
4. D. W. Wallace, C. Staudt-Bickel and W. J. Koros, *J. Membr. Sci.*, 2006, **278**, 92.
5. C. A. Angell, K. L. Ngai, G. B. McKenna, P. F. McMillan and S. W. Martin, *J. Appl. Phys.*, 2000, **88**, 3113.
6. L. C. E. Struik, *Physical Aging in Amorphous Polymers and Other Materials*, Elsevier, Amsterdam, 1978.
7. J. M. Hutchinson, *Prog. Polym. Sci.*, 1995, **20**, 703.
8. A. R. Berens and I. M. Hodge, *Polym. Eng. Sci.*, 1984, **24**, 1123.
9. Y. Kobayashi, W. Zheng, E. F. Meyer, J. McGervey, A. Jamieson and R. Simha., *Macromolecules*, 1989, **22**, 2302.
10. A. J. Hill, K. J. Heater and C. M. Agrawal, *J. Polym. Sci. Part B: Polym. Phys.*, 1990, **28**, 387.
11. I. M. Hodge, *Science*, 1995, **267**, 1945.
12. R. W. Baker, *Membrane Technology and Applications*, John Wiley and Sons, Chichester, 2004.
13. A. S. Stern, *J. Membr. Sci.*, 1994, **94**, 1.
14. W. J. Koros and G. K. Fleming, *J. Membr. Sci.*, 1993, **83**, 1.
15. J. M. S. Henis and M. K. Tripodi, *J. Membr. Sci.*, 1981, **8**, 233.
16. I. Pinnau, J. G. Wijmans, I. Blume, T. Kuroda and K. V. Peinemann, *J. Membr. Sci.*, 1988, **37**, 81.
17. J. M. S. Henis and M. K. Tripodi, *Science*, 1983, **220**, 11.
18. J. L. Keddie, R. A. L. Jones and R. A. Cory, *Europhys. Lett.*, 1994, **27**, 59.
19. B. Frank, A. P. Gast, T. P. Russell, H. R. Brown and C. Hawker, *Macromolecules*, 1996, **29**, 6531.
20. C. W. Frank, V. Rao, M. M. Despotopoulou, R. F. W. Pease, W. D. Hinsberg, R. D. Miller and J. F. Rabolt, *Science*, 1996, **273**, 912.
21. C. J. Ellison and J. M. Torkelson, *Nat. Mater.*, 2003, **2**, 695.
22. R. Ruiz, H. M. Kang, F. A. Detcheverry, E. Dobisz, D. S. Kercher, T. R. Albrecht, J. J. de Pablo and P. F. Nealey, *Science*, 2008, **321**, 936.
23. J. A. Forrest and K. Dalnoki-Veress, *Adv. Colloid Interface Sci.*, 2001, **94**, 167.

24. R. Inoue, T. Kanaya, K. Nishida, I. Tsukushi, M. T. F. Telling, B. J. Gabrys, M. Tyagi, C. Soles and W. I. Wu, *Phys. Rev. E*, 2009, **80**, 031802.

25. M. Y. Efremov, A. V. Kiyanova and P. F. Nealey, *Macromolecules*, 2008, **41**, 5978.

26. C. J. Ellison, S. D. Kim, D. B. Hall and J. M. Torkelson, *Eur. Phys. J. E*, 2002, **V8**, 155.

27. C. B. Roth and J. R. Dutcher, in *Soft Materials: Structure and Dynamics*, ed. J. R. Dutcher and A. G. Marangoni, Marcel Dekker, New York, 2005, pp. 1–38.

28. R. D. Priestley, L. J. Broadbelt and J. M. Torkelson, *Macromolecules*, 2005, **38**, 654.

29. T. Masuda, E. Isobe, T. Higashimura and K. Takada, *J. Am. Chem. Soc.*, 1983, **105**, 7473.

30. S. D. Kelman, S. Matteucci, C. W. Bielawski and B. D. Freeman, *Polymer*, 2007, **48**, 6881.

31. B. W. Rowe, B. D. Freeman and D. R. Paul, *Polymer*, 2009, **50**, 5565.

32. M. E. Rezac, P. H. Pfromm, L. M. Costello and W. J. Koros, *Ind. Eng. Chem. Res.*, 1993, **32**, 1921.

33. P. H. Pfromm and W. J. Koros, *Polymer*, 1995, **36**, 2379.

34. L. Baayens and S. L. Rosen, *J. Appl. Polym. Sci.*, 1972, **16**, 663.

35. P. H. Pfromm and W. J. Koros, *Polym. Mater. Sci. Eng.*, 1994, **71**, 401.

36. M. S. McCaig and D. R. Paul, *Polymer*, 2000, **41**, 629.

37. Y. Huang and D. R. Paul, *J. Membr. Sci.*, 2004, **244**, 167.

38. Y. Huang and D. R. Paul, *Macromolecules*, 2006, **39**, 1554.

39. J. H. Kim, W. J. Koros and D. R. Paul, *Polymer*, 2006, **47**, 3104.

40. Y. Huang, X. Wang and D. R. Paul, *J. Membr. Sci.*, 2006, **277**, 219.

41. A. R. Berens and I. M. Hodge, *Macromolecules*, 1982, **15**, 756.

42. I. M. Hodge and A. R. Berens, *Macromolecules*, 1982, **15**, 762.

43. Y. Huang and D. R. Paul, *Polymer*, 2004, **45**, 8377.

44. J. H. Kim, W. J. Koros and D. R. Paul, *Polymer*, 2006, **47**, 3094.

45. G. Reiter, G. Reiter, M. Hamieh, P. Damman, S. Sclavons, S. Gabriele, T. Vilmin and E. Raphaël, *Nat. Mater.*, 2005, **4**, 754.

46. J. M. S. Henis and M. K. Tripodi, *Sep. Sci. Technol.*, 1980, **15**, 1059.

47. T. C. Merkel, V. I. Bondar, K. Nagai, B. D. Freeman and I. Pinnau, *J. Polym. Sci. Part B: Polym. Phys.*, 2000, **38**, 415.

48. H. Richardson, I. Lopez-Garcia, M. Sferrazza and J. L. Keddie, *Phys. Rev. E*, 2004, **70**, 051805.

49. S. D. Kelman, B. W. Rowe, C. W. Bielawski, S. J. Pas, A. J. Hill, D. R. Paul and B.D. Freeman, *J. Membr. Sci.*, 2008, **320**, 123.

50. H. A. Lorentz, *The Theory of Electrons*, Dover Publications, Inc., New York, 1952.

51. D. W. V. Krevelen, *Properties of Polymers*, Elsevier, Amsterdam, 1990.

52. Y. Ichiraku, S. A. Stern and T. Nakagawa, *J. Membr. Sci.*, 1987, **34**, 5.

53. K. Nagai, B. D. Freeman and A. J. Hill, *J. Polym. Sci. Part B: Polym. Phys.*, 2000, **38**, 1222.

54. D. Cangialosi, H. Schut, A. van Veen and S. J. Picken, *Macromolecules*, 2003, **36**, 142.
55. G. Dlubek, D. Kilburn and M. A. Alam, *Macromol. Symp.*, 2004, **210**, 11.
56. P. E. Mallon, in *Positron & Positronium Chemistry*, ed. Y. C. Jean, P. E. Mallon and D. M. Schrader, World Scientific, New Jersey, 2003, pp. 253–280.
57. S. J. Tao, *J. Chem Phys.*, 1972, **56**, 5499.
58. M. Eldrup, D. Lightbody and J. N. Sherwood, *Chem. Phys.*, 1981, **63**, 51.
59. R. Suzuki *et al.*, *Jpn. J. Appl. Phys., Part 2*, 1991, **30**, L532.
60. J. Algers, R. Suzuki, T. Ohdaira and F. H. J. Maurer, *Polymer*, 2004, **45**, 4533.
61. S.-H. Huang, W. S. Hung, D. J. Liaw, C. L. Li, S. T. Kao, D. M. Wang, M. D. Guzman, C. C. Hu, Y. C. Jean, K. R. Lee and J. Y. Lai, *Macromolecules*, 2008, **41**, 6438.
62. S.-H. Huang, W. S. Hung, D. J. Liaw, H. A. Tsai, G. J. Jiang, K. R. Lee and J. Y. Lai, *Polymer*, 2010, **51**, 1370.
63. W.-S. Hung, M. De Guzman, S.-H. Huang, K.-R. Lee, Y. C. Jean and J.-Y. Lai, *Macromolecules*, 2010, **43**, 6127.
64. B. W. Rowe, B.D. Freeman and D.R. Paul, *Polymer*, 2009, **50**, 6149.
65. J. Algers, P. Sperr, W. Egger, G. Kögel and F. H. J. Maurer, *Phys. Rev. B*, 2003, **67**, 125404(1).
66. A. Serghei, H. Huth, C. Schick and F. Kremer, *Macromolecules*, 2008, **41**, 3636.
67. C. J. Ellison and J. M. Torkelson, *J. Polym. Sci. Part B: Polym. Phys.*, 2002, **40**, 2745.
68. C. L. Soles, J. F. Douglas, W.-l. Wu, H. Peng and D. W. Gidley, *Macromolecules*, 2004, **37**, 2890.
69. C. M. Agrawal, K. J. Heater and A. J. Hill, *J. Mater. Sci. Lett.*, 1989, **V8**, 1414.
70. A. J. Kovacs and J. M. Hutchinson, *J. Polym. Sci. Polym. Phys. Ed.*, 1979, **17**, 2031.
71. J. Hadac, *et al.*, *J. Non-Cryst. Solids*, 2007, **353**, 2681.
72. D. Cangialosi *et al.*, *Phys. Rev. B: Condens. Matter,* 2004, **70**, 224213(1).
73. D. J. Enscore, H. B. Hopfenberg, V. T. Stannett and A. R. Berens, *Polymer*, 1977, **18**, 1105.
74. A. H. Chan and D. R. Paul, *J. Appl. Polym. Sci.*, 1980, **25**, 971.
75. G. K. Fleming and W. J. Koros, *Macromolecules*, 1990, **23**, 1353.
76. S. M. Jordan, W. J. Koros and G. K. Fleming, *J. Membr. Sci.*, 1987, **30**, 191.
77. B. W. Rowe, B. D. Freeman and D. R. Paul, *Polymer*, 2010, **51**, 3784.
78. J. H. Kim, W. J. Koros and D. R. Paul, *J. Membr. Sci.*, 2006, **282**, 21.
79. J. H. Kim, W. J. Koros and D. R. Paul, *J. Membr. Sci.*, 2006, **282**, 32.
80. C. A. Scholes, G. Q. Chen, G. W. Stevens and S. E. Kentish, *J. Membr. Sci.*, 2010, **346**, 208.
81. M. Wessling, M. Lidon Lopez and H. Strathmann, *Sep. Purif. Technol.*, 2001, **24**, 223.

82. N. R. Horn and D. R. Paul, *Polymer*, 2011, **52**, 1619.

83. A. Q. Tool, *J. Res. Natl. Bur. Stand.*, (*U. S.*), 1945, **34**, 199.

84. A. J. Kovacs, J. M. Hutchinson and J. J. Aklonis, in *The Structure of Non-Crystalline Materials*, ed. P. H. Gaskell, Taylor and Francis, London, 1977, pp. 153–163.

85. R. W. Rendell, K. L. Ngai and D. J. Plazek, *J. Non-Cryst. Solids,* 1991, **131–133**, 442.

86. J. G. Curro, R. R. Lagasse and R. Simha, *Macromolecules*, 1982, **15**, 1621.

87. A. W. Thornton, K. M. Nairn, A. J. Hill, J. M. Hill and Y. Huang, *J. Membr. Sci.*, 2009, **338**, 38.

88. M. S. McCaig, D. R. Paul and J. W. Barlow, *Polymer*, 2000, **41**, 639.

89. J. Y. Park and D. R. Paul, *J. Membr. Sci.*, 1997, **125**, 23.

90. J. H. Kim, J. Jang and W. C. Zin, *Langmuir*, 2000, **16**, 4064.

91. L. Starannikova, V. Khodzhaeva and Y. Yampolskii, *J. Membr. Sci.*, 2004, **244**, 183.

92. P. M. Budd, N. B. McKeown, B. S. Ghanem, K. J. Msayib, D. Fritsch, L. Starannikova, N. Belov, O. Sanfirova, Y. Yampolskii and V. Shantarovich, *J. Membr. Sci.*, 2008, **325**, 851.

93. H. B. Park, C. H. Jung, Y. M. Lee, A. J. Hill, S. J. Pas, S. T. Mudie, E. Van Wagner, B. D. Freeman and D. J. Cookson, *Science*, 2007, **318**, 254.

94. A. J. Hill, S. J. Pas, T. J. Bastow, M. I. Burgar, K. Nagai, L. G. Toy and B. D. Freeman., *J. Membr. Sci.*, 2004, **243**, 37.

Recent High Performance Polymer Membranes for CO_2 Separation

S.H. HAN AND Y.M. LEE*

WCU Department of Energy Engineering, College of Engineering, Hanyang University, Seoul 133-791, Republic of Korea

4.1 Introduction

During the last century, the so-called polymer, a macromolecule composed of repeated covalent bonds, has grown from an undefined natural stuff to the most common material in the world due to mass production capabilities based on convenience for scale-up.[1,2] Therefore, polymer has widely been used for applications such as packaging, transport, living necessaries, medicines, electronics, chemical industries, *etc.*[3–8] For separation processes, polymer membranes have the advantages such as flexibility, easy fabrication, convenient tailoring and low prices in water purification, submerged bio-reactor, desalination, electrodialysis, ion exchange and gas permeation.[9,10] Polymeric membranes for gas separation have received much attention as one of the most promising processes for separation and purification in gas treatment.

Since A. Fick observed that gas molecules tend to pass through thin nitrocellulose films at different permeation rates for each gas, membrane-based gas separation has been studied extensively to look for better membrane materials and to apply this property to gas separation on an industrial scale.[9] Numerous scientists have studied membrane materials with respect to transport

Membrane Engineering for the Treatment of Gases, Volume 1:
Gas-separation Problems with Membranes
Edited by Enrico Drioli and Giuseppe Barbieri

mechanisms, gas permeation measurements, and fabrication of high flux asymmetric membranes by phase inversion processes. Sir Thomas Graham proposed a well-known 'solution-diffusion' mechanism to illustrate gas permeation through the membrane in dense polymer films.[9,10] Acknowledging the contribution of H. A. Daynes and R. M. Barrer, quantitative measurement for gas permeabilities by the so-called time-lag method was applied to a number of membrane materials for gas separation membranes.[9,11] S. Loeb and S. Sourirajan developed a novel asymmetric cellulose acetate membrane process enabling the reduction of effective membrane thickness into sub-micrometer by phase inversion process.[12,13] Based upon the high fluxes of asymmetric hollow fiber membrane, Monsanto Company (now merged to Air Products and Chemicals Inc.) began manufacturing commercial synthetic polymer membranes and processes for gas separation in the mid-1970s.[9,14]

Competing with conventional gas separation technologies such as adsorption, absorption and cryogenics, membrane-based gas separation has played an important role in gas separation industries. By using common polymer materials such as polydimethylsiloxane (PDMS), cellulose acetate (CA), polysulfone (PSF), polyethersulfone (PES), and polyimide (PI), gas separation membranes have been applied to various gas separation processes such as hydrogen recovery in petroleum and ammonia processes, nitrogen enrichment by air separation, recovery of volatile organic compounds (VOCs), and separation of carbon dioxide in natural gas resources and steel industries.[14] PDMS, one of the highly permeable rubbery polymers, has been used in VOC recovery based on high flux due to its superior sorption capability. While glassy PES and PSF have been mainly used for nitrogen enrichment from air, glassy CA and fluorinated poly-imides have been employed to natural gas calorie-up process and petrochemical industries. For natural gas separation, the scale of membrane plants reached up to 500–1000 MMscfd in hundred or more sites worldwide operated by Air Liquide (Medal^{TM} membrane), UOP (Separex^{TM} membrane) and Natco Group Inc., *etc.*[14,15] Although these polymer membranes have been widely spread in industrial processes containing hydrogen recovery, natural gas calorie-up and recovery of VOCs, however, these membrane materials and processes are estimated to be insufficient for the forthcoming enormously large-scale applications such as CO$_2$ separation for carbon capture and storage (CCS).[15–17]

Recently, global warming and severe climate changes provide further opportunities for large-scale applications, separation of flue gas or synthetic gas from coal-based power generation. Carbon dioxide, the most important greenhouse gas (GHG), is mostly emitted during production of electricity in power plants over 5000 sites worldwide.[17–19] Total CO$_2$ emission of flue gas composed of carbon dioxide, oxygen and balanced nitrogen in a 600 MW coal-combustion power plant reaches to 3 million tons of CO$_2$ per year, corresponding to an enormous flow rate of 500 N m^3 s^{-1}. The driving force of the gas mixture for gas separation membrane is very low, usually 0.1–0.3 bar. Therefore, it is essentially required to develop highly permeable even at low pressure and highly selective membranes to treat the enormous amount of post-combustion flue gas from pulverized carbon (PC) power plants. Moreover, pre-combustion process, an

Table 4.1 Comparison among three representative CO_2 separation processes

	Natural gas sweetening	*Flue gas separation*	*Was gas shift recovery*
Objective	Enrichment of methane for fuel	Capture of CO_2 against global warming	Recovery of hydrogen with CO_2 capture
Major constituents	CO_2: 20–40% CH_4: 60–80%	CO_2: 8–15% N_2: 60–70% H_2O: 15–25%	H_2: 50–60% CO_2: 40–45%
Minor constituents	Hydrocarbons, H_2S	O_2, SO_x, NO_x	CO, H_2S, NH_3
Temperature	10–100 °C	50–200 °C	200–600 °C
Pressure	High (30–50 bar)	Low (0.9–1.5 bar)	High (20–60 bar)

alternative process substituting the present power plants due to the most efficient processability and economical feasibility, needs high performance membranes with the high thermal stability as well as mechanical stabilities at high pressure.[16,18] Pre-combustion CO_2 capture process called as integrated gasification combined cycle (IGCC) is composed of coal gasification by steam reforming, desulfurization, water-gas-shift (WGS) reaction, and H_2/CO_2 separation.[20] For H_2/CO_2 separation, membrane materials have been considerably limited only to metal or inorganic materials due to the high operation temperature over 300 °C. However, thermally stable rigid-rod polymers have also potentialities as candidates of the gas separation membrane if the materials possess high permeabilities as well as thermal and chemical stabilities.[20,21] For the major CO_2 separation processes, conditions and gas components of three representative processes are summarized in Table 4.1.

Therefore, there is a strong requirement to develop high-performance polymer membranes with superior thermal, chemical, mechanical and long-term stabilities for CO_2 separation. Thus recently developed polymer materials are suggested to be the excellent candidates.[22,23] In this chapter, gas transport mechanism with respect to sorption and diffusion through polymeric membrane materials will be introduced followed by a recent literature review on the chemical structure, physical properties, gas permeabilities and selectivities of membrane materials as well as up-to-date high performance and highly permeable membranes especially for CO_2 separation.

4.2 Background

4.2.1 Solution-diffusion Mechanism for Gas Permeation

As Fick's first law was first developed by gas diffusion through a polymer film, gas transport through polymers is an intrinsic property of polymers which can be used for barrier, packaging, film and membrane, *etc.*[9,11] According to the law, gas flux depends upon the concentration difference as a driving force, thus the following relationship can be derived:

$$J = -D\frac{dc}{dx} = -D\frac{\Delta c}{\Delta x} \tag{4.1}$$

where J is gas flux (mol cm^{-2} s^{-1}), D is diffusivity of a gas molecule (cm^2 s^{-1}), x is the position (cm), and c is concentration (mol cm^{-3}).

In gaseous phase, concentration is proportional to the applied pressure, and the constant is defined as solubility by Henry's law. Therefore, the eqn (4.1) is arranged to the following:

$$c = S \cdot P \tag{4.2}$$

$$J = D \cdot S \cdot \frac{\Delta P}{l} \tag{4.3}$$

where S is solubility parameter [cm^3 cm^{-3} (cmHg)$^{-1}$], l is the thickness of membrane, and ΔP is pressure difference applied to the membrane. Therefore, permeability of a gas molecule is obtained as the product of diffusivity through membrane and solubility of gases. Also, selectivity of binary gas mixture between a membrane material can be represented as the ratio of permeabilities of two gas molecules:

$$Pe = D \cdot S \tag{4.4}$$

$$\alpha_{i/j} = \frac{Pe_i}{Pe_j} = \frac{D_i}{D_j} \frac{S_i}{S_j} \tag{4.5}$$

where Pe is permeability represented as Barrer [1 Barrer $= 10^{-10}$ cm^3(STP) cm cm^{-2} s^{-1} (cmHg)$^{-1}$], and $\alpha_{i/j}$ is selectivity in gas mixture i and j.

Sorption, a thermodynamic factor, occurs when a gas molecule in the bulk state adsorbs physically onto a membrane surface with respect to the concentration difference, while diffusion, a kinetic factor, is related to how fast a gas molecule can pass through the membrane.[10] Sorption and diffusion are activated energy processes, which are affected substantially by temperature, thus, the relationships for D, S and Pe are expressed as a function of temperature:

$$D = D_0 \exp(-E_d/RT) \tag{4.6}$$

$$S = S_0 \exp(-\Delta H_S/RT) \tag{4.7}$$

$$Pe = Pe_0 \exp(-E_P/RT) \tag{4.8}$$

where E_d is the activation energy of diffusion, ΔH_s is the heat of sorption, and $E_P = E_d + \Delta H_s$ is the activation energy of permeation. Since diffusion of gas molecules through polymers and sorption of gases on polymers are affected by the type of polymers, gas permeabilities are determined by the gas–polymer affinities. Therefore, two factors, thermodynamics and kinetics, are important for high permeation and separation performances simultaneously.

4.2.2 Trade-off Relationship in Gas Separation

While permeability of gas molecule is determined by diffusivity and solubility, selectivity of binary gas mixtures is ideally calculated by the ratio of the two gas permeabilities as seen in eqn (4.5). It has been well-known in gas separation membranes in several references that polymer membranes exhibit 'trade-off' relationships in that highly permeable polymers usually have relatively restricted selectivities because high gas permeability of one molecule indicates that the membrane possessed high diffusive pathways or sorption sites where other gases can pass through.[24–26] This trade-off relationship shows an upper bound where the plot of the selectivity versus the permeability of highly permeable gas molecule exhibits a limit for achieving the desired result for polymeric membranes. Robeson has illustrated the empirical relationships in a plot of log $\alpha_{i/j}$ versus log P_i for binary gas mixtures containing helium, hydrogen, oxygen, carbon dioxide, nitrogen, and methane.[24,27] As summarized in Table 4.2, these data correlate with the universal values of gas molecules predicted by theories, thus, the upper bounds are arranged by simple equation:

$$Pe_i = k \cdot \alpha_{i/j} \tag{4.9}$$

where k is a front factor equivalent to Pe_i where the selectivity α is 1. In theoretical assumption, Robeson and Freeman correlated the k and n values with kinetic and thermodynamic factors, difference of kinetic diameters as well as solubility constants of the two gases.[25,28,29] Therefore, the relationship between P and α in eqn (4.9) then gives:

$$\cdot \, \alpha_{i/j} = \frac{\beta_{i/j}}{Pe_i^{\lambda_{i/j}}} \tag{4.10}$$

Table 4.2 Tabulation of the values of the front factor k and the upper bound slope n[24,27]

	Upper bound introduced in 1991		*Upper bound revisited in 2008*	
Gas pair	k *(Barrer)*	n	k *(Barrer)*	n
O_2/N_2	389 224	−5.800	1 396 000	−5.666
CO_2/CH_4	1 073 700	−2.6264	5 369 140	−2.636
H_2/N_2	52 918	−1.5275	97 650	−1.4841
H_2/CH_4	18 500	−1.2112	27 200	−1.107
He/N_2	12 500	−1.0242	19 890	−1.017
He/CH_4	5002	−0.7857	19 800	−0.809
He/H_2	960	−4.9535	59 910	−4.864
CO_2/N_2	N/A	N/A	30 967 000	−2.888
N_2/CH_4	N/A	N/A	2570	−4.507
H_2/CO_2	1200	−1.9363	4515	−2.302
He/CO_2	705	−1.220	3760	−1.192
H_2/O_2	35 760	−2.277	N/A	N/A
He/O_2	4600	−1.295	N/A	N/A

Table 4.3 Calculated values of $\lambda_{i/j}$ and $\beta_{i/j}$ for binary gas mixtures[30]

Gas pair	$\lambda_{i/j}$	$\beta_{i/j}$				
		200 K	*250 K*	*300 K*	*350 K*	*400 K*
O$_2$/N$_2$	0.1364	1.70×10^0	8.03×10^{-1}	4.87×10^{-1}	3.41×10^{-1}	2.61×10^{-1}
H$_2$/N$_2$	0.5407	6.99×10^{-3}	2.15×10^{-3}	9.80×10^{-4}	5.59×10^{-4}	3.67×10^{-4}
CO$_2$/CH$_4$	0.3175	2.33×10^0	4.50×10^1	1.50×10^1	6.86×10^2	3.81×10^2
H$_2$/CO$_2$	0.3383	2.86×10^{-4}	4.93×10^{-4}	7.09×10^{-4}	9.18×10^{-4}	1.11×10^{-3}
CO$_2$/H$_2$	−0.2528	4.44×10^2	2.96×10^2	2.26×10^2	1.86×10^2	1.61×10^2
CO$_2$/N$_2$	0.1513	8.39×10^1	1.38×10^1	4.14×10^0	1.75×10^0	9.21×10^{-1}

$$\lambda_{i/j} = (d_j/d_i)^2 - 1 \tag{4.11}$$

$$\beta_{i/j} = \left(\frac{S_i}{S_j}\right)S_i^{\lambda_{i/j}}\exp\left\{-\lambda_{i/j}\left[b - f\left(\frac{1-a}{RT}\right)\right]\right\} \tag{4.12}$$

where $\lambda_{i/j}$ and $\beta_{i/j}$ are empirical parameters depending upon the gas pairs. These factors have advantages that they can provide theoretical limitations where polymer materials can reach for any kind of binary mixtures at a given temperature as well.[30] The calculated constants for prediction of upper bounds are listed in Table 4.3.

4.2.3 High Performance Polymer Membranes for Gas Separation

For conventional polymers affected significantly by the trade-off relationship, it is quite difficult to find high-performance polymers representing both high permeability and high selectivity.[31] Although several fluorinated polyimides show advanced performances, their gas permeabilities are still restrained due to their high intermolecular forces.[16,32] Several representative upper bound relationships for CO$_2$/CH$_4$, CO$_2$/N$_2$ and H$_2$/CO$_2$ are shown in Figures 4.1 to 4.3.

Based on the 'solution-diffusion' mechanism, separation of gas mixtures through polymer membranes can be considered from diffusivity-selectivity membranes and solubility-selectivity membranes. Diffusion-selective membranes are commonly utilized by glassy polymers for permanent gas mixtures whereas sorption-selective membranes are usually prepared by rubbery polymers for the separation of condensable gases or vapors.[14] Improving selectivity as well as gas permeabilities should be targeted to these factors. The best design of ideal membrane materials is to improve both diffusivity-selectivity and solubility-selectivity, but it is difficult to achieve those simultaneously. In this regard, recent high-performance membranes can be classified into two parts by their driving forces, highly diffusive membranes by glassy polymers and solubility-selective one mainly by rubbery polymers.

Glassy polymers possessing rigid structure and strong intermolecular forces have been generally known to show reduced gas permeabilities although low diffusion through the membrane enabled to separate gas molecules efficiently

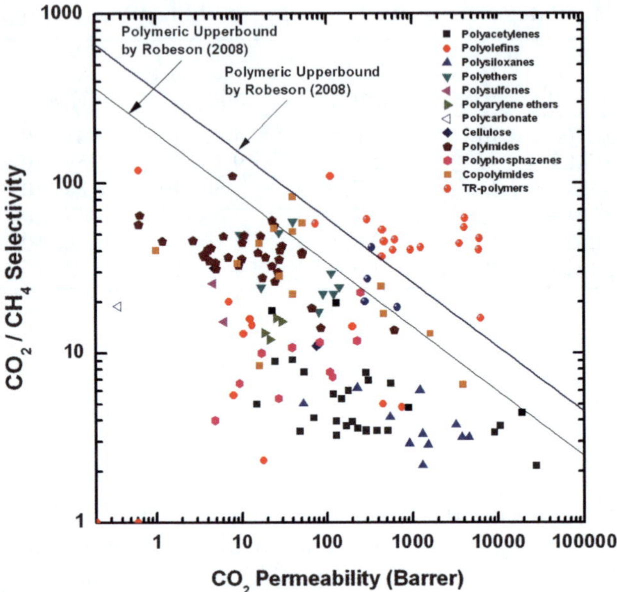

Figure 4.1 Trade-off relationship for CO_2/CH_4 separation.

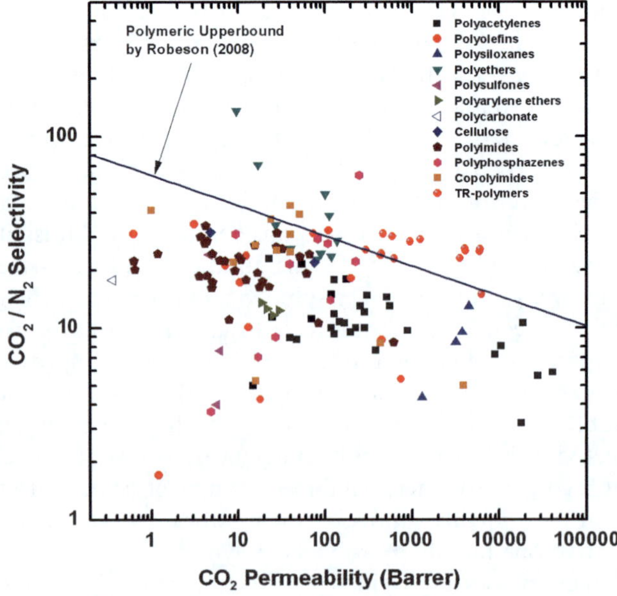

Figure 4.2 Trade-off relationship for CO_2/N_2 separation.

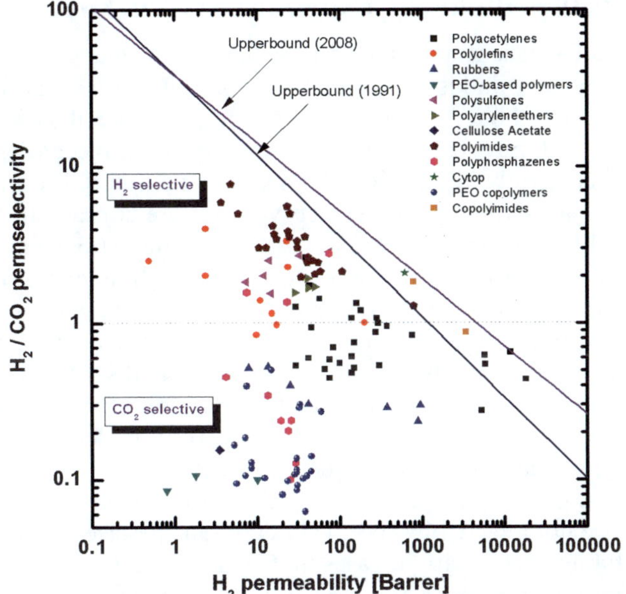

Figure 4.3 Trade-off relationship for H$_2$/CO$_2$ separation.

resulting in high diffusion selectivity.[9] However, glassy polymers containing distorted structure and the resulting increased free volume elements (FVE) were developed since T. Masuda *et al.* reported highly permeable glassy polymer, poly[1-(trimethylsilyl)-1-propyne] (PTMSP). Since PTMSP was developed in 1983, acetylene-based polymers such as poly(4-methyl-2-pentene) (PMP), poly(1-phenyl-1-propyne) (PPP) have been followed by the introduction of various substitution groups.[33] Amorphous fluorinated polymers, famous as the commercial names like AF 1600, AF 2400, Hyflon AD and Cytop, are candidate polymers for gas separation as well.[34–36] Recent development of polymers with intrinsic microporosities (PIMs) showed competitive performances resulted from the intrinsic FVE due to the bulky and distorted structures.[37,38] Thermally rearranged (TR) polymer membranes also presented extraordinary performances for gas separation. Chain rearrangement at the solid state improved the FVE in the polymer matrix followed by superior permselectivities as well as enormously high permeabilities.[23,39–41]

Sorption-enhanced polymer membranes are beneficial to obtain superior permeability as well as advanced selectivity especially for carbon dioxide, hydrocarbons and VOCs because condensable gas molecules are easier to be adsorbed on the polymer surface.[42] Recent research has focused on incorporating ethylene oxide into a polymer backbone, which has unique interaction with quadruple momentum of carbon dioxide; poly(ethylene oxide) (PEO), cross-linked PEO, PEO-based block copolymers such as poly(ethylene oxide-*b*-amide) (Pebax), poly(ethylene oxide-*b*-butylene terephthalate) (PEO-*b*-PBT).[43–49]

4.3   Sorption-enhanced Polymer Membranes

The solubility coefficient indicates the extent to which gas molecules are adsorbed in polymers, and is obtained by the concentration of the adsorbed gas per unit polymer volume and applied pressure, because it is dependent on condensability and physical interactions of the gas molecules for polymer membranes.[14] Generally, the concentrations as a function of pressure represent sorption isotherms with characteristic curves that are concave to the pressure axis. Governed by the relative polymer–penetrant interactions and the relative condensability of the penetrants, the sorption enthalpy implicates contributions resulting from the difference in cohesive forces operating between polymer segments before and after dissolving the penetrant as well as from cohesive van der Waals and electrostatic forces acting between the dissolved penetrant and the host polymer matrix.[50] Therefore, solubility-selectivity is substantially beneficial to separations of vapors and hydrocarbon molecules which have a relatively large size and are condensable.

Solubility coefficients are predicted by the critical temperature (T_c) of the gases if they are not affected by specific interactions of gases with the polymer. Thus, the solubility-selectivity of gases that do not interact with the polymer is primarily determined by the ratio of their critical temperatures.[22,51] Typically, permselectivity of rubbery polymers is governed by solubility-selectivity. In this regard, there are no large differences in permanent gas pairs nor in glassy polymers while, for vapor/gas or CO_2/permanent gas pairs, the solubility-selectivity is large enough to lead to a high permselectivity.

Lin and Freeman suggested material selection guidelines for membranes that remove CO_2 from gas mixtures.[22] Structure/property guidelines have been extensively explored between CO_2 and polymers containing various polar groups by surveying CO_2 and N_2 sorption. To achieve high CO_2 permeability and superior CO_2/light gas (*i.e.* CH_4, N_2 and H_2) selectivity, many polar groups such as ether (-O-), nitriles (-CN), carbonyls (-C=O-), acetates (-COO-), amides (-NHCO-) were introduced into the polymers. These polar groups contribute to high CO_2 solubility and CO_2/light gas solubility-selectivity, but polar groups in a polymer decrease CO_2 diffusion coefficients due to the strong affinity between CO_2 and the polar groups.

Silicon-based polymers such as polydimethylsiloxane (PDMS), which is known as the most permeable rubbery polymer, are some of the most investigated polymer membranes because of their high gas permeabilities owing to the chain flexibility, rotational mobility, and thus large free volume. However, gas selectivities are lower than those of common glassy polymers such as polyimides and polysulfones although siloxane polymers show much higher gas permeabilities than many other rubbery polymers. Physical and permeation properties of natural rubbers and siloxane polymers are summarized in Table 4.4. For instance, O_2/N_2 selectivity of PDMS reaches only to 2.1. On the other hand, these polysiloxane membranes are used in the separation of organic vapors from permanent gases, such as the recovery of high-value organic vapors, ethylene, propylene, gasoline, and vinyl chloride from industrial waste gas streams.

Table 4.4 Physical and permeation properties of basic rubbery polymers[52]

Polymer	Temperature (°C)	Permeability (Barrer)	$\alpha_{CO2/N2}$	References
Natural rubber	25	120	16.3	19, 49
Methyl rubber	25	6.9	15.8	52
Butyl rubber	25	4.7	16	52
Polybutadiene	24	119	10	53
Polyethylene (LDPE)	25	12.6	13	52
Poly(dimethyl siloxane)	35	3700	7.4	52

4.3.1 Poly(Ethylene Oxide) Membranes

Polyether membranes, which contain ether linkages possessing strong chemical interaction with carbon dioxide, represent very promising permeabilities as well as permselectivities as an efficient CO_2 separation membrane. So far, poly(ethylene oxide) (PEO) with ether oxygen has been reported to be the most useful groups to provide excellent CO_2 separation and permeation properties.[44,51] The transport behaviors of pristine PEO for various gas molecules have intensively investigated by Lin and Freeman. At 35 °C and infinite dilution, semi-crystalline PEO exhibits CO_2 permeability coefficient of 12 Barrers, and CO_2/H_2 and CO_2/N_2 pure gas selectivities of 6.7 and 48, respectively. In addition to good separation properties for quadrupolar–nonpolar gas pairs, PEO also shows interestingly high selectivity for olefins over paraffins, which is ascribed to favorable interaction between the polar ether groups in PEO and olefins.

Figure 4.4 presents permeability coefficients for gases at 35 °C as a function of upstream pressure. For the permanent gases (He, H_2, O_2, and N_2), permeability coefficients are usually independent of pressure, while the permeability coefficients of CO_2 and hydrocarbons such as CH_4 vary with the applied pressure. This behavior is consistent with gas permeation properties in rubbery polymers. The effect of temperature on permeation properties of PEO membranes was also investigated at 25, 35 and 45 °C. All penetrants exhibit higher permeability at higher temperature. It is also of interest to explore the difference between olefin and paraffin transport in these two polymers because of the importance of this separation to the petrochemical industry.[54] In amorphous PEO, olefins exhibit significantly higher permeability than their paraffin analogues. The pure gas C_3H_6/C_3H_8 selectivity at infinite dilution and 35 °C is 2.7.

However, pure PEO materials tend to crystallize easily, making it difficult to improve CO_2 separation and permeation properties, thus various cross-linked PEO membranes were developed to solve this problem. Hirayama reported poly(ethylene oxide) (PEO)-containing and cross-linked polymer films by mixtures of poly(ethylene glycol) dimethacrylate (DM), 2,2-bis(4-methacryloxy polyethoxy phenyl)propane and poly(ethylene glycol) methyl ether methacrylate (MM).[52] The films showed very high CO_2 permeability of 510 Barrer as well as high CO_2/N_2 selectivity of 36 for DM69/MM9(90/10) at 50 °C.

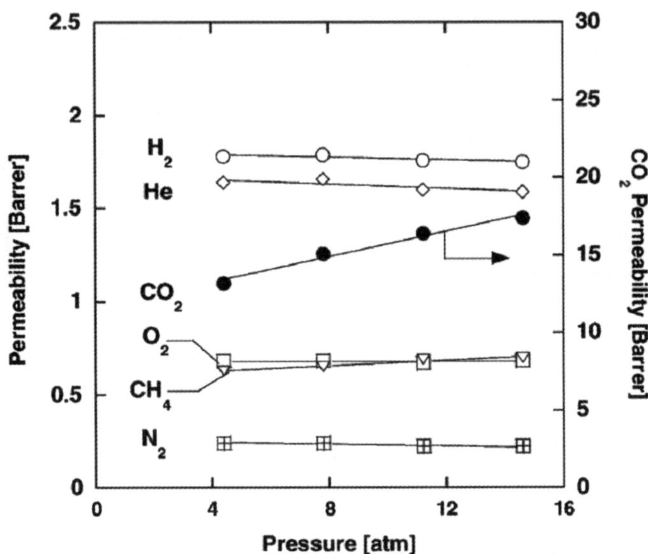

Figure 4.4 Permeability coefficients of penetrants in semi-crystalline PEO at 35 °C as a function of upstream pressure.[51]

The highest CO_2/N_2 selectivity of 69 was obtained at 25 °C for DM14/MM9(90/10) with P_{CO2} of 62 Barrer. It is notable that the CO_2/N_2 selectivity at 50 °C is still over 35 although sorption selectivity of CO_2/N_2 tends to decrease drastically at higher temperature (Table 4.5).

Cross-linked or grafted PEOs with mono- or diacrylate by UV curing were systematically investigated by Lin and Freeman (in Figure 4.5).[45,55–58] As the introduction of mono-functional acrylate provides more rotational free energies and thus prohibits crystallization of the ethylene oxide domains, PEGMEA copolymers represent higher free volume elements as well as high diffusion coefficients and gas permeabilities. As PEGMEA monomer content increases from 0 to 99 wt%, CO_2 permeability increases about four-fold, reaching 570 Barrers and CO_2/N_2 selectivity was retained from 52 to 41 at 35 °C.

Interestingly, the implication of methyl ether chain end groups in the monomer increases free volume and improves CO_2/H_2 separation performance of copolymers of PEGMEA and PEGDA as described in Figure 4.6. Moreover, it is notable that the effect of operation temperature is significant to the gas permeability and selectivity. The CO_2/H_2 pure gas selectivity of XLPEO reached a value of 40 at −20 °C (*i.e.* CO_2 permeability = 52 Barrers).[45] These reverse selective permeation properties are due to the increase of CO_2 solubility at reduced temperature and the resulting CO_2/H_2 solubility-selectivity. It indicates that solubility-selectivity can be strengthened by increasing the condensability of gas at lower temperatures, leading to high selectivity. However, considerations about economic process conditions (*i.e.* pressure and temperature) in practical applications must be considered.

Table 4.5 CO$_2$ permeabilities and permselectivies of cross-linked poly-(ethylene oxide)s

Cross-linker	Sample name	Temperature (°C)	CO$_2$ permeability (Barrer)	$\alpha_{CO2/N2}$	$\alpha_{CO2/CH4}$	Reference
No cross-linker	Poly(ethylene oxide) (PEO)	35	12	48	21	51
Dimethacrylate/ methacrylate	DM69/MM9	50	510	36	—	52
	DM14/MM23	25	240	65	—	
	DM14/MM9	25	62	69	—	
Diacrylate/ methyl ether acrylate	PEGDA/ PEGMEA(0)	35	112	52	13	22, 57
	PEGDA/ PEGMEA(50)	35	250	41	16	
	PEGDA/ PEGMEA(99)	35	570	41	13	
Cross-linked BPA-EDA/ PEGDMA	XLBPAEDA4	35	12	60	20	59
	BPAEDA4/ PEGMEA (50)	35	132	41	8.4	
	BPAEDA15/ PEGMEA (50)	35	300	51	19	
	BPAEDA15/ DGEEA (50)	35	270	46	19	
Dendritic imide	SPEO-III-1-ME (58 wt% of PEO)	35	55	50	—	60
	SPEO-II-3-AM (60 wt% of PEO)	35	40	55	—	
Amine- terminated PEO	Matrimid-PEO(20)	35	7.5	48	22	61
	Matrimid-PEO(50)	35	59	52	18	
	Matrimid-PEO(100)	35	115	52	17	

Richards *et al.* also studied cross-linked PEO based on bisphenol A ethoxylate diacrylate (BPA-EDA).[59] BPA-EDA was polymerized as a function of cross-linker molecular weight and copolymer composition. Copolymerization strategies involving the insertion of flexible PEG side chains along the network backbone proved effective in enhancing network free volume and increasing permeability. Copolymerization of the BPA-EDA cross-linkers with flexible PEG monomers was employed as a strategy to control cross-link density and to simultaneously increase the free volume of the membranes. Gas permeability of the cross-linked BPA-EDA membranes compared favorably with that reported for cross-linked PEGDA, with the cross-linked BPAEDA 15 network displaying CO$_2$ permeability at infinite dilution that was 40% higher than the value for cross-linked PEGDA 14. Here again, the insertion of methyl- or ethyl-terminated PEG branches *via* copolymerization proved effective in increasing the permeation properties of the membranes, with overall performance approaching or exceeding the literature upper bound for a number of CO$_2$/light gas pairs.

Yin *et al.* synthesized star-like poly(ethylene oxide)s by using anhydride-terminated hyperbranched polyimides as the central cores and poly(ethylene oxide)s (PEOs) as the linear arms.[60] Multi-armed star-like polymers are

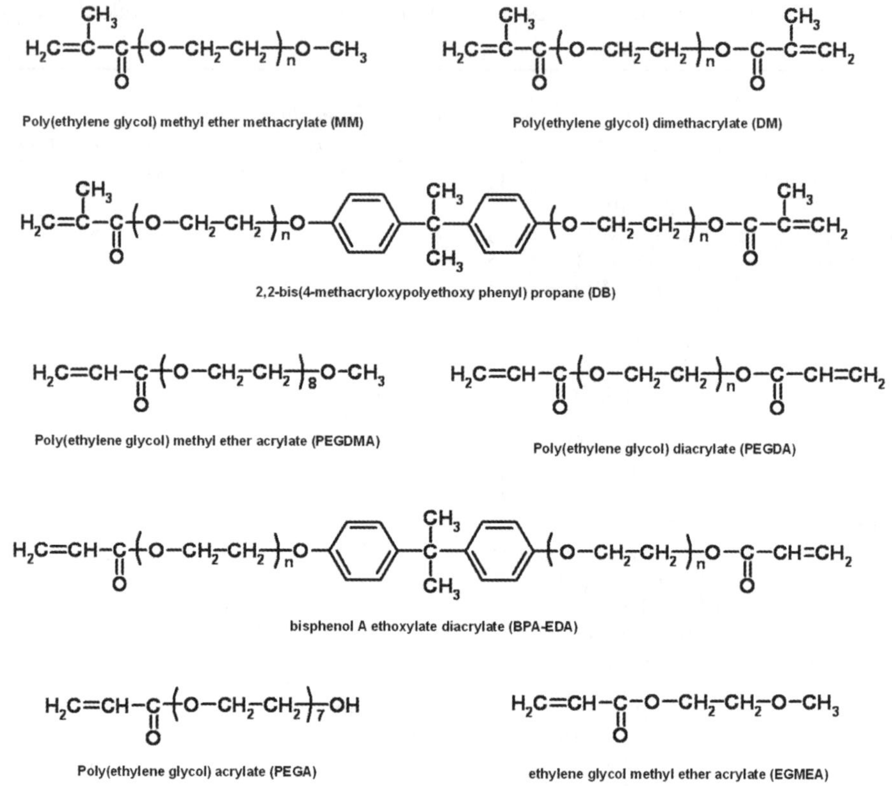

Poly(ethylene glycol) methyl ether methacrylate (MM)

Poly(ethylene glycol) dimethacrylate (DM)

2,2-bis(4-methacryloxypolyethoxy phenyl) propane (DB)

Poly(ethylene glycol) methyl ether acrylate (PEGDMA)

Poly(ethylene glycol) diacrylate (PEGDA)

bisphenol A ethoxylate diacrylate (BPA-EDA)

Poly(ethylene glycol) acrylate (PEGA)

ethylene glycol methyl ether acrylate (EGMEA)

Figure 4.5   Various mono- or diacrylate monomers for cross-linked poly(ethylene oxide)s.

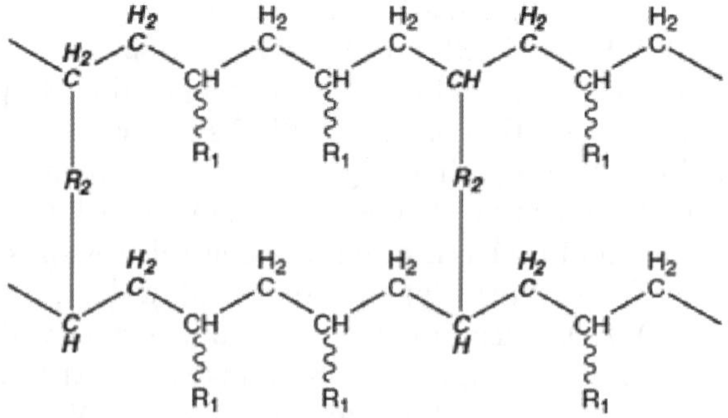

Figure 4.6   PEGDA/PEGMEA copolymer network. $R_1 = \text{-CO(OCH}_2\text{CH}_2)_8\text{OCH}_3$ and $R_2 = \text{-COO(CH}_2\text{CH}_2\text{O})_{14}\text{OC-}$.[45]

three-dimensional macromolecules, in which a large number of linear arms of similar molecular weight disperse from a central core. The three-dimensional microphase-separated structure in star-like polymers may be quite different from that of linear segmented ones, and the intermolecular packing of star-like polymers may also be different from the intersegmental packing of linear polymer chains. Owing to the ethylene oxide domains, the star-like PEO membranes were much more permeable to CO_2 (55 Barrer) than to N_2 (1.1 Barrer) and even to H_2 (9.1 Barrer), which is characteristic of PEO-based membranes. However, the permeation properties did not represent extra-ordinary results expected by the three-dimensional disorder probably due to the difference in the morphology.

Cross-linking of amine-terminated poly(ethylene oxide) in the imide domain was also tried by Zhao *et al.*[61,62] Poly(propylene glycol) block poly(ethylene glycol) block poly(propylene glycol) diamine (PPG/PEG/PPGDA), was employed to chemically cross-link Matrimid 5218 at room temperature. The effects of the cross-linking reaction on mechanical performance, gel content and H_2, CO_2, N_2 and CH_4 gas transport properties of the cross-linked Matrimid membranes were investigated. The presence of the PPG/PEG/PPGDA in the Matrimid greatly increased the solubility selectivity, which mainly contributes to high gas selectivity. In the Matrimid-PEO(100), the permeability of CO_2 increased 21 times over pristine Matrimid, and both the selectivities of CO_2/H_2 of 7.3 and CO_2/N_2 of 52.49 were also improved while the selectivity of CO_2/CH_4 dropped from 36 to 17. However, the mechanical properties of the cross-linked polyimide membranes were drastically diminished, which doubted that the break of stable imide domain into amide might result in long-term stabilities.

4.3.2 PEO-based Block Copolymer Membranes

Improvement of permselectivity by increasing solubility-selectivity in the polymer matrix have not been significant because molecules adsorbing favorably in the polymer act as plasticizers, resulting in a significant decrease of permselectivity by increasing the permeability coefficient of undesirable molecules. Thus, to mini-mize the effect of plasticization and to compensate the weak properties of PEO such as mechanical and thermal resistances while maintaining the advantages of poly(ethylene oxide), block copolymers are prepared to have a hard block com-posed of glassy polymer unit as well as a soft block of ethylene oxide unit.

Kim and Lee reported that poly(amide-*b*-ethylene oxide) (PEBAX could be a promising candidate to overcome this difficulty.[47] PEBAX is a block copolymer consisting of a polyamide (PA) block as a hard segment and a polyether (PE) block as a soft segment (Figure 4.7). A crystalline amide block in PEBAX functions as an impermeable phase, whereas an ether block acts as a permeable phase because of its high chain flexibility. Rezac and John reported diffusivity and solubility of water and methanol in PEBAX several research groups stu-died gas permeabilities of a wide range of activities near room tempera-ture.[47,63–65] In the removal of CO_2 from mixtures in syngas or flue gas, PEBAX

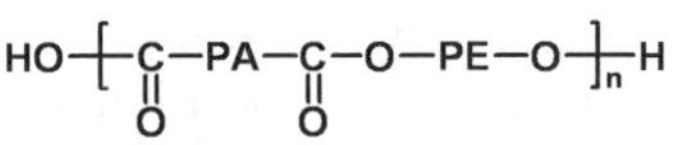

$$HO-\left[-\underset{O}{\underset{\|}{C}}-PA-\underset{O}{\underset{\|}{C}}-O-PE-O-\right]_n H$$

Figure 4.7 Structure of poly(amide-*b*-ethylene oxide) (PEBAX).

shows high selectivity in polar or quadrupolar/non-polar systems (*e.g.* CO_2/H_2 or CO_2/N_2). The sorption and permeation results indicate strong interactions between the polar gas CO_2 and the PE blocks in the copolymer. It is found that strong affinities of polar species to the PE block is attributable to the high permeability and permselectivity of polarizable gases through the PEBAX copolymer. Therefore, remarkably high selectivities of quadrupolar/non-polar gas, coupled with high CO_2 permeability coefficients, were observed. CO_2/H_2 selectivities as high as 9.8 and CO_2/N_2 selectivities as high as 56 were obtained in polymers with CO_2 permeability coefficients of approximately 220 Barrer. As the amount of polyether increases, permeability increases. CO_2/N_2 and CO_2/H_2 selectivities are higher in polymers with higher concentrations of polar groups. (Table 4.6) Because CO_2 is larger than H_2 and, therefore, has a lower diffusion coefficient than H_2, the weak size-sieving ability of the rubbery polyether phase, which is the locus of most of the gas permeation, also contributes to high CO_2/H_2 selectivity.

Along with amide unit, several glassy and rubbery groups are available for the fabrication of PEO-segmented copolymer membranes.[66,67] Polyurethanes and polyimides have been utilized because of easy preparation and better solution properties as well as superior mechanical properties and chemical stabilities. These polymers can be tailored by various structure and chain length. Yoshino and Okamoto synthesized PEO-PU, PEO-PA and PEO-PI from diverse chain length and compositions, and compared their structure–property relationships.[66] To obtain the moderate chain length of PEO, the phase separation of the PEO domains in the hard-segment polymer was investigated, showing that urethane and amide existed in large blocks in PEO region due to hydrogen bonding between the oxygen of PEO and the NH group of urethane or amide. Compared with PEO-PIs, PEO-PUs and PEO-PA had much smaller CO_2 permeabilities (as shown in Table 4.7) because of much smaller CO_2 diffusion coefficients and somewhat smaller CO_2 solubilities.

Polyimide containing poly(ethylene oxide) is a simple and convenient way to improve the gas separation properties. The copolymers are synthesized by the reaction of dianhydride and diamines with/without amine-terminated poly(ethylene oxide). Maya *et al.* and Munoz *et al.* reported thermal treatment effect of PEO-*b*-PI copolymer on the gas permeation properties.[68,69] Phase segregation of the polymer chain at higher treatment temperature to 300 °C contributed to improvement of gas permeability of the membrane. Molecular weight of poly(ethylene oxide) of 6000 was appropriate to induce microphase separation. CO_2 permeability was reached from 2.3 to 24 Barrer according to the PEO composition before thermal treatment, while it was enhanced by three to ten times higher permeability after thermal treatment at 300 °C without serious reduction in permselectivity.

These membranes are beneficial to separate CO_2/H_2 mixture as well. While conventional polyimides exhibited higher permeability for H_2 than that of CO_2, polyether-containing polyimides showed 'reverse selective' characteristics.[70] As increasing the chain length and molar composition in the block copolymer, CO_2 permeability as well as CO_2/H_2 selectivity improved simultaneously. As highly condensable CO_2 competed with H_2 for the sorption sites in hard segment, mixed

Table 4.6 Structure–property relationships of poly(amide-*b*-ethylene oxide) membranes for CO$_2$ separation

Copolymer	Soft segment	Hard segment	Temperature	CO$_2$ permeability (Barrer)	$\alpha_{CO2/N2}$	$\alpha_{CO2/H2}$	Reference
PEBAX 2533	Tetramethylene oxide 80 wt%	Nylon-12	35	180	18	3.5	64
PEBAX 4033	Tetramethylene oxide 53 wt%	Nylon-12	35	95	16	3	
PEBAX 4011	Ethylene oxide 57 wt%	Nylon-6	35	57	42	6.9	
PEBAX 1074	Ethylene oxide 55 wt%	Nylon-6	35	110	46	7.9	
PEBAX 3533	Tetramethylene oxide 70 wt%	Nylon-12	25	132	61	6.1	65
PEBAX 1657	Ethylene oxide 60 wt%	Nylon-6	25	122	71	—	
PEBAX 1657/TEOS 10wt%	Ethylene oxide 60 wt%	Nylon-6	25	154	72	—	47
PEBAX 1657/TEOS 19wt%	Ethylene oxide 60 wt%	Nylon-6	25	205	75	—	
PEBAX 1657/TEOS 27wt%	Ethylene oxide 60 wt%	Nylon-6	25	277	79	—	

Table 4.7 Structure–property relationships of poly (ethylene oxide)-segmented copolymer membranes for CO_2 separation

	Soft segment		Hard segment		Gas permeabilities and selectivities					
Copolymer	Type	Weight%	Type	Weight%	Temperature ($°C$)	Pe_{CO2} (Barrer)	$\alpha_{CO2/N2}$	$\alpha_{CO2/CH4}$	$\alpha_{CO2/H2}$	Reference
PEO-*b*-PU	PEG(75)	54	MDI-BPA	46	35	31	44	—	—	66
PEO-*b*-PU	PEG(85)	66	MDI-BPA	34	35	59	49	—	—	
PEO-*b*-PA	PEO3(80)	68	IPA-ODA	32	35	58	53	—	—	
PEO-*b*-PI	BPDA-PEO1(80)	47	BPDA-mDDS	53	35	3.8	58	—	—	
PEO-*b*-PI	BPDA-PEO4(80)	62	BPDA-ODA	38	35	117	51	—	—	
PEO-*b*-PI	PMDA-PEO2(75)	54	PMDA-ODA	46	35	40	54	—	—	
PEO-*b*-PI	PMDA-PEO4(80)	68	PMDA-pDDS	32	35	238	49	—	—	
PI-28	BPDA-DAPEG	28	BPDA-ODA	72	30	2.3	39	18	—	69
PI-28 (300 °C treat)	BPDA-DAPEG	28	BPDA-ODA	72	30	23	34	17	—	
PI-43	BPDA-DAPEG	43	BPDA-ODA	57	30	24	61	10	—	
PI-43 (300 °C treat)	BPDA-DAPEG	43	BPDA-ODA	57	30	77	54	18	—	
PI-68	BPDA-DAPEG	68	BPDA-ODA	32	30	16	53	15	—	
PI-68 (300 °C treat)	BPDA-DAPEG	68	BPDA-ODA	32	30	37	40	19	—	

PUU (PTMO/PDMS)	MDI-PTMO/PDMS				30	82	25	—	—	67
PUU (PEO/PDMS)	MDI-PEO/PDMS				30	624	11.5	—	—	
PEO-*b*-PBT	PEO (1K)	40	Butylene terephthalate	60	25	40	85	—	—	71
PEO-*b*-PBT	PEO (1 K)	64	Butylene terephthalate	36	25	105	55	—	—	
PEO-*b*-PBT	PEO (4 K)	75	Butylene terephthalate	25	25	650	—	—	—	
PEO-*b*-PA	PEO (1 K)	62	T6T6T	38	35	75	41	14	7.2	72
PEO-*b*-PA	PEO (2 K)	76	T6T6T	24	35	180	49	16	10	
PEO-*b*-PA	PEO (1 K)	78	Diamide TΦT	22	35	105	55	19	7.4	73
PEO-*b*-PA	PEO (1.5 K)	84	Diamide TΦT	16	35	126	53	18	8.2	
PEO-*b*-PA	PPO (2.2 K)	70	Diamide TΦT	30	35	418	27	8	6.1	
PEO-*b*-PA	PPO (4.2 K)	75	Diamide TΦT	25	35	520	25	7.5	6.2	
PEO-*b*-PTT	PEO (2 K)	70	Trimethylene terephthalate	30	30	183–200	51	—	10.2	74
PEO-*b*-PTT	PEO (3 K)	70	Trimethylene terephthalate	30	30	114	53	—	11.3	

gas permeation tests showed that the best reverse selective membranes had a CO_2 permeability of 179.3 Barrer and a CO_2/H_2 permselectivity of 22.7.

Park and Lee prepared mixed soft segment urethane urea membranes by using polysiloxane/polyether copolymer. Addition of PDMS into poly-(urethane-*b*-ether) membrane led to the phase separation in both hard segment (MBI/BD) and soft segment due to the difference in solubility parameters, and the dispersed PDMS phases served to provide more diffusive pathways in the polymer matrix.[67] Accordingly, the multi-block copolymer exhibited out-standing CO_2 permeability (984 Barrer), while the CO_2/N_2 selectivity dropped because the highly permeable siloxane domain enabled permeation of other light gases as well as CO_2.

The most impressive CO_2 separation performances in recent soft–hard segment copolymers have been reported by Wessling and coworkers.[71,75] As shown in Figure 4.8, poly(ethylene oxide)–poly(butylene terephthalate) (PEO-PBT) segmented multi-block copolymers allowed for such structural mod-ifications as the variation of the PEO segment length and amount of PEO. This enables a systematic study of the relation between the polymeric structure and the gas-transport properties. PBT forms the hard hydrophobic rigid crystalline phase: PEO forms the soft hydrophilic amorphous rubbery PEO segment. Depending on the composition of the block copolymer, up to five different phases may be present: two amorphous (PEO and PBT), two crystalline (PEO and PBT), and an interphase in which amorphous PBT is mixed with amorphous PEO. PEO-PBT block copolymers exhibit a high CO_2/N_2 selectivity ($\alpha \approx 60$ at 35 °C) due to the high solubility of CO_2 in the PEO phase. This selectivity is structure-dependent and is higher for block copoly-mers with larger amounts of PBT. With the PEO molecular weight of 1000, the CO_2 permeability is 130 Barrer and CO_2/N_2 selectivity is 60 at 25 °C.[71] At an increased temperature, copolymer containing high PEO length and composition, 4000PEO75PBT25 shows a drastic increase up to 650 Barrer as the PEO crystals start to melt. Moreover, these membranes exhibit outstanding water vapor permeability as well as water vapor/inert gas selectivity, due to the contribution of stagnant boundary layers at feed and permeate side.[75]

Exchange of the butylene terephthalate unit into a mono-disperse amide segment unit yielded microphase dispersed morphology, comprising a continuous PEO phase with dispersed crystallizable amide segments.[72,73] The CO_2 permeability increased strongly with PEO concentration, and this effect could partly be explained by the dispersed hard segment concentration and partly by the changing chain flexibility. The gas transport values were dependant on both the dispersed hard segment concentration and the polyether

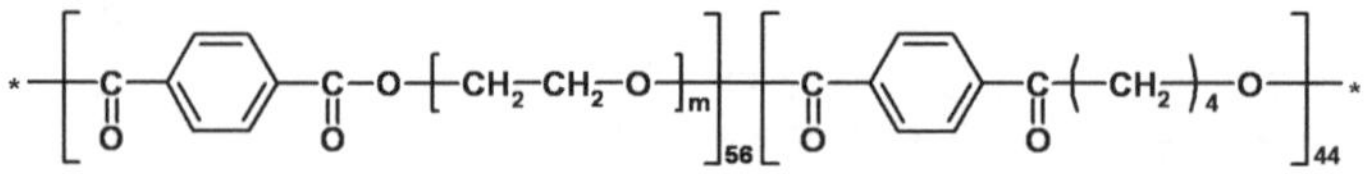

Figure 4.8 Structure of poly(ethylene oxide)–poly(butylene terephthalate) segmented multi-block copolymer.

segment length (length between cross-links). The gas permeation properties are summarized in Table 4.7.

Yave *et al.* investigated the effect of alkylene length in the terephthalate domain on the CO_2 separation performances.[74,76] PTT possesses several advantageous properties, including good tensile behavior, resilience, and outstanding elastic recovery. The interesting feature of this copolymer is that the exquisite combination of rigid (PTT) and flexible (PEO) segments allows manipulating its properties. The CO_2 permeability is 183–200 Barrer for membranes prepared from copolymers at the PEO molecular weight of 2000 and the content of 70 wt%.

These PEO-based copolymers show favorable CO_2 separations from flue gas and syngas. Copolymers enhanced the mechanical properties of PEO and controlled the crystallinity of PEO so that the ether linkage could maximize the adsorption of CO_2 in the free volume of polymer. The CO_2 permeabilties usually range from 50 to 200 Barrers (at most, 650 Barrer) while the CO_2/N_2 selectivities are over 40–50, the extent enough to enrich the CO_2 concentration to over 90% at a single stage from flue gas. On the other hand, the performances are significantly affected by the operation temperature mainly due to the solubility of CO_2 and the morphologies of the multi-block copolymers.

4.3.3 Dendrimer Membranes

Liquid membranes such as poly(amidoamine) (PAMAM) dendrimers are promising for highly selective separation of CO_2 from N_2 because amine-containing reagents are beneficial to adsorb acidic CO_2 due to the acid-base compatibility. The primary amine group -NH_2 is known to react strongly with CO_2 to produce a carbamate ion and a protonated base; apparently water is essential for this reaction:

$$CO_2 + NR_3 + H_2O \leftrightarrow NR_3H^+ + HCO_3^- \tag{4.13}$$

This type of facilitated transport membrane is favorable at a relatively low feed pressure in the presence of water because reactive carriers increase the CO_2 permeability drastically and CO_2 permeance significantly at low CO_2 partial pressures. Therefore, these liquid membranes have been studied by a number of researchers over the years.

Kovvali and Sirkar reported an excellent CO_2/N_2 selectivity for the viscous and non-volatile liquid PAMAM dendrimer, as an immobilized liquid membrane in a porous support membrane at atmospheric pressure.[77,78] Starburst polyamidoamine (PAMAM) generation 0 dendrimers having ethylene diamine (EDA) (as shown in Figure 4.9) as the core of the membrane liquid which will strongly facilitate CO_2 transport and efficiently block the transport of other gaseous species. The membranes composed of dendrimer and additives like glycerol in hydrophilic porous substrates function as a CO_2-selective molecular gate at low CO_2 partial pressures and high feed relative humidities. Although

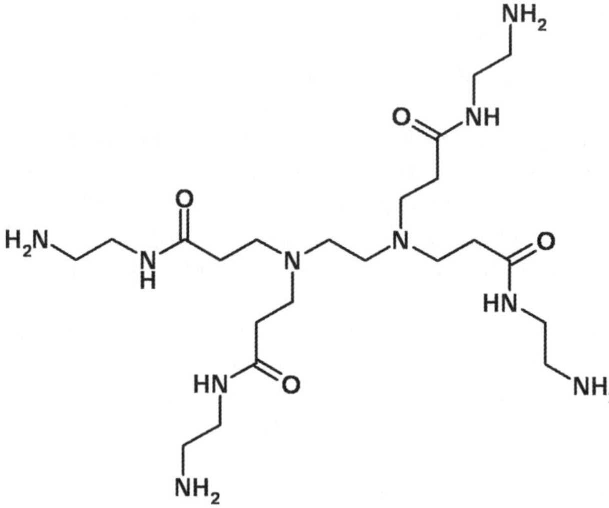

Figure 4.9 Schematic of the structure of the EDA core PAMAM dendrimer of generation 0.

considerably dependent on the CO_2 partial pressure difference, the CO_2/N_2 selectivities usually exceed 3000 at very low ($\sim 0.5\%$) feed CO_2 concentration while the CO_2 permeances do not reach over 10 GPU (1 GPU $= 10^{-6}$ cm^3(STP) cm^{-2} s^{-1} (cmHg)$^{-1}$.

Kazama and coworkers contribute to the higher flux of PAMAM membranes by several modifications.[79–82] Those include *in situ* modification method, allowing the interfacial precipitation of the dendrimer on the surface of commercially available porous hollow fiber substrates composed of poly(vinylidene difluoride) or polysulfone, UV cross-linking with PEGDMA for selective permeation of CO_2, and introduction of chitosan gutta layer and hyaluronic acid (HA) additives, which improved the CO_2 permeance of more than 110 GPU. Furthermore, they enlarged the membrane as a commercial scale due to the advantage that the *in situ* modification permitted direct post-treatment of commercial polysulfone ultrafiltration membrane modules.[83] Impregnation of the chitosan gutta layer gave birth to a thin PAMAM layer, thus the composite modules of 180 cm^2 and 0.4 m^2 membrane area exhibited the CO_2/N_2 selectivity of 110–170 and a CO_2 permeance of 20–29 GPU at humidified 5% CO_2/95% N_2 mixture with the long-term stability for more than 1000 h.

As PAMAM dendrimers are amine-terminated at each generation, those were utilized as cross-linkers disassociating imide segment into cross-linked amide unit. Xiao and Chung introduced the 0–2 generation of dendrimers in the 6FDA-durene polyimide so that they can control the length of cross-linking between the polyamide chains as well as the impregnation degree of dendritic amine.[84] Because the specific interactions of membranes with CO_2 increase, both CO_2 permeability and CO_2/N_2 selectivity are improved over polymeric upper bound with the modification time and the generations of PAMAM dendrimers.

In the meantime, other types of dendrimers are reported to enhance the permeation properties of PAMAM. Pentaerythrityl tetraethylenediamine (PETEDA) dendrimer was synthesized from pentaerythrityl tetrabromide and ethylenediamine.[85] The composite membranes for selectively permeating CO$_2$ were prepared by using PETEDA-PVA blend polymer as the active layer and polyethersulfone (PES) ultrafiltration membrane as the support layer. For pure gases, the membrane containing 78.6 wt% PETEDA and 21.4 wt% PVA in the blend has a CO$_2$ permeance of 81 GPU and CO$_2$/CH$_4$ selectivity of 52.

4.4 Diffusion-enhanced Membranes

Glassy polymers composed of rigid aromatic or conjugated aliphatic chain tend to present high diffusivity-selectivity for gas mixtures. Because diffusivity-selectivity means the ability of polymer materials to be selective for the size and shape of gas molecules, diffusivity-selectivity is relatively higher in glass polymers which possess rigid polymer chain and regular intersegmental distance than rubbery polymers.[14] On the other hand, common glassy polymers have relatively low diffusivity-selectivity due to the existing chain mobility of covalent bonds, in comparison with inorganic materials with a controlled molecular size. To improve the diffusivity-selectivity as well as diffusion coefficient of gas molecules in polymer membranes, it is efficient to demonstrate the well-tailored and narrow distribution of free volume elements between distorted structures of highly rigid glassy polymers so that the intermolecular chain packing can be disturbed, and thus small penentrants can be transported promptly.[9]

Recently developed highly permeable glassy polymers, therefore, are basically constructed by rigid main chain such as ladder-like or heteroaromatic structure, and also contains flexible or rotatable domains in the main chain or side chain. A number of poly(substituted acetylene) membranes has been reported by modifying the substitution group at side chain.[33,86,87] Also, alicyclic fluorinated polymers have been introduced as candidates.[34,35] In 2003, polymers with intrinsic microporosity (PIMs) were reported to have intrinsic cavities in the polymer matrix by constructing ladder-like polymers with torsional domain.[37,38] Moreover, thermally rearranged (TR) polymer membranes have been presented to have extraordinary gas selectivities and permeabilities by solid-state rearrangement of polyimide into rigid heteroaromatic polymers by creating micro-cavities.[39] Here, these highly permeable glassy polymers are reviewed in relation with their free volume elements characterized by density, intermolecular distance, sorption behavior and positron annihilation lifetime spectroscopy (PALS).

4.4.1 Substituted Polyacetylene-based Membranes

Since Masuda firstly reported a novel synthesis of poly(1-trimethylsilyl-1-propyne) in 1983, poly(substituted acetylene)s have considered as the most permeable polymers owing to their high free volume elements and the following superior diffusion coefficients.[33] In addition, this type of polymer has good solubility in organic solvents and thermal stability up to 200 °C as well as wide intermolecular distance and low densities, thus it has been widely studied for gas separation membrane regarding its diffusivity, sorption behaviors and permeation characteristics, *etc.*[88] Moreover, the most crucial advantages of this polymer are the potentials available to utilize various bulky side chains from mono- or di-substituted acetylene monomers.[43,86,89]

In particular, PTMSP is one of the most studied polymers for membrane applications because of its unique structure and excellent gas and vapor permeation properties. A large number of publications have appeared due to its unique structure and properties and, in particular, its extremely high gas permeability. High molecular weight PTMSP was obtained by introducing transition metal catalysts from groups 5, 6 and 8 in periodic table such as WCl_6-Ph_4Sn, $NbCl_5$ and $TaCl_5$.[33,88] Analogous to PTMSP, numerous mono-substituted and di-substituted acetylene polymers have been reported: poly(4-methyl-2-pentene) (PMP), poly(1-trimethylgermyl-1-propyne) (PTMGP), poly(1-*p-tert*-butylphenyl-2-phenylacetylene) (PTBPPA), and poly(1-triethylsilyl-1-propyne) (PTESP), *etc.* These polymers called as superglassy polymers were soluble in non-polar solvents such as hydrocarbons, toluene, and carbon tetrachloride, indicating ease of film preparation by solution casting methods. Nagai reported a detailed review on the synthesis of substituted poly(substituted acetylenes)s derived from various di-substituted monomers and the characterization of the membranes for gas permeations.[88]

Among the unique characteristics of these polymers, fractional free volume (FFV) of this kind of polymers is almost two times higher than FFVs of conventional glassy polymers like polysulfones, poly(phenylene oxide)s and polyimides; the fractional free volume (FFV) of PTMSP is up to 0.34. The free volume elements dispersed in this polymer is known to have two or three different radii of 0.35 and 0.55 nm.[90] Furthermore, it is believed that these free volumes are connected mutually, thus enabling gases or vapors to diffuse through polymer matrix. Other substituted acetylene polymers, analogous to PTMSP structures, represent still superior physical properties (glass transition temperature, densities, and free volumes, *etc.*) as compared to conventional polymers even though they do not reach to those of PTMSP.

The large size and amount of free volume elements of these polymers result in high gas and vapor permeation properties, permitting the materials to be applied to high-flux gas separation membranes.[91] PTMSP shows the most permeable properties compared to any kind of polymer although it has low selectivities to permanent/permanent gas mixtures; the oxygen permeability reached to 6100 Barrer, whereas the O_2/N_2 selectivity is only 1.8. The potential applications of PTMSP include oxygen enrichment from air applicable for use in car engines, respiration-aiding apparatuses, and combustion furnaces.[88] However, the physical aging of PTMSP prevents the commercial application where the permeability of fresh PTMSP drops to the oxygen permeability of 1800 Barrer after 1 month storage at room temperature.

4.4.2 Amorphous Fluoropolymer Membranes

Fluoropolymer is a kind of polymer which has been extensively investigated along with the synthesis of fluorine-containing monomers such as tetrafluoroethylene, trifluoroethylene and vinylidene fluoride, *etc.* Fluorinated polymers have many potential characteristics compared to common hydrocarbon polymers. Polytetrafluoroethylene (PTFE), known by the commercial trade name of Teflon, possesses extraordinary chemical, thermal, mechanical and electric properties. Whereas most PTFE- or PVDF-based fluoropolymers are either crystalline or semicrystalline, perfluoro-alicyclic Teflon AF polymers (Figure 4.10) have been reported to exhibit similar desirable electric, chemical, thermal and mechanical properties, compared to semicrystalline fluoropolymers.[34] Moreover, these amorphous fluoropolymers have unique physical properties, such as high glass transition temperatures and high free volume elements, similar to the most famous high free volume glassy polymer, PTMSP. These fluoropolymers can be dissolved in a few fluorinated solvents due to chain flexibility, high free volume fraction and low cohesive energy density (CED), which contributes to low activation energies for diffusion of small molecules as well as solubility of gases.[92]

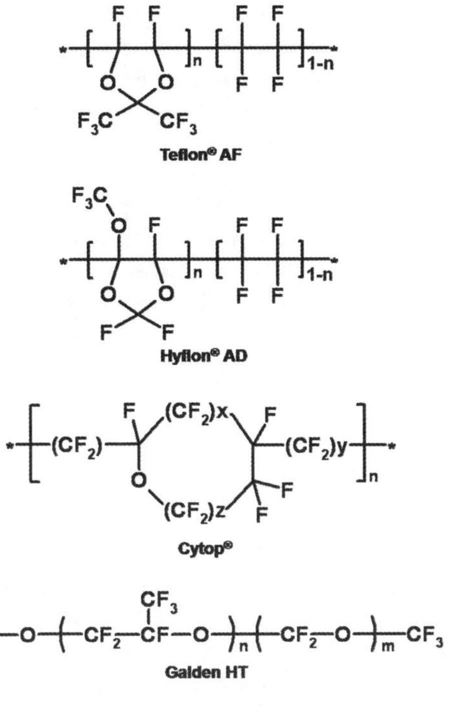

Figure 4.10 Structures of major perfluoropolymers and fluorinated solvent.

To prepare Teflon AF series, reactions of hexafluoroacetone (HFA) and ethylene oxide (EO) give rise to defluorinated 2,2-bis(trifluoromethyl)-4,5-difluoro-1,3-dioxole (BDD) monomer, followed by the copolymerization with tetrafluoroethylene (TFE). The physical properties of these amorphous copolymers are dependent upon the compositions in the copolymers as shown in Table 4.8. Currently, there are two commercial grades of Teflon AF polymers produced by DuPont, AF-2400 and AF-1600, which contain the composition of copolymers for which $n = 0.87$ and 0.65, respectively, where n is the molar ratio of the BDD monomers.[35]

Hyflon AD is an amorphous perfluorinated copolymer of tetrafluoroethylene (TFE) and 2,2,4-trifluoro-5-trifluoromethoxy-1,3-dioxole (TTD). Compared to the situation in the Teflon AF series, the reduced intermolecular distance due to the structural change result in moderate fractional free volume of 23%. Thus, the gas permeabilities are lower than those of Teflon AF membranes, allowing the increase of permselectivities. While this fluoropolymer has the advantage of solubility in fluorinated liquid, such as Galden HT 55 and HFE 7100, the permeation properties are affected by the kind of residual solvents as well.

Because the syntheses of these fluorinated polymers require the treatment of very toxic fluorine chemicals they are mainly produced by major chemical companies such as Solvay Solexis. Membrane application using the polymers has been considerably restricted. However, several recent studies report various transport properties such as gas permeation, diffusion and sorption, and free volume characteristics in these amorphous fluoropolymers.[34,93–96] Similar to PTMSP, these polymers are permeable to permanent gases. As summarized in Table 4.8, carbon dioxide and nitrogen permeability of a Teflon AF-2400 membrane is 2200 and 480 Barrers, respectively. Pinnau and Toy compared gas permeation properties of a AF-2400 flat-sheet membrane prepared by a solution-casting method to PTMSP membrane.[34] Although PTMSP and amorphous fluoropolymers are both ultra-high free volume glassy polymers, their gas permeation properties are different from each other. Permeabilities of small gas molecules in AF-2400 membranes are higher than those of large, condensable gases while PTMSP membranes exhibit the opposite permeation behaviors. This opposite phenomenon is based on the differences between the permeation properties of two polymers affected by both diffusivities and solubilities. Therefore, these amorphous fluorinated polymer membranes have potentials for the separation of small gas molecules such as O_2/N_2 separation, H_2 separation from ammonia process and CO_2 separation, *etc.*

Jansen, Drioli and co-workers investigate the transport properties of Hyflon AD membranes.[97–99] Because the molar weight of the solvents used for film casting of Hyflon is relatively high the residual solvents affect the membrane performances. Treatment of the membranes above their glass transition temperature allows regeneration of the reproducible permeability and permselectivity without plasticization by the residual solvents. As the smaller and packed alicyclic structure of the fluorinated monomer, the density of Hyflon AD membranes are higher than those of Teflon AF membranes and thus the

Table 4.8 Physical properties and gas transport properties of amorphous perfluorinated polymers

Polymer	Physical properties			Gas permeability (Barrers)					Permselectivity				
	$T_g(^\circ C)$	Density (g cm^{-3})	FFV	CO$_2$	H$_2$	O$_2$	N$_2$	CH$_4$	O$_2$/N$_2$	CO$_2$/N$_2$	CO$_2$/CH$_4$	H$_2$/CO$_2$	Reference
Teflon AF 2400	240	1.74	0.33	2200	2090	960	480	390	2.0	4.6	5.7	0.95	95
Teflon AF 1600	240	1.8	0.31	520	550	270	110	80	2.4	4.7	6.5	1.05	96
Hyflon AD 80	134	1.92	0.23	150	210	67	24	12	2.8	6.3	13	1.4	100
Hyflon AD 60	134	1.92	0.23	130	180	57	20	10	2.9	6.5	13	1.38	100
Cytop	108	2.03	0.21	35	59	16	5.0	2.0	3.2	7	18	1.69	100
PVDF	30	2.15	—	1.2	2.4	1.4	0.7	1.3	2.0	1.7	0.9	2.0	101

fractional free volume relatively diminishes. Therefore, the gas permeabilities of Hyflon AD membranes do not reach those of Teflon AF membranes while the permselectivies of permanent gas mixtures are higher.

4.4.3 Polymers with Intrinsic Microporosity

Whereas conventional glassy polymers have been known to be dense without micropores except free volume elements resulting from chain rotation in the matrix, polymers with intrinsic microporosity (PIMs) is a kind of microporous polymer containing microporous cavities due to the mixed conformation of highly stiff ladder-like domain and flexible benzodioxane structure as described in Figure 4.11. Their rigid but contorted molecular structures to interfere with chain packing can retain an efficient size of free volume elements for the permeation of small gas molecules and enable to grant high diffusion coefficients.[102]

Since the first synthesis of this kind of intrinsic microporous polymer was reported by Budd, McKeown and co-workers in 2004 and in 2006, several studies have been performed regarding their microporous characters, sorption, hydrogen storage and transport properties.[37,102] Up to now, more than 50 scientific papers have been published focusing mainly on the syntheses of new PIMs, the specific surface areas by structural modifications and the sorption capabilities of hydrogen as a promising candidate for hydrogen storage materials. PIMs have a large number of free volume elements by which they have high surface areas $(500–1000 \ m^2 \ g^{-1})$ with micro-cavity diameters in the range 0.6–0.8 nm as

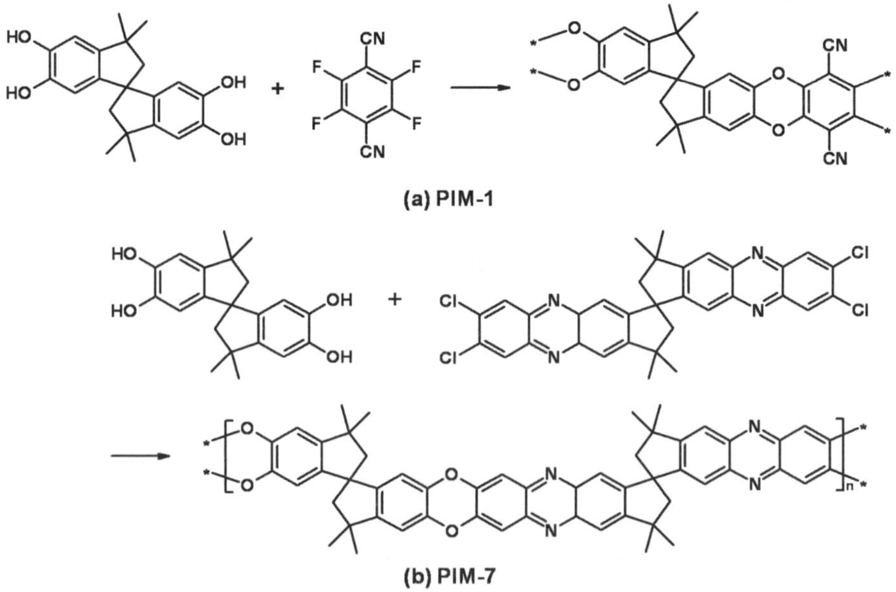

Figure 4.11 Synthesis of polymers with intrinsic microporosity: (a) PIM-1 and (b) PIM-7.

measured by positron annihilation lifetime spectroscopy (PALS).[103] PIMs have represented high gas permeation properties; O$_2$ permeabilities and O$_2$/N$_2$ selectivities of PIM-1 and PIM-7 are 370 Barrer, 4.0 and 190 Barrer, 4.5, respectively.[104] Gas permeabilities of these high free volume polymers can vary regarding the solvent during the syntheses and present physical aging and deformation of the free volume elements. In several references, CO$_2$ permeabilities of PIM-1 range from 2300 to 6500 Barrers, as summarized in Table 4.9.

Several structural modifications have been reported to prevent efficient space packing of the stiff polymer chains and consequently show intrinsic microporosity. Exchange of a biscatechol monomer into a bis(phenazine)-based tetrahalide monomer provides the potential diversity of polymers suitable for making PIMs.[105] Tetrahalide monomers containing the spirocenter allows the use of a greater number of biscatechol monomers, which would otherwise lead to linear polymers that pack efficiently in space to give a dense, non-microporous material. Catechol-containing PIM-7 to PIM-10 represent diverse microporous sorption behaviors; surface area of PIM-10 is 713 m^2 g^{-1}. Cardo-PIM-1 based on monomers containing fluorene moiety, a polycyclic aromatic hycrocarbon, showed CO$_2$ permeability of 430 Barrer and CO$_2$/N$_2$ selectivity of 33, respectively.

To regulate the free volume in PIMs, Du and Guiver introduced several bulky and rotational groups, and copolymerized those functional monomers. First, in the trifluoromethyl-containing aromatic tetrahalide the pendent phenylsulfone groups is relocated within the inter-chain free volume of the rigid and contorted ladder polymer, acting to reduce gas permeability and increase selectivity.[106] Therefore, gas permeabilities for oxygen and carbon dioxide decreased with increasing content of the trifluoromethylphenylsulfone versus the dinitrile monomer within the copolymer, while the selectivities of gases against nitrogen increased.

Second, in 5,5′,6,6′-tetrahydroxy-3,3,3′,3′-tetramethylspirobisindane (TTSBI) with tetrafluoroterephthalonitrile (TFTPN) and disulfone-based monomers: the incorporation of disulfonyl side groups into PIM structures increases selectivity.[107] The gas permeabilities and selectivities of DS-PIMs can be tuned by the size and nature of the pendant side chains and the copolymer ratio.

Third, in a series of copolymers from tetrahydroxy di-naphthyl, tetrahydroxy spirobisindane, and tetrafluoro terephthalonitrile monomers DNPIM-33 exhibits CO$_2$ permeability of 4646 Barrer and CO$_2$/N$_2$ permselectivity of 19.2, exceeding the polymeric upper bound.[108] This work demonstrates that significant improvements in properties may be obtained through development of copolymers with intrinsic microporosity (CoPIMs) that extends the spectrum of high-molecular-weight ladder structures of poly (dibenzodioxane)s. While bulky side chains in common glassy polymers enlarge the inter-chain distances and increase fractional free volume, these functional groups in PIM do not contribute to increment of torsional energies for the ladder-like main chain, but act as barriers for gas transport although the permselectivities are slightly improved.

Table 4.9 Physical properties and permeation properties of polymers with intrinsic microporosity (PIMs)

| | Physical properties | | | Gas permeability (Barrers) | | | | | Permselectivity | | | | |
Polymer	Density $(g\ cm^{-3})$	FFV	Surface area	CO_2	H_2	O_2	N_2	CH_4	O_2/N_2	CO_2/N_2	CO_2/CH_4	CO_2/H_2	Reference
PIM-1	1.09	0.26	860	2300	1300	370	92	125	4.0	25	18	1.8	104, 115
				3496	2332	786	238	360	3.3	15	10	1.5	103
				4030	1900	990	270	350	3.6	15	12	2.1	116
PIM-7			750	1100	860	190	42	62	4.5	26	18	1.3	104
Cardo-PIM-1			621	430	320	59	13	22	4.5	33	20	1.3	105
TFMPS-PIM1	1.214	0.22	—	731	—	156	33	—	4.7	22	—	—	106
TFMPS-PIM2	1.196	0.22	—	1476	—	308	75	—	4.1	20	—	—	
TFMPS-PIM3	1.156	0.24	—	2841	—	561	158	—	3.6	18	—	—	
TFMPS-PIM4	1.089	0.26	—	3616	—	737	217	—	3.4	17	—	—	
DS-PIM1-33	1.187	0.19	—	1408	—	322	88	—	3.7	16	—	—	107
DS-PIM2-33	1.198	0.19	—	1077	—	216	52	—	4.2	21	—	—	
DS-PIM3-33	1.162	0.21	—	2154	—	369	93	—	4.0	23	—	—	
DN-PIM-33	—	—		4646	2347	907	242	—	3.7	19	—	2.0	108
PIM-PI-1	1.15	0.232	680	1100	530	150	47	77	3.2	23	14	2.1	109, 110
PIM-PI-2	—	—	500	210	220	39	9	9	4.3	23	23	1.0	
PIM-PI-3	1.26	0.226	471	520	360	85	23	27	3.7	23	19	1.4	
PIM-PI-4	1.26	0.228	486	420	300	64	16	20	4.0	26	21	1.4	
PIM-PI-7	1.19	0.223	485	510	350	77	19	27	4.1	27	19	1.5	
PIM-PI-8i	—	—	683	3700	1600	545	160	260	3.4	23	14	2.3	
PIM-PI-8ii	1.14	0.231	—	2270	1020	320	100	170	3.2	23	13	2.2	

Polyimides constructed of the ladder-like structure are prepared by the reaction of benzodioxane-containing dianhydride and bulky diamines to acquire both high free volume of PIM and several superior stabilities of polyimides.[109,110] PIM-polyimides represent good solubilities for most polar aprotic solvents as well as extraordinary thermal stabilities, possessing the advantages of the two polymers. The gas permeabilities of PIM-like polyimides are the highest among the every reported polyimide membranes at close regions to upper bounds; PIM-PI-8ii have 490 Barrer of O$_2$ permeability and 3.5 of O$_2$/N$_2$ selectivity.[110] Note that the CH$_4$ permeability is higher than the N$_2$ permeability, different from the phenomena that permeation of N$_2$ is one to two times faster than that of CH$_4$ in usual glassy polymer, revealing that the high free volume characteristics governing PIM are dominant for the gas transport.

Without the introduction of ladder-like moiety induced from benzodioxane, polyimides derived from distorted backbone units have potentials to retain the intrinsic microporosity because the imide linkage itself is formed as very rigid and flat two-dimensional structure.[111–113] Similar to Cardo-PIM, these polymers are based on bifluorene units, which have a 90° kink that prevents space-efficient packing or crystallization of the otherwise stiff polymer chains. Weber and Thomas synthesized bi-, tri- and tetra-functional bifluorene and reacted with dianhydride, di(acid chloride) or trimesoyl chloride for contorted polyimide and polyamide. The most microporous polyimide represents the BET surface area of 982 m^2 g^{-1}, providing much potential for highly permeable gas separation membranes.

As a whole, McKeown and Budd present criteria for polymers with intrinsic microporosity by using more than 100 monomers.[114] With the analysis by BET sorption measurement, positron annihilation lifetime spectroscopy and atomistic computer simulation, they emphasized the importance of microporous materials in several applications.

4.4.4 Thermally Rearranged Polymer Membranes

A thermally rearranged (TR) polymer is a novel microporous material prepared by chain rearrangement of polyimide with *ortho*-functional groups in solid state.[23,39] As shown in Figure 4.12, conversion of aromatic polyimides containing functional groups into highly rigid heteroaromatic polybenzazoles or ladder-like polymers (*e.g.* polybenzoxazole (PBO), polybenzothiazole (PBT or PBZ), polybenzimidazole (PBI) and polypyrrolone (PPL)) at elevated temperature result in the evolution of micro-cavities with an intermediate cavity size, a narrow cavity size distribution and a shape reminiscent of bottlenecks connecting adjacent chambers, such as those found elegantly in nature in the form of ion channels and aquaporins.[39] The free volume elements in TR polymers are believed to three-dimensional networks of intermolecular micro-cavities, which are accessible for small gas molecules. This peculiarity of a free volume structure accounts for both the outstanding permeability of TR polymers with fast diffusion of gases and their still high permselectivity in the

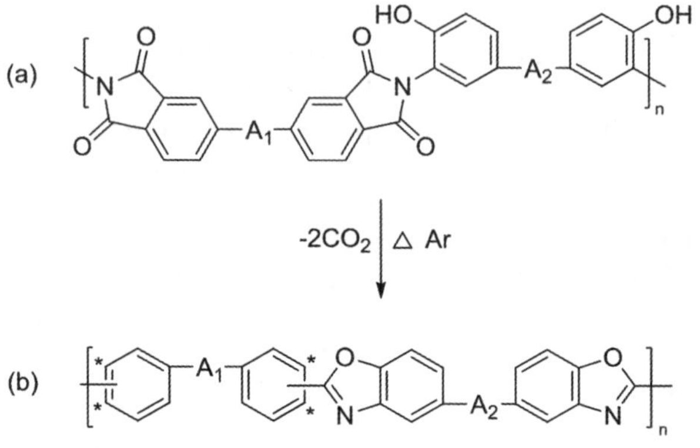

Figure 4.12 Mechanism of thermal rearrangement: (a) hydroxyl-containing polyimide[23] and (b) thermally rearranged polybenzoxazole (TR-PBO).

separation of small molecules. The increase of fractional free volume in the polymer matrix up to 30% is comparable to PTMSP, a well-known polymer with the highest gas permeability.[23,117]

The gas permeabilities of almost all TR polymer membranes are enhanced by at least two orders of magnitude over those of the original polymers and typical glassy polymers, confirming the presence of larger interconnected free volume elements. The gas permeabilities of TR polymer membranes are lower than those of PTMSP membrane, but gas selectivities of CO_2 separations (*e.g.* CO_2/CH_4, CO_2/N_2) are two to three times higher than for PTMSP. The CO_2 permeabilities reach up to 4000–5500 Barrer while the CO_2/CH_4 selectivities are between 20 and 40. Moreover, the additional advantage of TR polymers is that the permeabilities and permselectivies can be conveniently controlled by the thermal conversion ratio determined by thermal treatment times and temperatures.[23] All of the physical properties containing average intermolecular distance, density and fractional free volueme, and gas transport properties by single gas permeation measurement are summarized in Tables 4.10 and 4.11.

Copolymerization is a method to control the gas transport performances of copolymer membranes as well as to confirm the effect of thermal rearrangement. Poly(benzoxazole-*co*-imide) membranes were obtained from the thermal rearrangement of poly(hydroxyl imide-*co*-imide).[41] Size and distribution of free volume cavities created during thermal conversion could be easily controlled by varying HPI composition in the copolymer. CO_2 permeability of copolymer TR membranes varied from 2.7 Barrer [1 Barrer = 1 10^{-10} cm^3 (STP) cm cm^{-2} s^{-1} $(cmHg)^{-1}$] to 1014 Barrer depending on the membrane composition without a significant loss in selectivity. The copolymer TR membranes presented easily overcame the conventional polymeric upper bound limit, and improved the mechanical properties of TR polymer membrane.

Choi and Lee prepared poly(benzoxazole-*co*-pyrrolone) copolymers in various compositions from their precursors of polyimides containing both

Table 4.10 Polymer structure, chemical composition and physical properties of TR-polymer and copolymer membranes

Sample code			*Polymer structure and chemical compositions*		*Physical properties*			
			Dianhydride	*Diamine*	*d-spacing* ($\AA$)	*Density* (*g cm^{-3}*)	*FFV*	*Reference*
TR-PBO series	TR-1		6FDA	bisAPAF	6.0	1.293	0.26	23, 39
	TR-2		BPDA	bisAPAF	6.1	1.271	0.22	
	TR-3		ODPA	bisAPAF	6.1	1.304	0.21	
	TR-4		BTDA	bisAPAF	6.0	1.240	0.24	
	TR-5		PMDA	bisAPAF	7.0	1.362	0.19	
PAI_PPL _PBI	PAI (250)*		6FDA	DAB	6.00	1.475	1.124	117
	PPL (450)		6FDA	DAB	6.35	1.406	0.214	
	PBI (450)		6FDA	DAB	6.40	1.262	0.267	
	PBO		6FDA	bisAPAF	7.5	—	—	40
PBO-PPL copolymer	PBO-co-PPL 82		6FDA	bisAPAF 8 DAB 2	6.7	—	—	
	PBO-co-PPL 55		6FDA	bisAPAF 5 DAB 5	6.93	—	—	
	PBO-co-PPL 28		6FDA	bisAPAF 2 DAB 8	6.57	—	—	
	PPL 450		6FDA	DAB	7.35	—	—	
PBO-PI copolymer	PBO-PI 10:0		BPDA	bisAPAF	7.18	1.127	0.32	41
	PBO-PI 8:2		BPDA	bisAPAF 8 ODA 2				
	PBO-PI 5:5		BPDA	bisAPAF 5 ODA 5	6.45	1.353	0.156	
	PBO-PI 2:8		BPDA	bisAPAF 2 ODA 8				
	PBO-PI 0:10		BPDA	ODA	5.41	1.387	0.102	

*Parenthesis in sample code means the heat treatment temperature, in °C.

Table 4.11 Gas permeabilities and permselectivities of TR-polymer and copolymer membranes

Sample code		Permeability (Barrer)					Permselectivities				Reference
		H_2	CO_2	O_2	N_2	CH_4	O_2/N_2	CO_2/N_2	CO_2/CH_4	H_2/CO_2	
TR-PBO series	TR-1	2774	4045	747	156	73	4.8	26	55	0.69	23, 29
	TR-2	444	597	93	20	15	4.7	30	41	0.74	
	TR-3	91	73	14	2.3	1	6.1	31	58	1.25	
	TR-4	356	469	81	15	10	5.4	31	41	0.76	
	TR-5	635	952	148	34	23	4.4	28	46	0.67	
PAI_PPL_PBI	PAI (250)	36	24	3	0.7	0.3	4.1	35	92	1.5	117
	PPL (300)	207	73	23	4.0	2.0	5.9	19	35	2.8	
	PPL (350)	206	82	32	14.0	6.0	2.3	6	14	2.5	
	PPL (400)	248	126	34	7.0	4.0	5.2	19	32	2.0	
	PPL (450)	376	234	65	13.0	8.0	4.9	18	29	1.6	
	iPBI	44	11	3	0.3	0.1	8.3	34	93	3.9	
	PBI (450)	1779	1624	337	62.0	35	5.4	26	46	1.1	
PBO-PPL copolymer	PBO	4194	4201	1092	284	151	3.9	15	28	1	40
	PBO-co-PPL 82	1989	1874	421	94	50	4.5	20	37	1.1	
	PBO-co-PPL 55	2895	1805	475	85	46	5.6	21	39	1.6	
	PBO-co-PPL 28	1680	525	132	18	6.7	7.3	29	78	3.2	
	PPL 450	376	234	65	13	8.1	4.9	18	29	1.6	
PBO-PI copolymer	PBO-PI 10:0	1228	1014	220	48	41	4.6	21	24	—	41
	PBO-PI 8:2	623	389	90	18	14	4.9	21	27	—	
	PBO-PI 5:5	47	25	4.8	0.82	0.65	5.8	31	39	—	
	PBO-PI 2:8	38	11	2.2	0.4	0.3	5.7	29	38	—	
	PBO-PI 0:10	14	2.7	0.72	0.09	0.03	7.7	30	91	—	

hydroxyl and amino groups. All the copolymers showed higher gas permeabilities than their precursors, as well as higher gas selectivity than PBO or PPL homopolymers. PPL, which showed relatively high selectivity because of its well-packed structure, was used as the selective site. With increasing pyrrolone content, the permselectivity, specifically, the diffusion selectivity of the PBO-*co*-PPL copolymers, gradually increased. Thermally rearranged copolymerization of stiff and selective pyrrolone and high free-volume, highly permeable benzoxazole moieties were novel routes to enhance gas selectivity without significant losses in gas permeability.

Han and Lee modified TR-polypyrrolone into TR-PBI, which improved the highly permeable characteristics to small gas molecules as well as excellent molecular sieving properties.[117] Microporous characteristics were obtained by alkaline hydrolysis of polypyrrolone followed by simple thermal treatment. The TR-PBI membrane showed enhanced gas separation performance compared with the precursor TR-PPL membrane. Moreover, the transport properties of TR-PBI for small gas molecules at elevated temperature were much improved compared with those measured at room temperature, because the gas permeabilities increased without selectivity loss, while their rigid structures still allowed high diffusion-selectivity. For H$_2$/CO$_2$ separation, reduction of CO$_2$ permeability at elevated temperature resulted in the improvement of selectivity due to the significant drop of CO$_2$ sorption capability at high temperature. TR-PBI membrane displayed an advantageous transport performance for H$_2$/CO$_2$ at 120 °C surpassing the 2008 H$_2$/CO$_2$ upper bound with high H$_2$ permeabilities.

As a whole, the physical properties and free volume characterizations of the representative highly permeable glassy polymers such as PTMSP, PMP, AF 2400, AF 1600, PIM-1 and TR-PBOs are summarized and compared in Table 4.12. These polymers all represented bimodal micro-cavity distributions by positron annihilation lifetime spectroscopy (PALS). The larger cavities are efficient for higher gas diffusivity and permeability, while the size and intensities of smaller cavities are crucial for selective transport of gas molecules and evident for interconnections of free volume elements compared to conventional glassy polymers. The most permeable glassy polymers, PTMSP, have the largest τ_3 and τ_4 pores in terms of size, as well as intensity. Their smaller cavities have lifetimes of more than 1.7 ns, corresponding to cavity size of 5–6 Å, which is significantly larger than the kinetic diameter of most permanent gas molecules. AF 1600 and AF 2400 presented similar cavity characteristics with TR-PBOs. Note that the smaller cavities with lifetimes of 1.2–1.3 ns, corresponding to 3–4 cavity size are more selective to gas molecules, and highly permeable. The 3–4 size of cavities in the Teflon AFs and TR-PBOs membranes seems to contribute to the diffusion-selectivities and higher permeability of small penetrants over hydrocarbons.

The relationship between the high free volume elements and transport properties are supported by computer simulation research by Thornton, Hill and co-workers.[91,119,120] Gas diffusion behavior in different size, shape and composition of pores was studied, and it was revealed that the four

Table 4.12 Comparison of physical properties and free volume characterization of representative highly permeable polymers

Sample	Physical properties		Cavity characterization						Reference
	Density $(g\ cm^{-3})$	FFV	$\tau_3\ (ns)$	$I_3\ (\%)$	$\tau_4\ (ns)$	$I_4\ (\%)$	Cavity diameter $(\mathring{A}_3)$	Cavity diameter $(\mathring{A}_4)$	
PTMSP	0.75	0.34	1.7	7.4	8.8	34	5.1	12.4	103
PMP	0.78	0.28	2.3	8.4	7.6	29.4	6.2	11.6	
AF 2400	1.74	0.33	1.32	4.87	5.78	13.6	4.2	10.2	
AF 1600	1.8	0.31	1.2	5.51	5.11	17.3	3.9	9.6	
PIM-1	1.13	0.15	2.06	6.15	6.28	18.6	5.8	10.6	
TR-tPBO	1.17	0.28	1.2	5.0	5.26	6.0	4.0	9.7	118
TR-aPBO	1.38	0.22	1.1	7.1	3.9	12.7	3.5	8.4	
TR-cPBO	1.15	0.35	1.0	5.3	5.00	3.7	3.4	9.5	

representative glassy polymer membranes were optimized for selective transport of gas molecules.

4.5 Concluding Remarks

Membrane-based gas separation is one of the promising future technologies to reduce energy consumption and to battle with global warming. Polymer membrane materials have been used for industrial applications during the last several decades, and recently reported high performance polymers have enabled us to expand these novel materials for next-generation gas separation applications.

Based on 'solution-diffusion' mechanism, glassy polymers represent superior diffusivity selectivities as well as extraordinary diffusivities, while rubbery polymers have displayed high solubility selectivities. PEO and PEO-segmented block copolymers show more favorable results for CO$_2$ separations from flue gas and syngas. Copolymers improve mechanical properties of PEO and control the crystallinity of PEO so that the ether linkage can maximize the adsorption of CO$_2$ in the free volume. The CO$_2$ permeabilties of PEO containing polymers are usually ranged from 50 to 200 Barrers (at most 650 Barrer) while the CO$_2$/N$_2$ selectivities are over 40–50, the extent enough to enrich CO$_2$ concentration over 90% at a single stage from flue gas. On the other hand, the performances are significantly affected by the operation temperature mainly due to the solubility of CO$_2$ and the morphologies of the multi-block copolymers.

Poly(substituted acetylene)s such as PTMSP and PMP, amorphous fluoropolymers like Teflon AF and Hyflon AD, polymers with intrinsic microporosity, and thermally rearranged (TR) polymers are the candidate polymers for highly permeable glassy polymer membranes. The high free volume in glassy polymers contributes to enhanced diffusion and permeation of small gas molecules. The gas permeation performances of these highly permeable polymers even surpass upper bounds for CO$_2$/N$_2$, CO$_2$/CH$_4$ and H$_2$/CO$_2$ separations.

References

1. O. Vogl and G. D. Jaycox, *Prog. Polym. Sci.*, 1999, **24**, 3.
2. J. Scheirs, *Polymers*, John Wiley and Sons, 2003.
3. Y. Osada and J. Gong, *Prog. Polym. Sci.*, 1993, **18**, 187.
4. F. Ghebremichael, M. G. Kuzyk and H. S. Lackritz, *Prog. Polym. Sci.*, 1997, **22**, 1147.
5. G. Maier, *Prog. Polym. Sci.*, 2001, **26**, 3.
6. B. Adhikari and S. Majumdar, *Prog. Polym. Sci.*, 2004, **29**, 699.
7. K. Matyjaszewski and A. H. E. Müller, *Prog. Polym. Sci.*, 2006, **31**, 1039.
8. J. C. Scott, *Science*, 1997, **278**, 2071.
9. D. R. Paul and Y. P. Yampolskii, *Polymeric Gas Separation Membranes*, CRC Press, Florida, 1993.

10. M. Mulder, *Basic Principles of Membrane Technology*, 2nd ed., Kluwer Academic Publishers, Dordrecht, 2nd edn, 1996.
11. R. M. Barrer, *Nature*, 1937, **140**, 106.
12. S. Loeb and S. Sourirajan, in *Adv. Chem. Ser.*, American Chemical Society, 1962, p. 117.
13. H. Strathmann, P. Scheible and R. W. Baker, *J. Appl. Polym. Sci.*, 1971, **15**, 811.
14. R. W. Baker, *Membrane Technology and Applications*, 2nd ed., Wiley, New York, 2nd edn, 2004.
15. A. Brunetti, F. Scura, G. Barbieri and E. Drioli, *J. Membr. Sci.*, 2010, **359**, 115.
16. A. M. Kratochvil and W. J. Koros, *Macromolecules*, 2008, **41**, 7920.
17. H. Sijbesma, K. Nymeijer, R. van Marwijk, R. Heijboer, J. Potreck and M. Wessling, *J. Membr. Sci.*, 2008, **313**, 263.
18. C. E. Powell and G. G. Qiao, *J. Membr. Sci.*, 2006, **279**, 1.
19. E. M. Maya, D. M. Muñoz, J. G. de la Campa, J. de Abajo and A. E. Lozano, *Desalination*, 2006, **199**, 188.
20. J. R. Klaehn, C. J. Orme, E. S. Peterson, T. A. Luther, M. G. Jones, A. K. Wertsching and J. M. Urban-Klaehn, CO_2 Separation using thermally optimized membranes: A comprehensive project report (2000–2007), 2008.
21. B. T. Low, T. S. Chung, H. Chen, Y. C. Jean and K. P. Pramoda, *Macromolecules*, 2009, **42**, 7042.
22. H. Lin and B. D. Freeman, *J. Mol. Struct.*, 2005, **739**, 57.
23. H. B. Park, S. H. Han, C. H. Jung, Y. M. Lee and A. J. Hill, *J. Membr. Sci.*, 2010, **359**, 11.
24. L. M. Robeson, *J. Membr. Sci.*, 1991, **62**, 165.
25. B. D. Freeman, *Macromolecules*, 1999, **32**, 375.
26. M. L. Cecopieri-Gómez, J. Palacios-Alquisira and J. M. Domínguez, *J. Membr. Sci.*, 2007, **293**, 53.
27. L. M. Robeson, *J. Membr. Sci.*, 2008, **320**, 390.
28. M. M. Dal-Cin, A. Kumar and L. Layton, *J. Membr. Sci.*, 2008, **323**, 299.
29. L. M. Robeson, B. D. Freeman, D. R. Paul and B. W. Rowe, *J. Membr. Sci.*, 2009, **341**, 178.
30. B. W. Rowe, L. M. Robeson, B. D. Freeman and D. R. Paul, *J. Membr. Sci.*, 2010, **360**, 58.
31. R. W. Baker, *Ind. Eng. Chem. Res.*, 2002, **41**, 1393.
32. J. N. Israelachvili, *Intermolecular and Surface Forces*, Academic Press, London, 1992.
33. T. Masuda, E. Isobe, T. Higashimura and K. Takada, *J. Am. Chem. Soc.*, 1983, **105**, 7473.
34. I. Pinnau and L. G. Toy, *J. Membr. Sci.*, 1996, **109**, 125.
35. A. M. Polyakov, L. E. Starannikova and Y. P. Yampolskii, *J. Membr. Sci.*, 2003, **216**, 241.
36. G. Dlubek, J. Pionteck, K. Rätzke, J. Kruse and F. Faupel, *Macromolecules*, 2008, **41**, 6125.

37. P. M. Budd, B. S. Ghanem, S. Makhseed, N. B. McKeown, K. J. Msayib and C. E. Tattershall, *Chem. Commun.*, 2004, **10**, 230.
38. B. S. Ghanem, K. J. Msayib, N. B. McKeown, K. D. M. Harris, Z. Pan, P. M. Budd, A. Butler, J. Selbie, D. Book and A. Walton, *Chem. Commun.*, 2007, **1**, 67–69.
39. H. B. Park, C. H. Jung, Y. M. Lee, A. J. Hill, S. J. Pas, S. T. Mudie, E. Van Wagner, B. D. Freeman and D. J. Cookson, *Science*, 2007, **318**, 254.
40. J. I. Choi, C. H. Jung, S. H. Han, H. B. Park and Y. M. Lee, *J. Membr. Sci.*, 2010, **349**, 358.
41. C. H. Jung, J. E. Lee, S. H. Han, H. B. Park and Y. M. Lee, *J. Membr. Sci.*, 2010, **350**, 301.
42. A. Bos, I. G. M. Punt, M. Wessling and H. Strathmann, *J. Polym. Sci. Polym. Phys*, 1998, **36**, 1547.
43. A. Morisato, B. D. Freeman, I. Pinnau and C. G. Casillas, *J. Polym. Sci., Part B: Polym. Phys.*, 1996, **34**, 1925.
44. H. Lin and B. D. Freeman, *Macromolecules*, 2006, **39**, 3568.
45. H. Lin, E. Van Wagner, B. D. Freeman, L. G. Toy and R. P. Gupta, *Science*, 2006, **311**, 639.
46. C. P. Ribeiro and B. D. Freeman, *J. Polym. Sci., Part B: Polym. Phys.*, 2010, **48**, 456.
47. J. H. Kim and Y. M. Lee, *J. Membr. Sci.*, 2001, **193**, 209.
48. S. R. Reijerkerk, M. H. Knoef, K. Nijmeijer and M. Wessling, *J. Membr. Sci.*, 2010, **352**, 126.
49. W. Yave, A. Car, J. Wind and K. V. Peinemann, *Nanotechnology*, 2010, **21**, 295.
50. N. F. A. van der Vegt, V. A. Kusuma and B. D. Freeman, *Macromolecules*, 2010, **43**, 1473.
51. H. Lin and B. D. Freeman, *J. Membr. Sci.*, 2004, **239**, 105.
52. Y. Hirayama, Y. Kase, N. Tanihara, Y. Sumiyama, Y. Kusuki and K. Haraya, *J. Membr. Sci.*, 1999, **160**, 87.
53. L. K. Massey, *Permeability Properties of Plastics and Elastomers: A Guide to Packaging and Barrier Materials*, Plastics Design Library, Norwich, NY, 2nd edn, 2003.
54. R. B. Eldrige, *Ind. Eng. Chem. Res.*, 1993, **32**, 2208.
55. H. Lin, T. Kai, B. D. Freeman, S. Kalakkunnath and D. S. Kalika, *Macromolecules*, 2005, **38**, 8381.
56. H. Lin, E. Van Wagner, R. Raharjo, B. D. Freeman and I. Roman, *Adv. Mater.*, 2006, **18**, 39.
57. H. Lin, E. V. Wagner, J. S. Swinnea, B. D. Freeman, S. J. Pas, A. J. Hill, S. Kalakkunnath and D. S. Kalika, *J. Membr. Sci.*, 2006, **276**, 145.
58. S. Kelman, H. Lin, E. S. Sanders and B. D. Freeman, *J. Membr. Sci.*, 2007, **305**, 57.
59. J. J. Richards, M. K. Danquah, S. Kalakkunnath, D. S. Kalika, V. A. Kusuma, S. T. Matteucci and B. D. Freeman, *Chem. Eng. Sci.*, 2009, **64**, 4707.

60. Y. Yin, L. Yang, M. Yoshino, J. Fang, K. Tanaka, H. Kita and K. I. Okamoto, *Polym. J.*, 2004, **36**, 294.

61. H. Y. Zhao, Y. M. Cao, X. L. Ding, M. Q. Zhou, J. H. Liu and Q. Yuan, *J. Membr. Sci.*, 2008, **320**, 179.

62. H. Y. Zhao, Y. M. Cao, X. L. Ding, M. Q. Zhou and Q. Yuan, *J. Membr. Sci.*, 2008, **323**, 176.

63. M. E. Rezac, T. John and P. H. Pfromm, *J. Appl. Polym. Sci.*, 1997, **65**, 1983.

64. V. I. Bondar, B. D. Freeman and I. Pinnau, *J. Polym. Sci., Part B: Polym. Phys.*, 2000, **38**, 2051.

65. J. H. Kim, S. Y. Ha and Y. M. Lee, *J. Membr. Sci.*, 2001, **190**, 179.

66. M. Yoshino, K. Ito, H. Kita and K. I. Okamoto, *J. Polym. Sci., Part B: Polym. Phys.*, 2000, **38**, 1707.

67. H. B. Park, C. K. Kim and Y. M. Lee, *J. Membr. Sci.*, 2002, **204**, 257.

68. E. M. Maya, D. M. Muñoz, J. G. de la Campa, J. de Abajo and A. E. Lozano, *Desalination*, 2006, **199**, 188.

69. D. M. Muñoz, E. M. Maya, J. de Abajo, J. G. de la Campa and A. E. Lozano, *J. Membr. Sci.*, 2008, **323**, 53.

70. H. Chen, Y. Xiao and T. S. Chung, *Polymer*, 2010, **51**, 4077.

71. S. J. Metz, M. H. V. Mulder and M. Wessling, *Macromolecules*, 2004, **37**, 4590.

72. D. Husken, T. Visser, M. Wessling and R. J. Gaymans, *J. Membr. Sci.*, 2010, **346**, 194.

73. S. R. Reijerkerk, A. Arun, R. J. Gaymans, K. Nijmeijer and M. Wessling, *J. Membr. Sci.*, 2010, **359**, 54.

74. W. Yave, A. Szymczyk, N. Yave and Z. Roslaniec, *J. Membr. Sci.*, 2010, **362**, 407.

75. S. J. Metz, W. J. C. Van De Ven, J. Potreck, M. H. V. Mulder and M. Wessling, *J. Membr. Sci.*, 2005, **251**, 29.

76. A. Szymczyk, *Eur. Polym. J.*, 2009, **45**, 2653.

77. A. S. Kovvali, H. Chen and K. K. Sirkar, *J. Am. Chem. Soc.*, 2000, **122**, 7594.

78. A. S. Kovvali and K. K. Sirkar, *Ind. Eng. Chem. Res.*, 2001, **40**, 2502.

79. S. Duan, T. Kouketsu, S. Kazama and K. Yamada, *J. Membr. Sci.*, 2006, **283**, 2.

80. T. Kouketsu, S. Duan, T. Kai, S. Kazama and K. Yamada, *J. Membr. Sci.*, 2007, **287**, 51.

81. S. Duan, F. A. Chowdhury, T. Kai, S. Kazama and Y. Fujioka, *Desalination*, 2008, **234**, 278.

82. I. Taniguchi, S. Duan, S. Kazama and Y. Fujioka, *J. Membr. Sci.*, 2008, **322**, 277.

83. T. Kai, T. Kouketsu, S. Duan, S. Kazama and K. Yamada, *Sep. Purif. Technol.*, 2008, **63**, 524.

84. Y. Xiao, T. S. Chung and M. L. Chung, *Langmuir*, 2004, **20**, 8230.

85. Z. Wang, M. Li, Y. Cai, J. Wang and S. Wang, *J. Membr. Sci.*, 2007, **290**, 250.

86. T. Nakagawa, T. Saito, S. Asakawa and Y. Saito, *Gas Sep. Purif.*, 1988, **2**, 3.

87. I. Pinnau, Z. He and A. Morisato, *J. Membr. Sci.*, 2004, **241**, 363.

88. K. Nagai, T. Masuda, T. Nakagawa, B. D. Freeman and I. Pinnau, *Prog. Polym. Sci.*, 2001, **26**, 721.

89. A. Fukui, K. Hattori, Y. Hu, M. Shiotsuki, F. Sanda and T. Masuda, *Polymer*, 2009, **50**, 4159.

90. K. De Sitter, P. Winberg, J. D'Haen, C. Dotremont, R. Leysen, J. A. Martens, S. Mullens, F. H. J. Maurer and I. F. J. Vankelecom, *J. Membr. Sci.*, 2006, **278**, 83.

91. A. W. Thornton, K. M. Nairn, A. J. Hill and J. M. Hill, *J. Membr. Sci.*, 2009, **338**, 29.

92. H. B. Park, Y. M. Lee, in *Advanced Membrane Technology and Applications*, ed. N. N. Li, A. G. Fane, W. S. Winston and T. Matsuura, John Wiley & Sons, Hoboken, NJ, 2008, p. 633.

93. A. Y. Alentiev, Y. P. Yampolskii, V. P. Shantarovich, S. M. Nemser and N. A. Platé, *J. Membr. Sci.*, 1997, **126**, 123.

94. T. C. Merkel, V. Bondar, K. Nagai and B. D. Freeman, *Macromolecules*, 1999, **32**, 370.

95. T. C. Merkel, V. Bondar, K. Nagai, B. D. Freeman and Y. P. Yampolskii, *Gas Sorption, Macromolecules*, 1999, **32**, 8427.

96. A. Y. Alentiev, V. P. Shantarovich, T. C. Merkel, V. I. Bondar, B. D. Freeman and Y. P. Yampolskii, *Macromolecules*, 2002, **35**, 9513.

97. J. C. Jansen, F. Tasselli, E. Tocci and E. Drioli, *Desalination*, 2006, **192**, 207.

98. J. C. Jansen, M. Macchione and E. Drioli, *J. Membr. Sci.*, 2007, **287**, 132.

99. M. Macchione, J. C. Jansen, G. De Luca, E. Tocci, M. Longeri and E. Drioli, *Polymer*, 2007, **48**, 2619.

100. Y. Yampolskii, I. Pinnau and B. D. Freeman, *Materials Science of Membranes for Gas and Vapor Separation*, John Wiley & Sons, Hoboken, NJ, 2006.

101. C. J. Orme, M. L. Stone, M. T. Benson and E. S. Peterson, *Sep. Sci. Technol.*, 2003, **38**, 3225.

102. N. B. McKeown, B. Gahnem, K. J. Msayib, P. M. Budd, C. E. Tattershall, K. Mahmood, S. Tan, D. Book, H. W. Langmi and A. Walton, *Angew. Chem. Int. Ed.*, 2006, **45**, 1804.

103. C. L. Staiger, S. J. Pas, A. J. Hill and C. J. Cornelius, *Chem. Mater.*, 2008, **20**, 2606.

104. P. M. Budd, N. B. McKeown and D. Fritsch, *J. Mater. Chem.*, 2005, **15**, 1977.

105. B. S. Ghanem, N. B. McKeown, P. M. Budd and D. Fritsch, *Macromolecules*, 2008, **41**, 1640.

106. N. Du, G. P. Robertson, J. Song, I. Pinnau, S. Thomas and M. D. Guiver, *Macromolecules*, 2008, **41**, 9656.

107. N. Du, G. P. Robertson, I. Pinnau and M. D. Guiver, *Macromolecules*, 2009, **42**, 6023.

108. N. Du, G. P. Robertson, I. Pinnau, S. Thomas and M. D. Guiver, *Macromol. Rapid Commun.*, 2009, **30**, 584.
109. B. S. Ghanem, N. B. McKeown, P. M. Budd, J. D. Selbie and D. Fritsch, *Adv. Mater.*, 2008, **20**, 2766.
110. B. S. Ghanem, N. B. McKeown, P. M. Budd, N. M. Al-Harbi, D. Fritsch, K. Heinrich, L. Starannikova, A. Tokarev and Y. Yampolskii, *Macromolecules*, 2009, **42**, 7881.
111. J. Weber, Q. Su, M. Antonietti and A. Thomas, *Macromol. Rapid Commun.*, 2007, **28**, 1871.
112. J. Weber, M. Antonietti and A. Thomas, *Macromolecules*, 2008, **41**, 2880.
113. N. Ritter, M. Antonietti, A. Thomas, I. Senkovska, S. Kaskel and J. Weber, *Macromolecules*, 2009, **42**, 8017.
114. N. B. McKeown and P. M. Budd, *Macromolecules*, 2010, **43**, 5163.
115. P. M. Budd, E. S. Elabas, B. S. Ghanem, S. Makhseed, N. B. McKeown, K. J. Msayib, C. E. Tattershall and D. Wang, *Adv. Mater.*, 2004, **16**, 456.
116. J. Song, N. Du, Y. Dai, G. P. Robertson, M. D. Guiver, S. Thomas and I. Pinnau, *Macromolecules*, 2008, **41**, 7411.
117. S. H. Han, J. E. Lee, K.-J. Lee, H. B. Park and Y. M. Lee, *J. Membr. Sci.*, 2010, **357**, 143.
118. S. H. Han, N. Misdan, S. Kim, C. M. Doherty, A. J. Hill and Y. M. Lee, *Macromolecules*, 2010, **43**, 7657.
119. A. W. Thornton, T. Hilder, A. J. Hill and J. M. Hill, *J. Membr. Sci.*, 2009, **336**, 101.
120. A. W. Thornton, K. M. Nairn, A. J. Hill, J. M. Hill and Y. Huang, *J. Membr. Sci.*, 2009, **338**, 38.

Design of Membrane Modules for Gas Separations

M. SCHOLZ,[a] M. WESSLING[a] AND J. BALSTER*[b]

[a] RWTH Aachen University, Chemische Verfahrenstechnik (CVT), 52064 Aachen, Germany; [b] Evonik Degussa GmbH, Process Technology & Engineering, Paul-Baumann Straße 1, 45764 Marl, Germany

5.1 Introduction

In the last three decades, membrane-based gas separation has become a well-established unit operation for a variety of important applications. Gas separation membranes operate in the chemical industry in hydrogen separation, monomer recovery, the enrichment of nitrogen from air, and natural gas treatment. Recovery of hydrogen from the purge gas of an ammonia reactor was established around 1980. Vinylchloride monomer recovery from the PVC production process was started up about 10 years later. Recently, the recovery of propylene from polypropylene purge bins became feasible. The removal of sour gases from natural gas as well as biogas operates reliably, today.[1-4]

Gas separation is possible even with the two extreme types of membrane considered, *i.e.* porous and non-porous. The transport mechanisms through these two types of membrane, however, are completely different.[1] Gas separation is performed using membranes based on three general transport mechanisms: Knudsen diffusion, solution-diffusion, molecular sieving.[2] Industrially relevant are solution-diffusion based membranes.

In gas separation processes a high pressure on the feed side, a low pressure on the permeate side or a combination of both can be used to create the driving force.

Membrane Engineering for the Treatment of Gases, Volume 1:
Gas-separation Problems with Membranes
Edited by Enrico Drioli and Giuseppe Barbieri
© Royal Society of Chemistry 2011
Published by the Royal Society of Chemistry, www.rsc.org

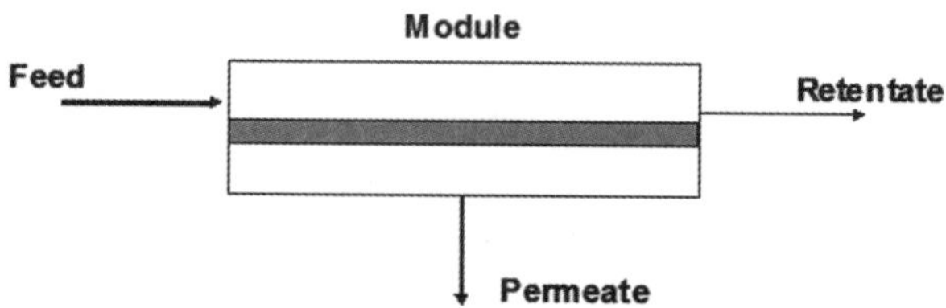

Figure 5.1 Schematic drawing of a module.

Since the driving force of permeation is a difference in chemical potential it is also possible to use a sweep stream with low concentrations of the target components in order to increase the driving force. The product can be the retentate or the permeate stream depending on the components present in the system.

In order to apply membranes on a technical scale, large membrane areas are normally required. The smallest unit into which the membrane area is packed is called a module and is the central part of a membrane installation.[1]

Figure 5.1 shows a feed stream entering the module at a certain composition and with a certain flow rate. Because the membrane has the ability to transport one component faster than another, both the feed composition and the flow rate inside the module will change along the length of the module. The feed stream is separated during the passage through the module into a permeate stream and a retentate stream. The permeate stream is the fraction of the feed stream which passes through the membrane whereas the retentate stream is the fraction retained.

Designs of the membrane module and of the operational conditions have to be chosen carefully for each separation process. In order to fully exploit the maximum separation property of the membrane, the following general requirements need to be fulfilled:[1,5]

- Good mechanical, thermal and chemical stability
- Good flow distribution (no dead zones, no channeling)
- High packing density
- Low pressure drop
- Possibility of cleaning
- Ease of maintenance and operation
- Cheap manufacturing
- Compactness of the system scale
- Possibility of membrane replacement.

Depending on the process and application different characteristics are more important than others. Because the cleaning ability is of less importance in gas separation, the main interest of module design is a high packing density. High ratios of membrane area to module volume are required which result in low manufacturing costs.[5]

The three major module types for gas separation processes are plate-and-frame, spiral wound, and hollow fiber.[2] In general, a system does not consist of just one single module but of a number of modules arranged together. In fact, each technical application has its own system design based on the specific requirements. The choice of module configuration, as well as the arrangement of the modules in a system, is based solely on economic considerations with the correct engineering parameters being employed to achieve this.[1]

5.2   Membrane Modules

A number of module designs are possible and all are based on two types of membrane geometry: (i) flat sheet membranes and (ii) capillary fibers. Typical dimensions are shown in Figure 5.2.

Plate-and-frame as well as spiral-wound modules house flat membranes. Capillary and hollow fibers are packed in larger bundles in a pressure vessel.[1] Commercially relevant are flat membranes and hollow fibers. However, new developments in perovskite membranes for oxygen transport are based on tubular membranes as well.[6]

5.2.1   Plate-and-frame Modules

Plate-and-frame modules were the earliest version of membranes modules and are today still used in ultra-filtration and pervaporation processes. There is only one plate-and-frame configuration used in solution-diffusion membranes.[7] This design may be relevant in the future for flat perovskite membranes.[8–10] The most important elements of the module construction are:[11]

- Flat membrane
- Membrane supporting plate/spacer
- Feed distribution plate.

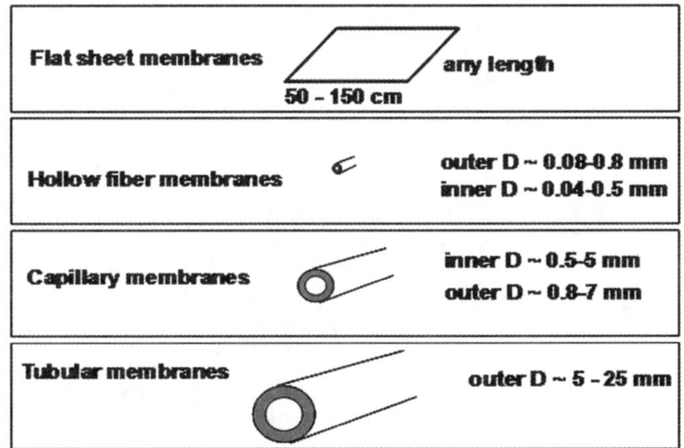

Figure 5.2   Available types of membranes.

In a plate-and-frame module configuration (Figure 5.3) sets of two membranes are placed in a sandwich-like fashion with their feed sides facing each other. In each feed and permeate compartment, a suitable spacer is placed. A plate-and-frame stack is build up by the number of membrane sets needed for a given membrane area equipped with sealing rings and two end plates. The packing density (membrane surface per module volume) of such modules is low[1] and about 100–400 m^2 m^{-3}.

A schematic flow path in a plate-and-frame module is illustrated in Figure 5.4. In order to reduce channeling (tendency to by-pass parts of the membrane along a short cut) and to establish a uniform flow distribution baffles have been introduced.[1]

The advantages of plate-and-frame modules are:[11]

- Exchange ability of single membranes
- Low sensitivity to particulate blocking of the feed channels
- Usage of flat membranes without the usage of glue.

The disadvantages are:

- The need for several sealings
- Pressure drop
- Low packing density.

The plate-and-frame module is highly effective in pervaporation applications, but is much less popular than spiral wound and hollow fiber modules for gas

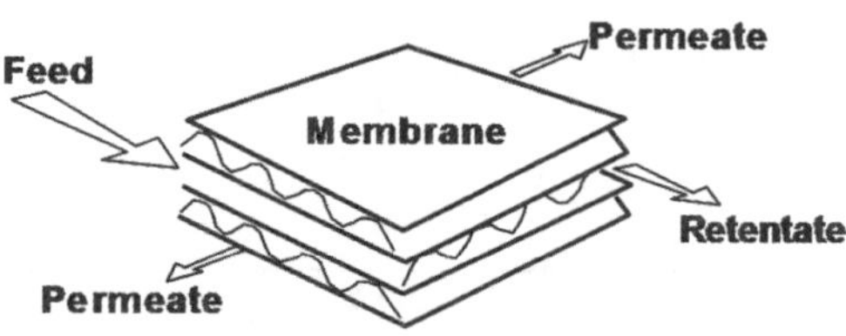

Figure 5.3 Drawing of a plate-and-frame module.

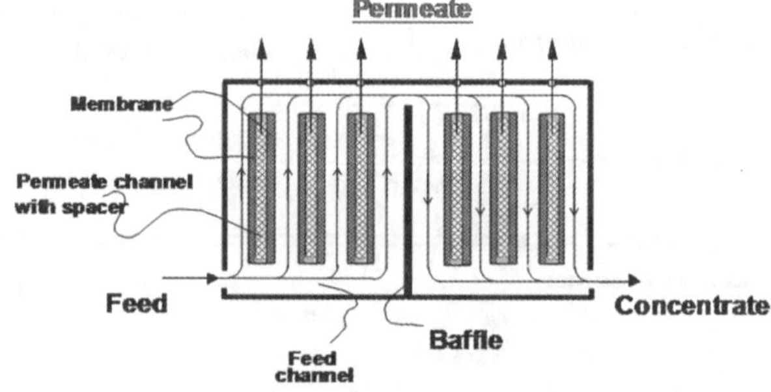

Figure 5.4 Schematic flow path in a plate-and-frame-module.

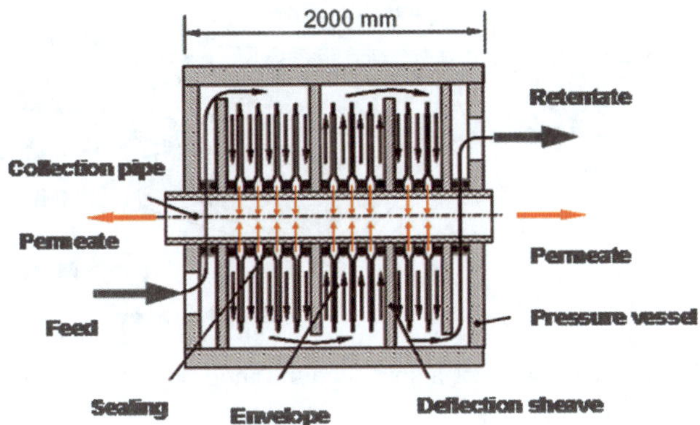

Figure 5.5 GKSS envelope module.

separation applications due to the lower membrane area per unit volume that can be accommodated.[3] The only commercial application for a plate-and-frame device in gas separation is the oxygen enrichment from air for small scale medical applications[14,15] and for organic vapor recovery as commercialized by Borsig, with a special custom design gas separation module.

A futher development of the plate-and-frame module is the envelope module. In the GKSS module (Figure 5.5) flat sheet membranes are designed in the envelope type that enables membranes exchange easily. These modules are applicable in nanofiltration, vapor permeation and gas permeation processes and are suited for high operation pressure up 120 bars. The flat sheet membrane module can be used with sweep gas or vacuum operation on the permeate side. The feed gas flows around the membrane envelopes and the permeate is collected over a central collecting pipe, which lead to low pressure losses. The flexible distance between the membrane envelopes makes it possible to create relative constant retentate flows.[11]

5.2.2 Spiral-wound Modules

A more important module type for membrane applications is the spiral-wound module. Initially it was developed for reverse osmosis applications and is nowadays also used in ultra-filtration and gas permeation applications.[11] The spiral-wound module is the next logical step from a flat membrane. It is in fact a plate-and-frame system wrapped around a central collection pipe, similar to a sandwich roll. Feed and permeate-side spacer material is glued along three edges to build a membrane envelope. A feed-side spacer is separating the top layer of two flat membranes simultaneously acting as a turbulence promoter (Figure 5.6).[1]

The feed flows in axial direction over the envelopes built up by two membrane stacks. The permeate is collected on the inside of the envelope and spirals into the central permeate collection pipe. The packing density of this module

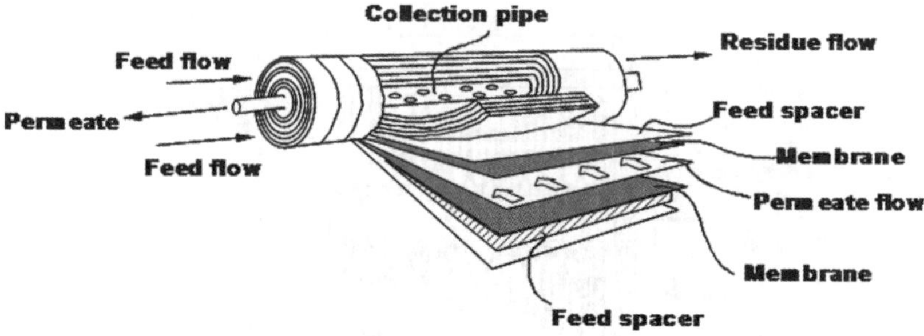

Figure 5.6   Schematic drawing of a spiral-wound module.

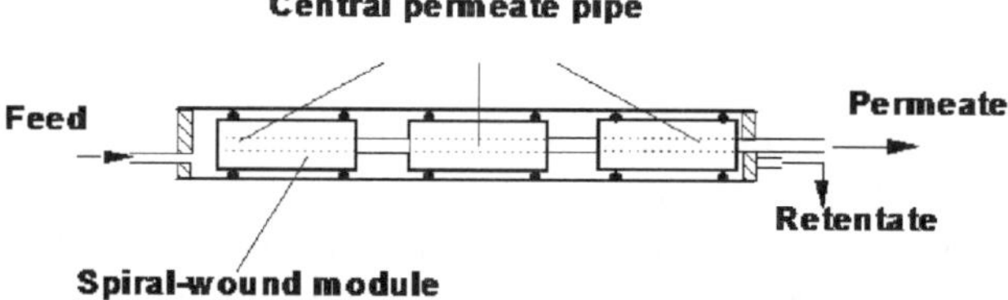

Figure 5.7   Schematic of a pressure vessel containing three spiral-wound modules in series.

(300–1000 m^2 m^{-3}) is greater than that of the plate-and-frame module but strongly depends on the channel height, which in turn is determined by the thickness of the permeate and feed-side spacer material.[1] The spacer is not only used to keep the distance between the membranes, but it also enhances the mass transfer at a minimum pressure drop.[11] Net-type spacers are common for spiral-wound modules. Important characteristics determining the module performance influenced by this type of spacer are:

- Filament diameter
- Hydrodynamic angle
- Mesh size
- Voidage
- Position to flow (transverse).

The analysis of pressure drop within spacer-filled channels can be found in Da Costa *et al.*[12] Recently a double helix spacer with enhanced mass transfer characteristics combined with a reduced pressure drop was developed at RWTH Aachen University.[13]

Several (up to six) spiral-wound modules are assembled in one pressure vessel and are connected in series via the central permeate tubes (Figure 5.7).[1,11] New vessel designs currently emerge that addresses the huge gas flows to be potentially treated in post combustion CO_2 capture.[16]

The advantages of spiral-wound modules are:[6]

- Simple, cost effective construction
- Relatively high packing density/membrane area to volume ratio (up to 1000 m^2 m^{-3})
- Good mass transfer due to feed spacer.

The disadvantages are:

- Long permeate path
- Difficult to clean.

In total up to 20% of the commercial gas separation membranes are being formed into spiral wound modules.[4]

5.2.3 Hollow Fiber Modules

Hollow fibers consist of a porous, non-selective support layer (about 200 μm) and an active layer (<40 nm). The active layer is the actual membrane, but due to its small thickness it must be supported by a thicker layer in order to obtain mechanical strength, to withstand the pressure difference between feed to permeate side.

The hollow fiber membrane module assembles a shell-and-tube heat exchanger. It consists of a large number of hollow fibers assembled in a module, as shown schematically in Figure 5.8. The free ends of the fibers are potted with agents such as epoxy resins, polyurethanes, silicone rubber, thermoplastics, thermosets or inorganic cements. In some cases they can also be fused by heating.

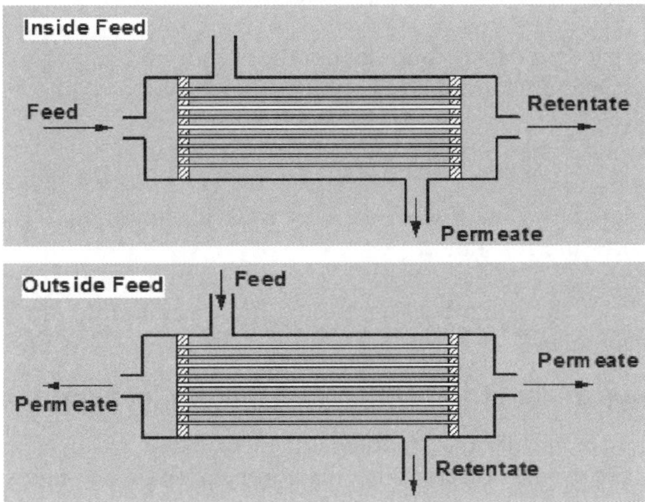

Figure 5.8   Schematic of a hollow fiber module with the feed on the inside of the fiber (bore side feed) and the feed on the outside of the fibers (shell side feed).

Table 5.1   Properties of hollow fiber modules

Property	Description
Structure	Self-supporting
Active layer	Inner/outer diameter
Feed	Lumen/shell side
Inner diameter	40–250 μm
Outer diameter	80–400 μm
Packing density	$<10\,000$ m^2 m^{-3}
Allowed pressure	100 bar shell side; 15 bar lumen side
Applications	GP, RO, DL

The hollow fiber membranes are self-supporting. Two types of module arrangement can be distinguished: (i) where the feed passes through the bore of the hollow fiber (lumen) whereas the permeate is collected on the outside of the hollow fibers (Figure 5.8, 'Inside Feed'); and (ii) where the feed enters the module on the shell side of the hollow fibers (external) and the permeate passes into the fiber bore (Figure 5.8, 'Outside Feed'). The choice between the two concepts is mainly based on the application where parameters such as pressure, pressure drop, type of membrane available, *etc.* are important. Depending on the concept chosen, asymmetric hollow fibers are used with their skin on the inside or on the outside.[1]

The difference between hollow fiber modules and capillary modules is a matter of dimensions while the module concepts are the same. The hollow fiber module is the configuration with the highest packing density,[1] which can attain values of 30 000 m^2 m^{-3}.

The typical properties and dimensions of hollow fiber membrane modules are summarized in Table 5.1.

The advantages of hollow fiber modules are:[11]

- Highest packing density and membrane area to module volume ratio
- Cheapest manufacturing costs.

The disadvantages are:

- Mostly laminar flows (increased mass transfer limitations)
- Lower pressure resistance.

Up to 80% of the commercial gas separation membranes are formed into hollow fiber modules.[4]

5.2.4   Comparison of the Different Module Configurations

The choice of the membrane module is mainly determined by economic considerations. The production costs in a membrane production plant are sensitive to the produced volume. In a hollow fiber spinning plant operating continuously, the membrane costs are in a range of 2–5 \$ per m^2 of membrane area. An equivalent of spiral-wound modules[4] would cost 10–100 \$ per m^2.

Table 5.2 Features and applications of the different gas separation modules[4]

Configuration	Features	Typical applications	Applicants

High-pressure
Shell-side feed
Hollow fibers
	Cross-flow
	Usually no fouling
	    problem
		H_2 recovery in
		    refineries
		CO_2 removal from
		    natural gas
			Medal
			Cynara

Low-pressure
Bore side feed
Hollow fibers
	Counter-current flow
	Pretreatment required
	Good feed flow
	    distribution
		Nitrogen enrichment
		    from air
		Oxygen enrichment
		    from air
		Dehydration of air
			Medal
			Air Products
			Parker

Spiral-wound
    modules
	Cross-flow
	Usually no fouling
	    problem
	Wide range of
	    membranes can be used
		CO_2 removal from
		    natural gas
		Vapor/gas separations
			MTR

However, costs are just one factor which must be considered and the proper choice of the membrane module depends on the application.[1] In gas separation plants, especially in refinery and petrochemical operations, the module costs are only 10–25% of the total costs, so that significant reductions in membrane costs might not markedly change the cost of the complete plant.[4]

Spiral-wound modules are used where pressure drop has to be considered and when counter current flow is not needed to maximize separation efficiency. Higher pressure applications involving costly pressure vessels and piping make the hollow fiber modules more favorable because this reduces the component costs of the system by as much as a factor of ten in some cases.[2] An overview of the features and applications of the different module types is given in Table 5.2.

5.3 Operation of Gas Separation Hollow Fiber Membrane Modules

The operation of hollow fiber membrane modules is influenced by various parameters on different scales. In order to design a membrane module the flow patterns (module scale) as well as the site of the inlet and the active membrane layer (fiber scale) have to be specified. These specifications affect the module performance due to an influence on the driving force of permeation and on the membrane area.

5.3.1 Flow within the Fiber (Lumen-side Feed, Shell-side Feed)

In general, hollow fibers consist of a porous support as well as an active membrane layer. They can be fed through the lumen or the shell side which is indicated in Figure 5.9. Dependent on the separation problem the active layer can be sited on the outer or on the inner diameter and three technical relevant configurations exist:

- Feed on the shell side with the active membrane layer on the outer fiber diameter
- Feed on the lumen side and the active membrane layer on the outer fiber diameter
- Feed on the lumen side with the active membrane layer on the inner fiber diameter (Figure 5.9).

The correct fiber configuration has to be chosen carefully. When high pressures are applied the feed should enter the shell side of the module since the mechanical stability of lumen side feed is lower compared to shell side feed. This is caused by a difference in compression strength and tensile strength. The mechanical stability of hollow fibers and the maximum applicable pressure can be determined using eqn (5.1) and eqn (5.2):

$$P_i = \frac{\sigma_t \cdot 2 \cdot s}{d_i} \tag{5.1}$$

$$P_o = \frac{\sigma_c \cdot 2 \cdot s}{d_o} \tag{5.2}$$

where d_i is the inner fiber diameter, d_o the outer fiber diameter, P the applicable pressure within and outside the fiber, σ_t the tensile strength and σ_c the compression strength, s is the thickness of the porous support and the active layer whereas the thickness of the active layer is much smaller than that of the

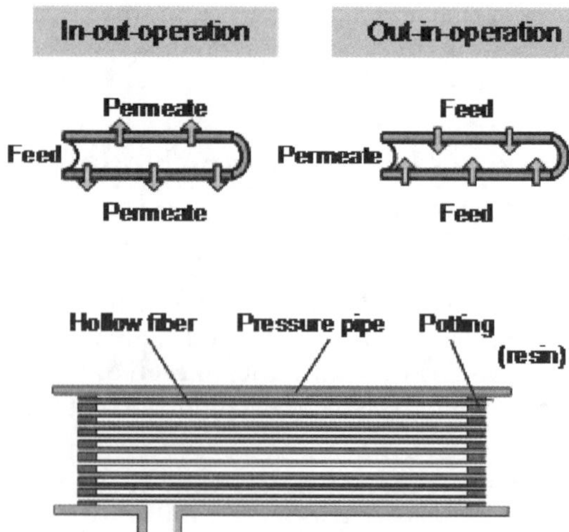

Figure 5.9   Lumen-side feed as well as shell-side feed is illustrated. In the lower part a general scheme of a hollow fiber membrane module is presented.

support. The maximal allowed tensile and compression strength are tabulated for various materials, so that the tolerable pressure can be determined. Typically the compression strength for polymeric membrane materials is in the order of 100 N mm^{-2} and the tensile strength is in the order of 50 N mm^{-2}.

Drawbacks of a shell side feed are channeling and bypass effects. This can be avoided by feeding the hollow fiber through the lumen side. By locating the feed at the lumen side, the site of the active membrane layer has to be reconsidered. In fact the membrane area is higher when the active layer is sited at the outer fiber diameter which results in high membrane area to volume ratios. In this case the fluid has to pass through the porous support layer before it arrives at the active layer. In liquid applications, where diffusion coefficients are low, it can have a large contribution to the mass transfer resistance. In gaseous application the impact is lower, but for high permeances concentration polarization can still play a crucial role. Most commercial gas permeation hollow fiber modules have the active membrane layer sited on the outer fiber diameter. In general there is a trade-off between higher membrane area and concentration polarization on the feed side. In the case of highly permeable membranes, where concentration polarization dominates the active layer should be located on the inner diameter. Another advantage of lumen side feed is that a more even flow distribution of the feed gas can be realized and the tolerance to non-uniform packing is higher. Furthermore, the requirement of mechanical strength of the module casing is reduced due to low pressures on the shell side.[17]

The flow on the shell side is often considered as plug flow. Lemanski and Lipscomb[18] presented concentration profiles within the module which indicate that the flow on the shell side is not homogenous. In particular the concentration of the flow close to the shell side inlet and outlet where the fluid flows perpendicular to the fibers differs in radial direction. Hence the module performance is affected due to driving force reduction in regions where the fluid flows not in counter-current flow, which is caused by different residence times of fluid taking different routes through the module and due to variations in fluid velocities which influence the mass transfer in the boundary layer.

In particular, randomly packed fibers lead to channeling on the shell side due to local high packing densities, which is the ratio of area occupied by the fibers to the area of the shell compartment.[19] Channeling has a dramatic impact on the mass transfer, particularly when the flow is fully developed. Therefore a regular fiber packing is essential[20] which can be realized by a square or triangular fiber packing.

In regions with high packing density, the mass transfer is higher due to higher residence times. Bao and Lipscomb[20] found that the flow in the low packing regions dominate the mass transfer as higher flows will decrease overall mass transfer coefficients.

5.3.2 Operational Modes

As mentioned earlier membrane modules generally operate in three-end mode where the feed is split into one retentate and one permeate stream. It is also

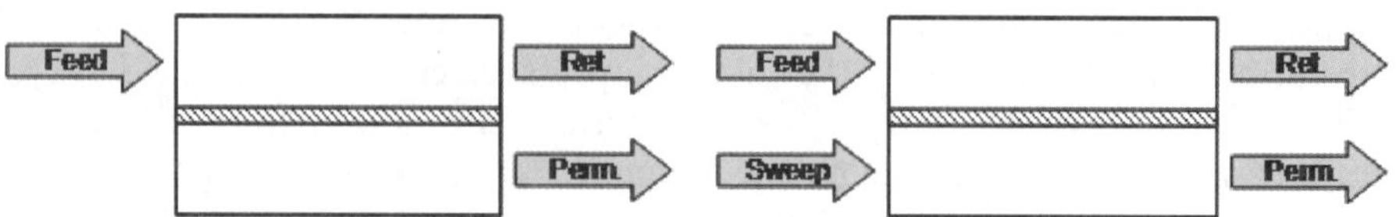

Figure 5.10   A 3-end as well as 4-end module is presented, where the 4-end module is operated using a sweep stream on the permeate side.

possible to use a sweep stream at the 'dead end' of the permeate side. These modules are referred to as four-end modules, which are illustrated in the Figure 5.10.

The sweep gas can be an inert gas or a fraction of the retentate stream which is recycled to the permeate inlet. The sweep gas is used in order to enhance the module performance by reducing the partial pressure of the faster permeating components on the permeate side which increases the driving force. Using a fraction of the product stream reduces the recovery of the process whereas the purity of the product can be increased. Hence four-end modules with sweep gases are applied when high product purities are required. Sweep gas operation is also implemented when driving force generation by high pressures on the feed side or low pressures on the permeate side is not economically feasible. Apparently, the permeate gas can generally not be used in further process steps, since it is contaminated with large amounts of sweep gas.

5.3.3   Flow Patterns

Membrane modules can be operated in five different flow patterns. The flow pattern describes the state in which the feed and the permeate stream flow through the module. It can have an influence on the model performance when the concentration on the feed side changes significantly and when the impact of the porous support layer can be neglected[11] which is true for almost all gas permeation applications. In this section we focus on three-end modules since they are common in industrial application. The distinct flow modes in which three-end modules can be operated are:

- Co-current
- Counter-current
- Cross-flow
- Free flow permeate
- Ideal mixed feed and permeate side.

The flow modes are illustrated in Figure 5.11.

In co-current flow the feed and the permeate flow parallel in the same direction. The quality of the locally produced permeate along the module is higher than the quality of the permeate in the bulk stream of the permeate

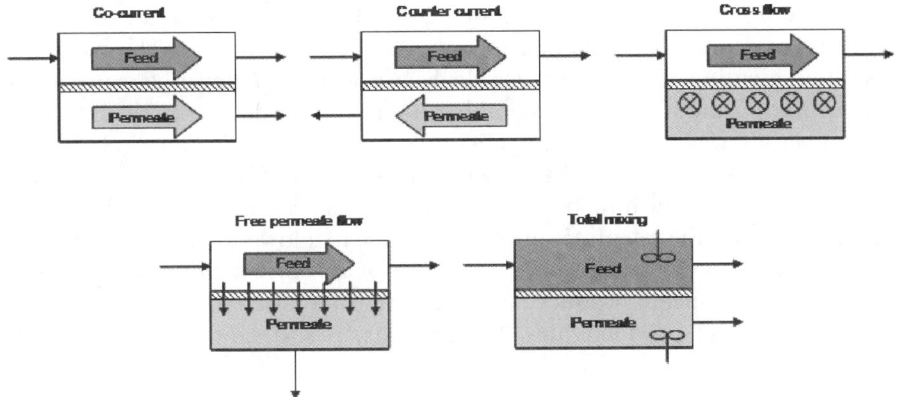

Figure 5.11   The five different flow patterns applied in gas permeation modules.

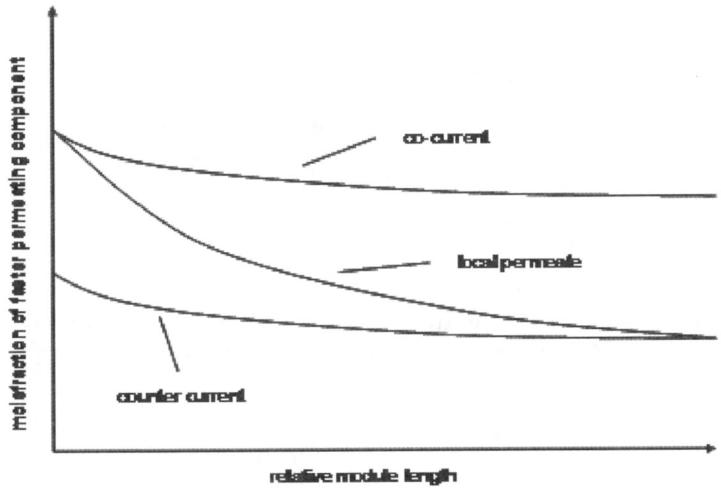

Figure 5.12   Driving force of permeation for co-current, counter-current and locally produced permeate.[5]

chamber. This results in a reduced driving force compared to free flow permeate, which is illustrated in Figure 5.12.

In counter-current flow the feed and the permeate flow parallel but in the opposite direction. As indicated by Figure 5.12 the driving force in counter current flow is higher compared to co-current and free permeate flow. In most applications, in particular those with retentate products, counter current flow is utilized since the higher driving force is favorable,[17] which is demonstrated by experiments.

Cross-flow is realized in spiral wound modules, in which the permeate flows perpendicular to the feed flow. However, there are also counter-current flow spiral wound modules known.[21] Free flow permeate is implemented by orthogonal flow of permeate to the feed fluid. Free flow permeate and ideal mixed module compartments are of theoretical importance but should not be applied in a real membrane module due to low module performance. The modes of operation mentioned previously are ideal cases which can never be realized in a real membrane module where a combination of this modes is observed.

Another alternative is to feed the module by the shell side and discharge the permeate on both ends of the lumen side. Feng and Ivory found that it is superior to co-current flow but not as favorable as counter current flow when the product is on the retentate side.[17] This is due to a lower pressure build up on the permeate side and a negative influence of co-current flow like indicated above. For a permeate product they found that the discharge from both fiber ends yields in high productivities. These results are just valid for shell side feed operation.

5.4 Mathematical Description of the Performance of a Gas Separation Module

5.4.1 Characteristic Numbers

In order to characterize the module performance several characteristic parameters can be specified.

5.4.1.1 Selectivity

The selectivity, S, is defined as ratio of compositions of a target component to another component in the product to the ratio of compositions of a target component to another component of the inlet shown by eqn (5.3):

$$S_{i/j} = \frac{y_i/y_j}{x_i/x_j} \tag{5.3}$$

In gas permeation, selectivity is often referred as the permeability of one component to the permeability of the second component:

$$S_{i/j} = \frac{Pe_i}{Pe_j} \tag{5.4}$$

5.4.1.2 Purity

The purity specifies the mole fraction of the target component in the product. For example, in biogas, which is a multi-component mixture, the purity refers to the methane content in the product.

5.4.1.3 Recovery

The recovery is the stream of the target component entering the membrane module to the stream of target component in the product stream. Unfortunately, a trade-off exists between purity and recovery. The sufficient level of purity has to be chosen carefully since it will determine process economics:

$$\text{recovery} = \frac{\dot{\text{Product}}_i}{\dot{F}_i} \tag{5.5}$$

5.4.1.4   Stage Cut and Pressure Ratio

The module performance is also determined by operation conditions which are the pressure ratio as well as the stage cut. The pressure ratio is the ratio of feed pressure to permeate pressure and represents the driving force of permeation (see eqn (5.7)). The stage cut is the ratio of permeate stream to the feed stream (see eqn (5.6)). High retentate purities require high stage cuts thus the stage cut has an impact on product recovery:

$$\theta = \frac{\dot{P}}{\dot{F}} \tag{5.6}$$

$$\varphi = \frac{p^{\text{Feed}}}{p^{\text{Permeate}}} \tag{5.7}$$

5.4.1.5   Membrane Performance

Since the feed stream is split into one retentate and one permeate stream, the product can be either the retentate or the permeate. Considering retentate side products every product purity can be achieved. The purity of permeate side products is limited. It is strongly dependent on membrane selectivity (indicated

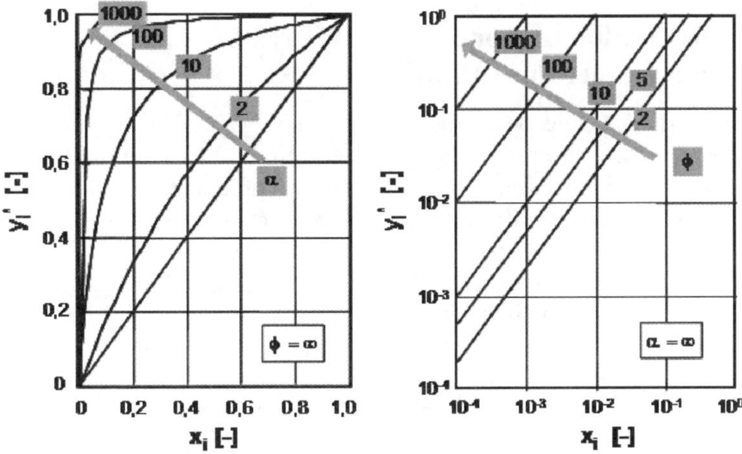

Figure 5.13   Permeate mole fraction of component i dependent on the mole fraction of i in the feed and the applied pressure ratio (left). Permeate mole fraction of component i dependent on the mole fraction of i in the feed and the selectivity (right).

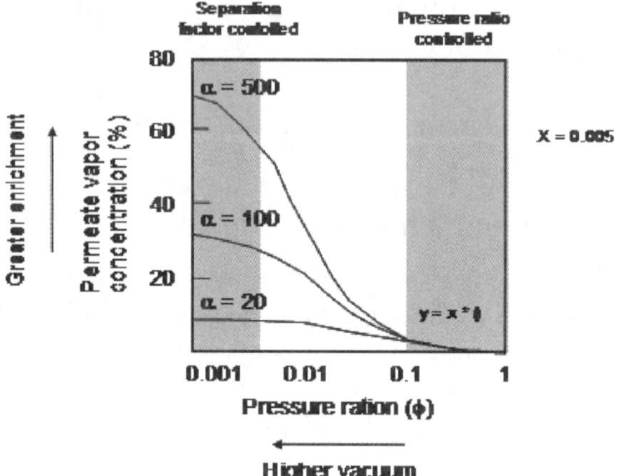

Figure 5.14 Infinitesimal element of a membrane module, where n indicates the mole stream and H is the enthaly stream.

by α) as well as applied pressure ratio (indicated by φ) which is illustrated in Figures 5.13 and 5.14. The figures show that, theoretically, a pure permeate product can be obtain by applying an infinite pressure ratio and using an infinite selective membrane. In industrial applications the pressure ratio is limited since it is the main contributor to the operating costs. The membrane selectivities depend on the permeating components and generally less than 1000.[22] An increased selectivity is coupled with a reduced permeability so that the required membrane area will increase, which result in increased investment as well as operating costs.

5.4.2 Description of Concentration, Pressure and Temperature Profiles

To describe the module performance mass, energy and momentum balances have to be solved. The mass balance is determined for each component in the gas mixture. The balances calculated for the feed and permeate side are coupled using a mass transfer equation which depends on the operating conditions as well as on membrane characteristics. Assuming Henry sorption, non-coupling of permeate fluxes and equality of the chemical potential at the surface of the active membrane layer a simplified mass transfer equation can be used:

$$\dot{n}_i = \frac{Pe_i \cdot \left(x \cdot P^{\text{Feed}} - y \cdot P^{\text{Permeate}}\right) \cdot A}{thickness} \tag{5.8}$$

where Pe is the permeability, A is the active membrane area which is multiplied by the partial pressure difference between feed and permeate side. Detailed information on the derivation of eqn (5.8) can be found in Melin and Rautenbach.[11] The permeability must be estimated by experimental investigations.

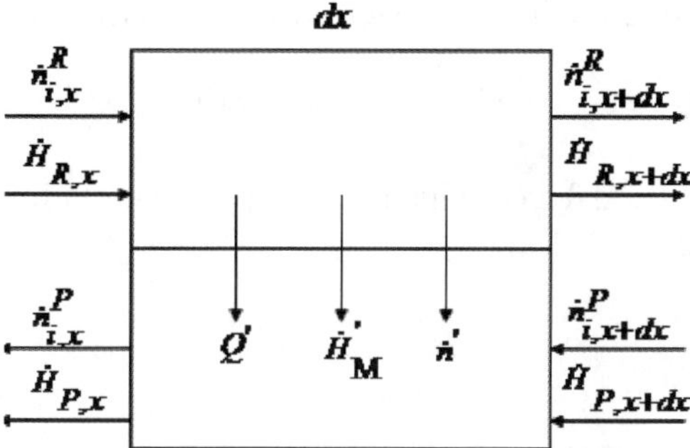

Figure 5.15 This diagram shows the permeate concentration dependent on pressure ratio and selectivity. Two regions can be identified in which one indicates the pressure ratio controlled range and the other the selectivity controlled range.

Demonstrating the conservation equations we will focus on counter current flow since it is of greater importance compared to co-current flow. The derivation of co-current operation is obtained analogously. A differential description of the conservation relations is required to approximate the module performance. The streams necessary to derive the balances are illustrated in Figure 5.15. The list of symbols provides an overview over the abbreviations used in the following section.

Deriving the mass balance for each component on the feed side results in:

$$0 = \dot{n}_i'' \cdot W + \frac{d\dot{n}_i^{\mathrm{R}}}{dx} \tag{5.9}$$

where W is the membrane width. Similarly we obtain the mass balance for the permeate side:

$$0 = \dot{n}_i'' \cdot W + \frac{d\dot{n}_i^{\mathrm{P}}}{dx} \tag{5.10}$$

5.4.3 Energy Balance

The energy balance can be determined similarly. Two phenomena can be observed when energy is transferred from feed to permeate side. On the one hand enthalpy is transferred through the membrane due the permeation. On the other hand this will induce a temperature difference on both sides of the membrane so that a heat flux can be observed. Especially when gases with high Joule–Thomson coefficients permeating through the membrane the energy balance must be incorporated as temperature will influence permeability.

The energy balances for the feed and permeate side are:

$$0 = Q'' \cdot W + \dot{H}_{\mathrm{M}}'' \cdot W + \frac{d\dot{H}_{\mathrm{R}}}{dx} \tag{5.11}$$

$$0 = Q'' \cdot W + \dot{H}_{\mathrm{M}}'' \cdot W + \frac{d\dot{H}_{\mathrm{P}}}{dx} \tag{5.12}$$

where Q is the conductive heat transferred to the permeate side:

$$Q'' = k \cdot (T_{\mathrm{R}} - T_{\mathrm{P}}) \tag{5.13}$$

H_{M} is the enthalpy transferred which is the specific enthalpy multiplied by the permeating stream:

$$\dot{H}''_{\mathrm{M}} = \dot{n}'' \cdot h_{\mathrm{M}} \tag{5.14}$$

5.4.4 Pressure Losses

Pressure losses on the feed and permeate side will influence the module performance drastically (see eqn (5.8)). Pressure losses can be calculated by eqn (5.15):

$$\frac{dP}{dx} = \frac{1}{2} \lambda \frac{\rho}{d} v^2 \tag{5.15}$$

The friction factor is estimated by assuming laminar flow using eqn (5.16):

$$\lambda = \frac{64}{\mathrm{Re}} = \frac{64 \cdot \eta}{d \cdot v \cdot \rho} \tag{5.16}$$

On the shell side, pressure losses can be estimated using eqns (5.15) and (5.16) when applying the hydraulic diameter instead of the inner fiber diameter.

In order to determine the concentration, pressure and temperature profiles these set of coupled differential equations must be solved numerically, which can be done by conventional software.

5.5 Non-ideal Construction of Membrane Modules and the Influence of Non-idealities of Defect-free Dense Hollow Fiber Membranes

Besides non-ideal operation, non-ideal module construction influences the module performance as well. Module performance refers to the recovery, which is the ratio of a product stream to the feed stream of the target compound, for a given product purity. The objective of manufacturing process is to produce uniform, defect free hollow fibers. Unfortunately, real membrane fibers are not uniform since small changes in production conditions as well as manufacturing tolerances result in different fiber properties. Variations in fiber dimensions (diameters, length, and membrane thickness) and fiber properties (permeability) as well as manufacturing defects (blocked fibers, pinholes) have an impact on the module performance. In particular, the variation of fiber diameters, which is one of the major drawbacks in module performance, was well investigated by Lemanski and Lipscomb[23] and Rautenbach *et al.*[24] Among various investigations Struck[5] has analysed the effects of blocked fibers and pinholes. In particular, at higher product purities a minimum in variation of fiber properties is required. Therefore the objective of manufacturing hollow fiber membranes is to produce fibers with equal quality and reduce variations in properties as well as avoid defects during manufacturing process.

5.5.1 Influence of Fiber Diameter Variation

As shown in various studies[5,23,24] the variation of fiber diameter (inner as well as outer diameter) should be avoided. Apparently the variation in fiber diameter has two effects, which are a variation of membrane area as well as a variation in flow rate in every single fiber. The membrane area of a single fiber is linearly dependent on the fiber diameter:

$$A_{\text{Membrane}} = \pi \cdot d \cdot L \tag{5.17}$$

The flow rate through a fiber[23] is influenced by the diameter by the power of 4. Applying the Hagen–Poiseuille law, which is valid for laminar flow in hollow fibers, results in:

$$\dot{V} = \frac{\pi}{128} \frac{d^4}{\eta} \frac{dP}{dz} \tag{5.18}$$

If the fiber diameter varies either effect must taken into account. In case of a fiber diameter distribution within a module, small fibers could consume the product. If the whole feed stream entering a fiber permeates through the membrane, the pressure within the fibers will drop under the pressure of the retentate pressure, leaving larger fibers so that a backflow into the small fiber is induced. This results in a loss of product and will decrease the module recovery. The reduced pressure on the feed side also decreases the mass transfer through the membrane, which is indicated by eqn (5.19):

$$\dot{n} = \frac{Pe_i \cdot \left(x \cdot P^{\text{Feed}} - y \cdot P^{\text{Permeate}}\right) \cdot A}{\textit{thickness}} \tag{5.19}$$

This will influence the quality of the permeate produced by single fibers.

5.5.2 Influence of Variation in Membrane Thickness

Intrinsic permeability is a material property which is independent of membrane dimensions. A crucial factor in the design and application of membrane modules is permeance. In contrast to the intrinsic permeability the permeance is determined by material properties as well as thickness of the active membrane layer. Hence variations in membrane thickness result in variations of permeances though the selectivity is not affected, since the permeance will change for all permeating components:

$$Q = \frac{S \cdot D}{\delta} \qquad (5.20)$$

As indicated by the eqn (5.4), the permeance depends on the solubility S of the permeating molecule in the polymeric membrane, the diffusion coefficient, D, through the membrane material as well as the membrane thickness, δ.

The quality of permeate produced by any fiber is highly dependent on the permeance, and the permeate produced by fibers with different permeances will differ. Similar to a variation in fiber diameter, variations in permeance result in fibers in which the entering gas permeates completely through the membrane, so that product is consumed. Lemanski and Lipscomb,[23] and Struck and co-workers,[5,24] found that the effect of different permeances is similar to a variation in fiber diameter and should be avoided.

5.5.3   Influence of Variation in Fiber Length

The active fiber length within a membrane module can vary due to capillary forces during the potting process as well as due to fibers which are twisted to enhance the mass transfer on the shell side. Variations in fiber length result in a variation in membrane area of each fiber. Struck *et al.*[5,24] showed that deviations in fiber length have only an insignificant impact on module performance. He demonstrated that if 50% of the fibers are 10% shorter than the residual fibers, the module performance is not influenced significantly. Further, he revealed that if a fiber fraction is 25% shorter than the residual fibers, the impact on module performance cannot be neglected. However, this case is uncommon in technical applications of straight hollow fiber modules. Some module suppliers tend to braid hollow fibers to enhance mixing on the shell side. In this particular case the variation of fiber length shall be considered and the module must be designed carefully.

5.5.4   Influence of Membrane Defects

Dense membranes are essential for proper module operation. Due to the manufacturing process defects which are pinholes, broken fibers as well as poor sealed adhesive joints can influence the module performance.

In general, the flow-through defects are not selective and uncontrolled which is detrimental to module performance and operation. The loss through a defect is negative since product is lost and the recovery is lower than in the case of defect free fibers, which are illustrated in Figure 5.16. The unselective flow will also influence the permeate concentration. Two cases have to be distinguished where in the first case the defect is close to the retentate outlet and in the latter

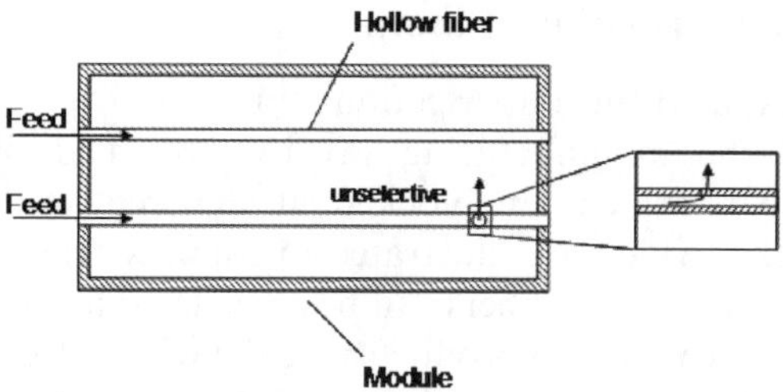

Figure 5.16 Membrane module with a defect which causes unselective flow through the membrane.

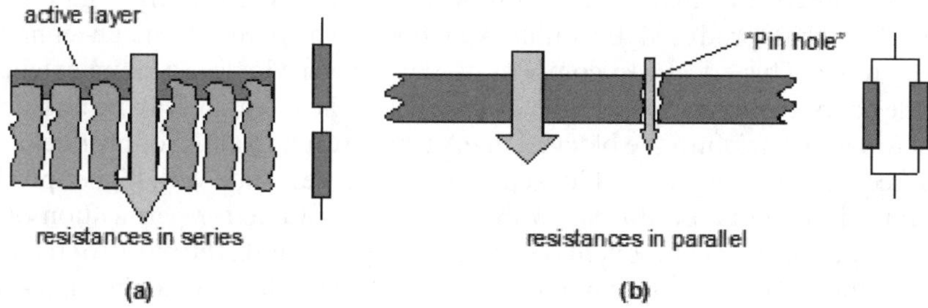

Figure 5.17 Mass transfer resistance model for defect free (a) as well as for a defect (b) membrane.

case the defect is close to the feed inlet. If the flow through a defect is close to the retentate outlet, the major fraction is contributed by the retentate stream. This results in a reduced recovery and the permeate concentration is influenced since the unselective flow acts like a sweep stream. In the case of a defect close to the inlet, the recovery is reduced, but the permeate concentration is not significantly influenced.

Struck[5] demonstrated that pinholes have significant negative influences on the membrane performance and that pinholes should be avoided. The flow through defects can be modeled using the resistance model proposed by Ismail and Shilton.[25] The model is illustrated in Figure 5.17 and is similar to electric resistance models. In defect free membranes two mass transfer resistances occur, which are the resistance in the active layer and the resistance in the porous support. Compared to the active layer the resistance of the porous support can be neglected in many applications. Ismail and Shilton[25] also found that defects in membrane material are generally less than 100 Å. Thus it is reasonable to assume Knudsen flow within the defects:[26]

$$\dot{n}_k'' = \frac{4\varepsilon d_{\text{pore}}}{3\tau\sqrt{2\pi\Re T M_k}} \cdot \frac{\Delta P_k}{\delta} \tag{5.21}$$

The uncontrolled flow through a defect can be in the same order of magnitude as the stream permeating through the membrane which has a tremendous effect on the module performance.

5.5.5   Influence of Blocked Fibers

Blocked fibers result from imperfections in adhesive area, crushing and impurities during the manufacturing process. Blocked fibers can be distinguished into fibers which are blocked at the inlet and those which are blocked at the outlet, which are illustrated in Figures 5.18 and 5.19.

In the latter case the blocked fibers can be considered as fibers with a recovery of zero in case of a retentate side product since the whole stream flowing into the fiber will permeate through the membrane. The concentration of the fast permeating component will be lower compared to the concentration in an unblocked fiber. Since the permeate mixes on the shell side with the permeate of unblocked fibers, the mixed concentration of the faster permeating component is lower than that produced by unblocked fibers which results in an enhanced driving force. This effect can compensate the recovery losses to some extent.

If the fibers are blocked at the inlet, a fraction of the retentate leaving unblocked fibers will flow back into the blocked fibers which result in a loss of product and a lower recovery. In the case of blocked inlet fibers a permeate with higher product concentration will be produced, so that the driving force for permeation of the preferred permeating component is enhanced. Furthermore, the purity of retentate stream is limited in contrast to unblocked fibers. In a module containing blocked as

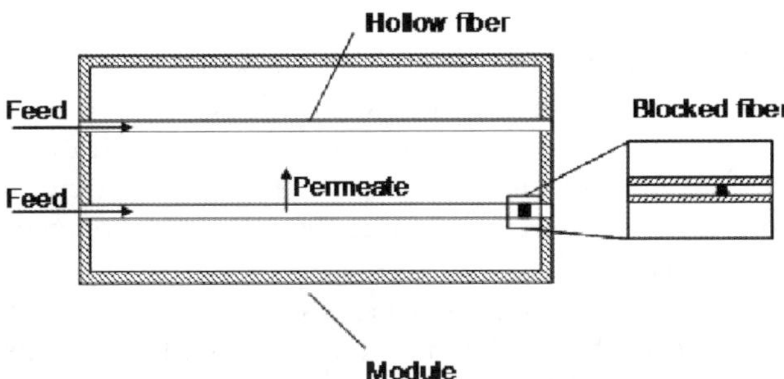

Figure 5.18   Membrane module with blocked fibers at the retentate outlet. The blocked fibers can be assumed as fibers with a recovery of 0.

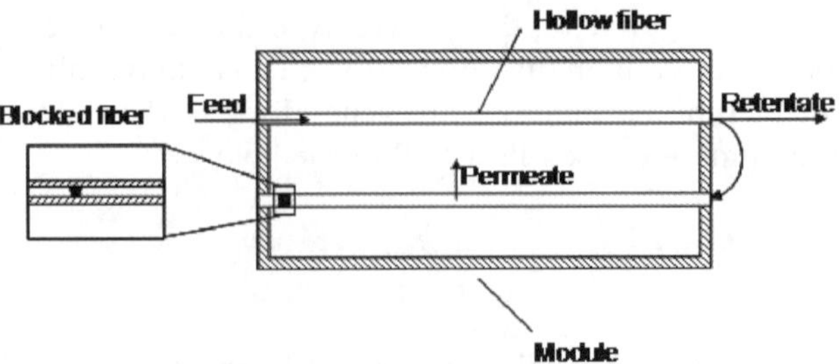

Figure 5.19   Membrane module with blocked fibers at the feed inlet. The blocked fibers consume product, so that the recovery decreases.

well as unblocked fibers the recovery of unblocked fibers cannot be reduced in any order since recovery is lost through the blocked fibers.

In order to evaluate if the fibers in a membrane module are blocked or have any defects Feng and Ivory[26] proposed the following procedure. The shell side of the module should be connected to a pure gas supply at a constant pressure where one end of the lumen side is closed (Figure 5.18). Then permeate flow rate measurements are performed. In the second step the first end is closed and the second end should be opened. If all fibers are unblocked then the module performance should not differ in both cases. The whole procedure should be repeated for various pressures in order to obtain a linear dependence of the flow rate from the pressure which indicates that there is no internal leakage caused by defects in the membrane.

5.6 Concluding Remarks

When membranes shall be applied in processes or any other applications, they must be assembled in so called modules. The three major module types for gas separation processes are plate and frame, spiral wound and hollow fiber modules. These designs differ in packing density, manufacturing cost and their range of application. The plate and frame modules, which is highly effective in pervaporation applications, is much less popular than the spiral wound and hollow fiber modules due to the lower area per volume ratio that can be accommodated. The majority of gas separation membranes are arranged in hollow fiber modules due to their high area per volume ratio. This can be achieved by a high packing density which is required since most membranes have low permeabilities and so high membrane areas are necessary. Hollow fibers are easy to manufacture and manufacturing costs are quiet low. In a hollow fiber module the fibers are in parallel and they are placed in a shell. The fibers are connected to the shell by a potting resin which seals a module's shell side. It allows use of the modules with shell or bore-side feed.

The objective of designing hollow fiber modules is to minimize non-ideal effects like pressure losses and concentration polarization. Usually, there is trade-off between reducing pressure losses and reducing concentration polarization. The effect of co-current, counter current and cross-flow configurations on the mass transfer was elaborated.

In order to obtain an optimal separation performance various operational as well as design parameters have to be considered carefully. Here the flow through the module, which is usually counter current flow, the location of the feed and active layer are the most important parameters. During module manufacturing various detrimental effects can occur, which reduce the module performance, *e.g.* variations in fiber dimensions and properties (diameter, length, membrane thickness, permeability) and defects (pinholes, blocked fibers). In order to ensure a proper operation of the membrane module these effects have to be avoided.

5.7 List of Symbols

A	Active membrane area
A_{Membrane}	Membrane area
d	Diameter
D	Diffusivity
d_{Pore}	Pore diameter
$\dot{F}$	Feed stream
$\dot{F}_{\text{i}}$	Feed stream of component i
h_{M}	Specific enthalpy
$\dot{H}_{\text{M}}''$	Enthalpy
k	Heat transfer coefficient
L	Length
M_k	Molar mass of component k
$\dot{n}$	Mole stream
$\dot{n}_{\text{i}}$	Stream of component i
$\dot{n}_{\text{i}}''$	Flux of component i
$\dot{n}_k''$	Flux of component k
Pe	Permeability
$\dot{P}$	Permeate stream
p^{Feed}	Feed pressure
Pe_i, Pe_j	Permeability of components i and j
δP_k	Partial pressure difference
p^{Permeate}	Permeate pressure
Q	Permeance
Q''	Heat flux transferred by conduction
$\Re$	Universal gas constant
Re	Reynolds number
S	Solubility
$S_{i/j}$	Selectivity
T	Temperature
T_{R}	Feed temperature
T_{P}	Permeate temperature
$\dot{V}$	Volume stream
W	Membrane width
x	Mole fraction retentate
y	Mole fraction permeate
δ	Thickness of active membrane layer
ε	Porosity
η	Dynamic viscosity
θ	Stage cut
λ	Friction factor
v	Velocity
π	pi
ρ	Density
τ	Tortuosity
φ	Pressure ratio

References

1. M. Mulder, *Basic Principles of Membrane Technology*, Kluwer Academic Publishers, Dordrecht, 2nd edn, 1997.
2. W. J. Koros and G. K Fleming, *J. Membr. Sci.*, 1993, **83**, 1.
3. R. W. Baker, E. L. Cussler, W. Eykamp, W. J. Koros and H. Strathmann, *Membrane Separation Systems: Recent Developments and Future Directions*, Noyes Data Corp., Park Ridge, NJ, 1991.
4. R. W. Baker, *Ind. Eng. Chem. Res.*, 2002, **41**, 1393.
5. A. Struck, PhD thesis, RWTH Aachen University, 1998.
6. J. Caro, K. J. Caspary, Ch. Hamel, B. Hoting, P. Kölsch, B. Langanke, K. Nassauer, Th. Schiestel, A. Schmidt, R. Schomäcker, A. Seidel-Morgenstern, E. Tsotsas, I. Voigt, H. Wang, R. Warsitz, St. Werth and A. Wolf, *Ind. Eng. Chem. Res.*, 2007, **46**, 2286.
7. K. Ohlrogge and K. Ebert, *Membranen - Grundlagen, Verfahren und industrielle Anwendungen*, Wiley-VCH, Weinheim, 2006.
8. G. Etchegoyen, T. Chartier and P. Del-Gallo, An architectural approach to the oxygen permeability of a La0.6Sr0.4Fe0.9Ga0.1O3-δ perovskite membrane, *J. Eur. Ceram. Soc.*, 2006, **26**, 2807.
9. L. Vratsanos *et al.*, International Pittsburgh Coal Conference, 20–23 September 2009.
10. J. Repasky *et al.*, ITM Syngas: Ceramic Membrane Technology for Lower Cost Conversion of Natural Gas, AIChE Spring National Meeting San Antonio TX, 21–25 March, 2010.
11. T. Melin and R. Rautenbach, *Membranverfahren - Grundlagen der Modul- und Anlagenauslegung*, Springer, Berlin, 2007.
12. A. Da Costa, A. Fane and D. Wiley, *J. Membr. Sci.*, 1994, **87**, 79.
13. C. Fritzmann, K. Ronzheimer and T. Melin, *IDA World Congress on Desalination and Water Reuse*, 7–12 November 2009, Dubai, UAE.
14. T. Kondo, *New Developments in Gas Separation Technology*, Toray Research Center, Inc., 3-3-7 Sonoyama, Otsu, Siga, 520, Japan, 1992.
15. R. H. Blackmer and J. H. Hedman, *US Patent* 4 174 955, 1979.
16. R. W. Baker, J. G. Wijmans, T. C. Merkel, H. Lin, R. Daniels and S. Thompson, *US Patent* 2010236404, 2010.
17. X. Feng and J. Ivory, *J. Membr. Sci.*, 2000.
18. J. Lemanski and G. G. Lipscomb, *J. Membr. Sci.*, 2001, **195**, 215.
19. L. Bao and G. Lipscomb, *J. Membr. Sci.*, 2002, **204**, 207.
20. L. Bao and G. Lipscomb, *J. Membr. Sci.*, 2002, **57**, 125.
21. T. Hofmann, M. Wessling and R. W. Baker, US Patent 5 711 882, 1998.
22. L. Robeson, *J. Membr. Sci.*, 2008, **320**, 390.
23. J. Lemanski and G. G. Lipscomb, *J. Membr. Sci.*, 2000, **167**, 241.
24. R. Rautenbach, A. Struck and M. F. M. Roks, A variation in fiber properties affects the performance of defect-free hollow fiber membrane modules for air separation, *J. Membr. Sci.*, 1998, **150**, 31.
25. A. F. Ismail and S. J. Shilton, *J. Teknologi.*, 2000, **32**(F), 93.
26. X. Feng and J. Ivory, *J. Membr. Sci.*, 2000, **176**, 197.

Gas/Vapor Permeation Applications in the Hydrocarbon-processing Industry

ARNAUD BAUDOT

IFP Energies nouvelles, Rond-point de l'échangeur de Solaize, BP 3, 69360 Solaize, France

6.1 Natural and Biogas Membrane Processing

Though raw natural gas is mainly composed of methane (generally from 75% to 90%), it also contains undesired components such as:

- Acid gaseous impurities such as carbon dioxide or hydrogen sulfide that should be removed in order to prevent pipeline corrosion
- Condensable compounds (higher C2+ hydrocarbons and water) that should be removed in order to prevent condensation troubles, hydrate formation or corrosion when flowing natural gas in pipelines
- Inert gases, such as nitrogen, that lower the calorific value of natural gas.

6.1.1 Membrane Suppliers

Since the beginning of the 1980s, gas permeation has been primarily applied in the natural gas processing industry to carbon dioxide removal. The very first field trials were carried in the beginning of the 1980s with cellulose acetate-based permeators by small companies such as Separex,[1] Envirogenics

Membrane Engineering for the Treatment of Gases, Volume 1:
Gas-separation Problems with Membranes
Edited by Enrico Drioli and Giuseppe Barbieri
© Royal Society of Chemistry 2011
Published by the Royal Society of Chemistry, www.rsc.org

or Grace Membrane Systems (a joint venture between W.R. Grace and Bend Research).[2] One of the first large-scale membrane-based carbon dioxide tertiary recovery project was launched in 1983 in Texas with two Cynara (Dow) permeation units[3] that were designed to handle up to 94 000 N m^3 h^{-1}.

At the end of the 1980s large producers of industrial gases entered the permeation membrane business:

- Air Products and Chemicals Inc. incorporated Separex in 1988. Air Liquide and DuPont created a dedicated joint venture, *i.e.* Medal (MEmbrane Dupont Air Liquide), in 1989 which became a fully owned subsidiary of Air Liquide the following year.[4]
- Permea Inc. (belonging formerly to Monsanto) became a subsidiary of Air Products and Chemicals Inc. in 1991.[4]
- In 1996 Medal launched a new generation of hollow-fiber permeation membranes, based on polyimides.[5]

The last move concerning producers of permeation membranes was the entry of large oil and gas engineering companies since the mid-1990s:

- Separex, formerly owned by Hoechst, was sold to UOP in 1994.[6]
- Grace Membrane Systems became Kvaerner Membrane Systems in 1994.
- Cynara was acquired by Natco in 1998.
- ABB signed a cooperation agreement with MTR Inc. in 2002.

In 2002, the estimated annual sales of membranes for the treatment of natural gas was estimated to be close to 30 million USD per year.[7] All the main current membrane providers for natural gas treatment are listed in Table 6.1.

Table 6.1 Main suppliers of membrane technology for natural gas treatment

Company (group/country)	Membrane material	Module type	CO$_2$	H$_2$O/VOC
Producers of industrial gases				
Permea (Air Products U.S.)	Polysulfone	HF	×	—
Medal (Air Liquide Fr.)	Polyimide/polyaramide	HF	×	—
Oil and gas engineering				
Kvaerner (U.K./N.)	Cellulose acetate	SW	×	—
Separex (UOP U.S.)	Cellulose acetate	SW	×	—
Cynara (Natco U.S.)	Cellulose acetate	HF	×	—
MTR (ABB U.S.)	Perfluoropolymers, silicone rubber	SW	—	×
Borsig (D.)	Rubbery polymers	SW	—	×
Others				
Ube Industries (J.P.)	Polyimide	HF	×	—
GKSS (D.)	Silicone/PEBA	SW	—	×

HF, hollow fibers; SW, spiral wound.[7]
×, supplier provides membranes for CO$_2$ or H$_2$O/VOC removal.

The estimated cost for the fabrication of hollow-fiber membranes modules ranged from 2 to 10 USD m^{-2} while it is comprised between 10 and 100 USD m^{-2} for spiral-wound membrane modules.[7,8] The selling price of the modules is generally between two and three times higher than the production cost of the membrane itself.[7] To be more precise, the selling price of a hollow fiber module was given at 108 USD m^{-2} in 2002[9] while it was equal to 160 USD m^{-2} for a spiral-wound module in 1993.[10,11]

6.1.2   Membrane-based Acid Gas Removal

6.1.2.1   *Carbon Dioxide Removal from Natural Gas*

Gas permeation constitutes the latest carbon dioxide extraction technology applied to natural gas treatment and, as such, had to gain its place against (or with) well-proven absorption-based sour gas removal processes.

Indeed, membranes offer several strong advantages over conventional solvent columns: compacity (which makes membranes particularly well suited for off-shore operations[12]), low investment costs, low maintenance costs, at least with single-stage configurations that do not operate compressors. The current level of commercial membrane CO_2/CH_4 selectivity is between 12 and 25 in field conditions.[7,13–15] As a matter of fact, the relatively moderate values of CO_2/CH_4 selectivity offered by the current commercial polymer membranes result in the field in a partial loss of the treated methane in the low-pressure permeate. Indeed, methane loss in the low-pressure permeate generally varies between 10 and 25%, on a single-stage configuration (Figure 6.1).

Depending on the price of natural gas, methane loss to the permeate can represent from 33 to 80% of the whole operating cost of a membrane unit, not taking into account the amortisation costs.[6,10,11,16] Process optimisation calculations led to the introduction of two-stage configurations, wherein the first stage permeate was pressurized and fed to a second permeation stage in order to recover a part of the methane. This generally leads to a significant improvement of the overall methane recovery (higher than 95%), though introducing extra cost due to the inter-stage compression unit (Figure 6.2).

Several studies have been carried out since the adoption of membranes by natural gas operators, aiming at comparing membrane-based carbon dioxide removal operations with conventional absorption operations (Table 6.2).

Several trends can be issued from these studies, depending on the various considered gas treatment scenario.

For moderate natural gas flows (lower than 25 000–30 000 N m^3 h^{-1})[20,21] containing high carbon dioxide concentrations it appears that membranes are more profitable than absorption processes. In 1995, it was shown that commercial permeation membranes were more competitive than DEA/MDEA-based absorption when the carbon dioxide content was higher than 20%.[22] Another study realized by Kellogs & Co.[21] sets this limit at approximately 16% for a gas price of 1.5 USD per MMBtu.

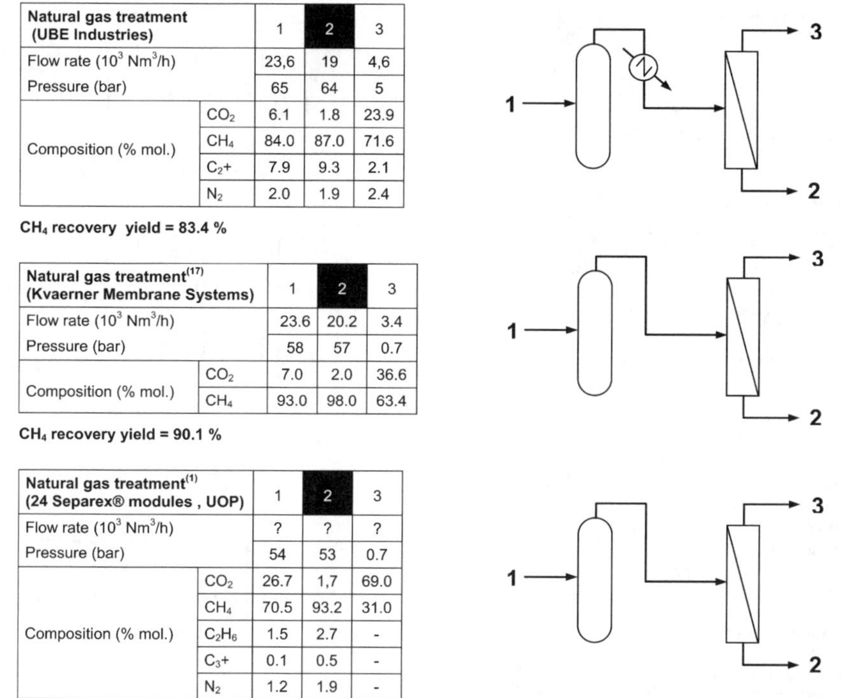

Natural gas treatment (UBE Industries)		1	2	3
Flow rate (10^3 Nm^3/h)		23,6	19	4,6
Pressure (bar)		65	64	5
Composition (% mol.)	CO$_2$	6.1	1.8	23.9
	CH$_4$	84.0	87.0	71.6
	C$_2$+	7.9	9.3	2.1
	N$_2$	2.0	1.9	2.4

CH$_4$ recovery yield = 83.4 %

Natural gas treatment[17] (Kvaerner Membrane Systems)		1	2	3
Flow rate (10^3 Nm^3/h)		23.6	20.2	3.4
Pressure (bar)		58	57	0.7
Composition (% mol.)	CO$_2$	7.0	2.0	36.6
	CH$_4$	93.0	98.0	63.4

CH$_4$ recovery yield = 90.1 %

Natural gas treatment[1] (24 Separex® modules , UOP)		1	2	3
Flow rate (10^3 Nm^3/h)		?	?	?
Pressure (bar)		54	53	0.7
Composition (% mol.)	CO$_2$	26.7	1,7	69.0
	CH$_4$	70.5	93.2	31.0
	C$_2$H$_6$	1.5	2.7	-
	C$_3$+	0.1	0.5	-
	N$_2$	1.2	1.9	-

Figure 6.1 Examples of single-stage membrane-based carbon dioxide removal operations.

For high natural gas flows with low carbon dioxide concentrations absorption processes prove to be more profitable than gas permeation modules.

For high natural gas flows combined with high carbon dioxide content an association of membranes and absorption seems to be the most profitable option.[21] Membrane modules, located before the absorption column, carry out a bulk removal of carbon dioxide, while the absorption columns allow the final polishing of the natural gas. This configuration leads to significant investment cost reductions on the absorption unit.

6.1.2.2 Other Related Permeation-based Carbon Dioxide Removal Operations

6.1.2.2.1 Enhanced Oil Recovery. This technique aims at increasing the oil fields recovery yield as a result of high pressure injection of gaseous carbon dioxide. This leads to the maintaining of high pressures in the reservoir and to an improvement of oil displacement because of several physical effects: decrease of oil viscosity due to carbon dioxide solubilisation, stripping effects and modification of multiphase equilibria due to CO$_2$ solubilization in the reservoir

Natural gas treatment (UBE Industries)		1	2	3
Flow rate (10^3 Nm^3/h)		118	103	22
Pressure (bar)		58	57	0,3
Composition (% mol.)	CO$_2$	9.6	1.9	61.0
	CH$_4$	90.0	97.6	38.9
	C$_2$+	0.3	0.4	0,0
	N$_2$	0.1	0.1	0.1

CH$_4$ (C2+) recovery yield = 94.4 % (87 %)

Natural gas treatment[6] (UOP LLC)		1	2	3
Flow rate (10^3 Nm^3/h)		35.4	31.3	4.1
Pressure (bar)		63	62	0,3
Composition (% mol.)	CO$_2$	11.0	1.9	81.1
	CH$_4$	86.0	95.2	18.7
	C$_2$+	2.1	2.3	0.1
	N$_2$	0.6	0.6	0.1

CH$_4$ (C2+) recovery yield = 97.8 % (96.7 %)

Natural gas treatment[17] (Kvaerner Membrane Systems)		1	2	3	4	5
Flow rate (10^3 Nm^3/h)		23,6	22,2	1,5	3.7	2.2
Pressure (bar)		58	57	0.7	0.7	58
Composition (% mol.)	CO$_2$	7.0	2.0	81.1	36.6	7.0
	CH$_4$	93.0	98.0	18.9	63.4	93.0

CH$_4$ recovery yield = 98.5 %

Natural gas treatment[18] (Modules Separex®. UOP LLC)		1	2	3	4	5	6
Flow rate (10^3 Nm^3/h)		17.1	30	12.6	16.9	12.7	4.1
Pressure (bar)		66.5	66.5	65.5	18.2	17.2	1.35
Composition (% mol.)	CO$_2$	10	11.5	2	16	13.7	23
	CH$_4$	64.9	64.2	86.5	48.8	63.2	4.8
	C$_2$+	5	4.2	8.6	2.4	3.1	0.1
	H$_2$S	20	20	3	32.7	20	71.6
	H$_2$O	0.1	0.1	0.01	0.1	0.01	0.5

CH$_4$ recovery yield = 98 %

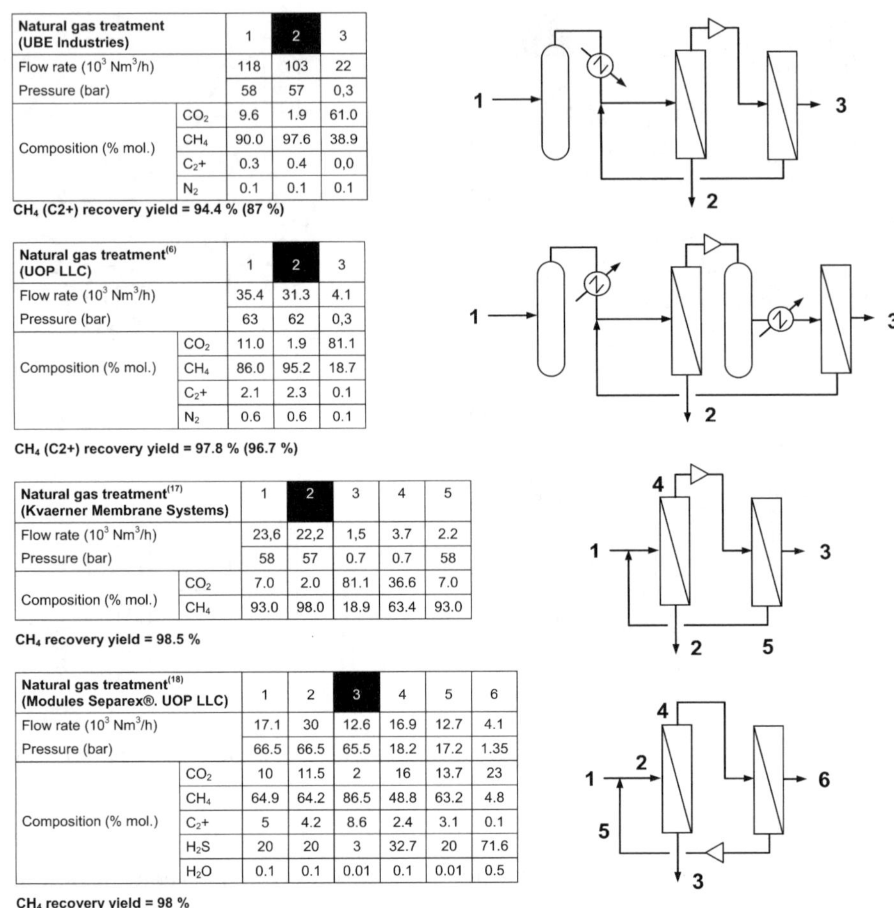

Figure 6.2 Examples of two-stage membrane-based carbon dioxide removal operations.

aqueous phase. When the oil reaches the surface, carbon dioxide is purged with the associated gas. Since the volumes of concerned carbon dioxide are very large (from 140 to 280 N m^3 per barrel extracted),[23] it is necessary to separate carbon dioxide from the hydrocarbon gaseous phase in order to have it recycled to the reservoir (after pressurization).

Since the quantity of CO$_2$ dissolved in oil varies over a large range of values during the whole production of the reservoir (levels from 40% up to 90% can be observed after a few years of production),[12] it is rather difficult to design a cost-effective CO$_2$ removal unit that could operate during the entire life of the reservoir. Therefore, modular CO$_2$ removal units such as membrane operations offer much flexibility for such an operation (Table 6.3).

The first large scale EOR project operating membranes has been carried out in the Sacroc oil field located in western Texas. Carbon dioxide injection was

Table 6.2 Comparison of relative investment costs (CAPEX) and operating costs (OPEX) of membrane-based, solvent-based and hydrid processes for acid gas removal from natural gas

Operating conditions	Economic study by	Year	CAPEX ref. OPEX ref.	1-stage mem.	2-stage mem.	Mem. +DEA	DEA	MDEA	Selexol
35 400 m^3 h^{-1} 11% CO$_2$, 63 bar	UOP[6]	1995	— 1.438 M$/a		— 1		— 0.99		
43 900 m^3 h^{-1} 10% CO$_2$, 49 bar	Grace (Kvaerner)[16]	1988	3.3 M$ 1.54 M$/a		1 1		1.68 1.20		
11 800 m^3 h^{-1} 12.2% CO$_2$	Perry Gas Co.[19]	1983		1 1		2.77 0.76	3.85 0.66		
	J. Brown Eng.[17]	1988		1 1	1.43 0.85	4.6 0.44	6.36 0.087	6.25 0.068	6.18 0.13

Table 6.3 Comparison of relative investment costs (CAPEX) and operating costs (OPEX) of membrane-based, solvent-based and hydrid processes for carbon dioxide capture (EOR applications)

Operating conditions	Study realised by	Year	CAPEX ref. OPEX ref.	Cryo.[*] + Mem.	Mem.	Mem. +DEA	DEA	Cryo.[*]	TEA	KCO_3 chaud
175 000 m^3 h^{-1}	Amoco[25]	1982	47 M\$			1	2.2	1.56	1.48	
90% CO_2,			20.4 M\$/a			1	1.76	1.39	1.12	
18 bar										
118 000 m^3 h^{-1}	Permea[26]	1983	16.1 M\$	1.50		1				1.31
80% CO_2			6.5 M\$/a	1.31		1				1.82
190 000 m^3 h^{-1}	Fluor Eng.[27,28]	1984			1				0.63	0.63[**]
40% CO_2					1				0.56	0.68[**]

[*]cryo.: cryogenic distillation (Ryan–Holmes process[2]).
[**]Hot potassium carbonate + membrane.

launched in 1972 (at a flow range up to 240 000 N m^3 h^{-1}). Three CO$_2$-recovery units were initially installed: two hot potassium carbonate-based absorption units (operatred Sun Explo with a 190 000 N m^3 h^{-1} capacity leading to a reduction of carbon dioxide from 24% to 0.5%; and Chevron, with 54 000 N m^3 h^{-1} lowering CO$_2$ content to 1%) and an amine (MEA)-based absorption unit (operated by Monsanto, with a 20 000 N m^3 h^{-1} capacity). At the end of the 1970s, realising that the level of CO$_2$ was increasing at a higher rate than expected, Chevron contacted Cynara in order to install and operate two gas permeation units upstream of the hot potassium carbonate absorption columns. The treatment capacities of the membrane units were, respectively, 60 000 N m^3 h^{-1} at 35 bar feed pressure prior to the Sun unit and 25 000 N m^3 h^{-1} at 33 bar feed pressure prior to the Chevron unit. Cynara chose to operate a single stage configuration, the membrane module being installed in parallel. The membrane operation proved to be highly flexible and reliable, as the membrane modules remained in operation twice longer than the expected lifetime (more than 5 years instead of 2–3 years) without any decrease in selectivity.[24] A slight decrease in membrane permeability has nevertheless been observed throughout the operation, and it was necessary to add more modules in order to maintain the production level of the membrane operation.[3]

Since then, other membrane companies have seen their products involved in other EOR projects:

- Kvaerner Membrane System[20]
  - Dallas Production Inc. (Texas) treated 200 000 N m^3 of gases containing up to 25% of carbon dioxide in 1994 during a feasibility study that lasted for 18 months. After this test, the membrane unit was in operation treating 1200 N m^3 h^{-1}
  - Hydrocarbon Operating, Inc. (Texas) processed 590 N m^3 h^{-1} at 51 bar feed pressure in 1994 on a pilot skid composed of two tubular membrane modules
- A Medal membrane unit was operated at the end of the 1990s to process 14 200 N m^3 h^{-1} at 56 bar, 48% CO$_2$, reducing the retentate CO$_2$ concentration down to 6%
- Cynara[2]
  - Amoco, Texas (1994): Cynara membranes were used to process 35 000 up to 120 000 N m^3 h^{-1} of gas containing 80% CO$_2$
  - Mobil, Salt Creek (1992): Cynara membranes were used to process 76 000 up to 120 000 N m^3 h^{-1} of containing 70% CO$_2$
  - Shell, Texas: Cynara membranes were used to process 13 000 N m^3 h^{-1} of gas containing 70% CO$_2$ (Figure 6.3).

6.1.2.2.2 Biogas Purification. Biogas is issued from the anaerobic digestion of domestic, agricultural or sewage wastes. The composition of biogas is variable, ranging from 25 to 75% of methane, 15 to 40% of carbon dioxide, from a few hundreds ppm up to 50% of nitrogen, oxygen, hydrogen sulfide,

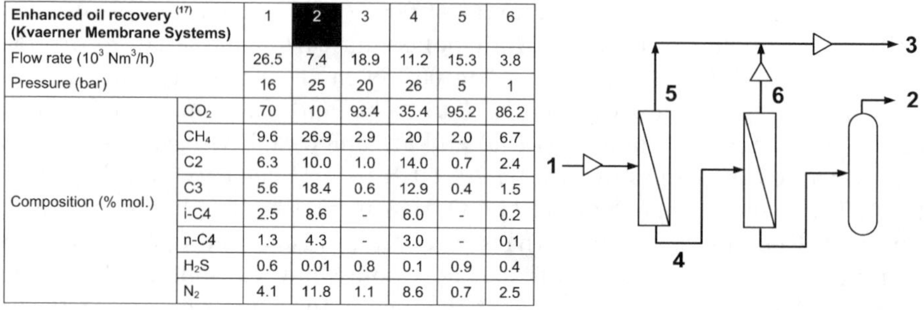

Enhanced oil recovery [17] (Kvaerner Membrane Systems)		1	2	3	4	5	6
Flow rate (10^3 Nm^3/h)		26.5	7.4	18.9	11.2	15.3	3.8
Pressure (bar)		16	25	20	26	5	1
Composition (% mol.)	CO_2	70	10	93.4	35.4	95.2	86.2
	CH_4	9.6	26.9	2.9	20	2.0	6.7
	C2	6.3	10.0	1.0	14.0	0.7	2.4
	C3	5.6	18.4	0.6	12.9	0.4	1.5
	i-C4	2.5	8.6	-	6.0	-	0.2
	n-C4	1.3	4.3	-	3.0	-	0.1
	H_2S	0.6	0.01	0.8	0.1	0.9	0.4
	N_2	4.1	11.8	1.1	8.6	0.7	2.5

CH_4 (C2+) recovery yield = 77,8 % (73 %)

Figure 6.3 Permeation application applied to EOR.

ammonia, water vapor, VOCs (such as halogenated or aromatic compounds). Biogas production flows are usually moderate,[34] ranging generally between 40 and 4000 N m^3 h^{-1}. After appropriate treatment, it is feasible to reinject (bio)methane for domestic consumption through the local gas network.

First attempts to value biogas with membranes were carried out at the beginning of the 1980s (Figure 6.4). For this application, single-stage configurations were studied with membranes from Envirogenics,[18] Permea,[23] Separex,[1,30] (which led to a biomethane recovery yield close to 70%), as well as two-stage configurations from Permea[18] or Membratek-Envig[31] (which raised the biomethane recovery yield to 80%). As an example, on can cite a pilot plant operated in 1996 by the Danish companies Naturscan and Bioscan close the biogas production plant located at Funen (Denmark). Carbon dioxide was extracted through low-pressure feed (8 bar) Membratek membrane modules supplied by the South African company Weir Envig. The biogas, containing between 33 and 43% (v/v) carbon dioxide in the feed, was purified through a two-stage recirculating membrane system. The recovery yield in biomethane reached values up to 90% while it contained relatively moderate amounts of carbon dioxide (less than 3% v/v).

6.1.2.3   Future Directions for CO_2 Removal Membranes Applied to Natural Gas Treatment

In a recent review by UOP,[33] the relative technical limitations of CO_2 removal polymer membranes were listed as follows.

6.1.2.3.1   Sensitivity Towards Chemicals.   The polymer may be sensitive to some chemicals (hydrocarbons, dehydration solvents such as glycols) or even water when in excess. For instance, the presence of excess liquid water can lead to the collapse of dried cellulose acetate membranes. That is why a pre-treatment of natural gas, such as a dew-pointing operation, is sometimes required before a membrane-based carbon dioxide removal operation.

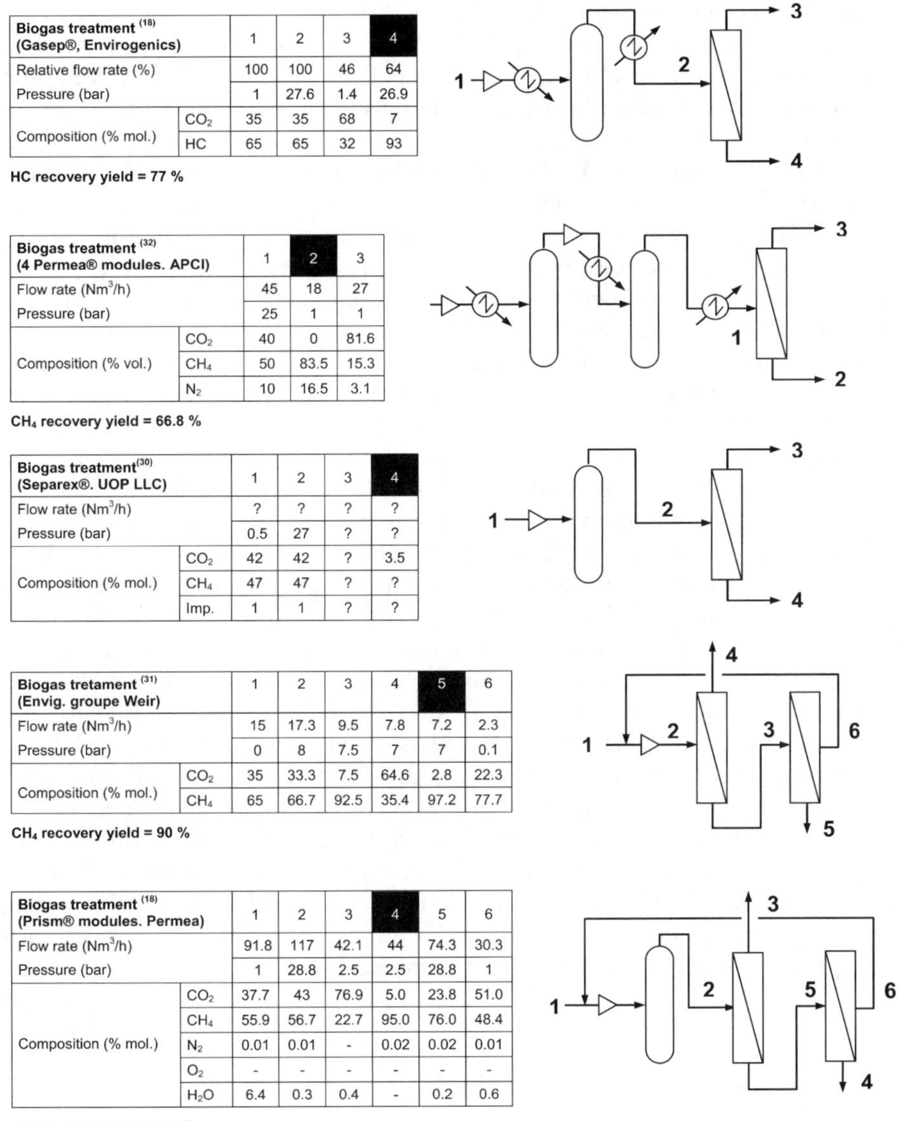

Biogas treatment [18] (Gasep®, Envirogenics)		1	2	3	4
Relative flow rate (%)		100	100	46	64
Pressure (bar)		1	27.6	1.4	26.9
Composition (% mol.)	CO_2	35	35	68	7
	HC	65	65	32	93

HC recovery yield = 77 %

Biogas treatment [32] (4 Permea® modules. APCI)		1	2	3
Flow rate (Nm^3/h)		45	18	27
Pressure (bar)		25	1	1
Composition (% vol.)	CO_2	40	0	81.6
	CH_4	50	83.5	15.3
	N_2	10	16.5	3.1

CH_4 recovery yield = 66.8 %

Biogas treatment [30] (Separex®. UOP LLC)		1	2	3	4
Flow rate (Nm^3/h)		?	?	?	?
Pressure (bar)		0.5	27	?	?
Composition (% mol.)	CO_2	42	42	?	3.5
	CH_4	47	47	?	?
	Imp.	1	1	?	?

Biogas tretament [31] (Envig. groupe Weir)		1	2	3	4	5	6
Flow rate (Nm^3/h)		15	17.3	9.5	7.8	7.2	2.3
Pressure (bar)		0	8	7.5	7	7	0.1
Composition (% mol.)	CO_2	35	33.3	7.5	64.6	2.8	22.3
	CH_4	65	66.7	92.5	35.4	97.2	77.7

CH_4 recovery yield = 90 %

Biogas treatment [18] (Prism® modules. Permea)		1	2	3	4	5	6
Flow rate (Nm^3/h)		91.8	117	42.1	44	74.3	30.3
Pressure (bar)		1	28.8	2.5	2.5	28.8	1
Composition (% mol.)	CO_2	37.7	43	76.9	5.0	23.8	51.0
	CH_4	55.9	56.7	22.7	95.0	76.0	48.4
	N_2	0.01	0.01	-	0.02	0.02	0.01
	O_2	-	-	-	-	-	-
	H_2O	6.4	0.3	0.4	-	0.2	0.6

CH_4 recovery yield = 84 %

Figure 6.4 Examples of permeation-based carbon dioxide removal from biogas.

6.1.2.3.2 Long-term Compaction of the Selective Layer of the Membranes.

The compaction is due to (i) the mechanical compression induced by the trans-membrane pressure difference, and (ii) a relaxation of the polymer chains together with the mechanical compression. The induced densification of the polymer selective layer can lead to a significant reduction of the permeability of the membrane after several years of operation.

6.1.2.3.3 Plasticization by CO_2. Plasticization by CO_2 can lead to a significant loss of separation selectivity when the membrane is operated in field conditions, especially at high CO_2 partial pressures.[34–38] Indeed, the sorption of carbon dioxide leads to the swelling of the selective layer of the membrane and consequently to a diminution of its molecular sieving properties. That is why, in field conditions, the carbon dioxide/methane selectivity of commercial membranes is rather limited, between 12 and 25.[45–48] The operation of membranes leads thus to an unavoidable loss of a part of the processed methane in the low-pressure permeate. In an environment of ever-growing prices of fossil energies, this compels the gas operator to deploy technical solutions in order to limit the methane loss (addition of a second-stage recovery unit, use of permeate as fuel gas) which raises the operating costs of membranes operating. This aspect constitutes the main bottleneck for a wider acceptance of membranes by gas operators. Yet, improved performances targets have been issued, as, for instance, it is estimated that the diffusion of membranes for carbon dioxide removal membrane would be booming if the effective CO_2/CH_4 selectivity of permeation membranes could reach values as high as 50.[13] Several R&D routes have been studied in parallel in order to reach this goal: cross-linking of the polymer network constitutive of the selective layer[35,38–47] by various physical (irradiation, thermal) or chemical means, or developing new inorganic membrane materials, based for instance on narrow-pore molecular sieves. As an illustration, since the mid-2000s, the Japanese company NGK Insulators has been conducting an ambitious R&D program on DDR zeolite membranes[48–52] while Shell Global Solutions has been funding an extensive R&D program at the University of Colorado on SAPO membranes for almost 10 years.[53–57] Since those membranes offer high selectivities on a laboratory scale (from 60 to 200 for DDR membranes,[58] 20 to 120 for SAPO membranes[59,60]), R&D on this topic is presently focused on scale-up[60] and issues related to the resistance of membrane materials towards contaminants.[59]

6.1.3 Removal of Hydrogen Sulfide

Three main types of polymer membrane have been only tested on a laboratory scale or pilot scale in order to remove hydrogen sulfide from natural gas.

6.1.3.1 *Membranes Based on Rubbery Polymers Selective Layers*

This type of polymer material generally offers permability values for low molecular weight components. The selectivity of this type of polymer is then essentially based on the affinity (*i.e.* the sorption coefficient) difference between several components to be separated. In the literature, three main polymer families can be identified.

6.1.3.1.1 Apolar Rubbery Polymers, such as Silicones (PDMS and PTMSP).
The affinity of these materials goes for the most condensable components.
As a general rule, the H_2S/CH_4 selectivity of these materials is rather modest,
as methane diffuses more rapidly than hydrogen sulfide (Table 6.4).[61]

6.1.3.1.2 Apolar Rubbery Polymers Impregnated with Polar Solvents. Two
patents have been issued by UOP in the mid-1980s about the use of PEG- or
glycerol-impregnated silicone membranes. The addition of polar solvents,
which display a high affinity for hydrogen sulfide, leads to a very significant
increase of the selectivity of these membrane materials towards hydrogen
sulfide (Table 6.4). Though this type of membranes was offering high pure
compounds selectivities, no data was provided by the author about real
separation performances of this type of membranes.[62,63]

6.1.3.1.3 Polar Rubbery Polymers. These type of polymer have mainly been
studied by academics teams and display high values of H_2S/CH_4 selectivities
associated with high H_2S permeabilities. To the authors knowledge, though,
those materials have not been studied on larger scale than in a laboratory.[64,65]

6.1.3.2   Membranes with a Selective Layer made of Block Polymers

This type of membrane materials has been principally studied by MTR Inc.
which has developed commercial spiral-wound membranes with a selective
layer made of various polyether-*block*-amides (commercialized under the
trademark PEBAX® by Arkema). These types of polymers are composed of
rigid segments (amide groups) and soft segments (polyether groups) and proved
to offer high performances for the separation of hydrogen sulfide and methane
under industrial conditions (high feed pressure, high carbon dioxide content).[66,67] Tests under industrial conditions were conducted on a production
facilities operated by Shell and located in west Texas.[66,68] Based on those pilot-
plant experiments, the economics of the association of such membranes and
Sulfatreat® process (sulfur removal process based on the chemistry of iron)
were compared with those of a conventional amine-based sour gas removal
process. It was shown that the PEBAX®-based membranes/Sulfatreat process
association was leading to significant reductions in operation costs (from 20%
to 40%) over the amine process alone for flow gases containing few percent of
H_2S and lower than 140 000 N m^3 h^{-1} while a PEBAX® membrane/amine
coupling was more valuable than a standalone amine process for gas flows
containing more than 5% H_2S and higher than 23 000 N m^3 h^{-1}.

6.1.3.3   Membranes with a Selective Layer made of Glassy Polymers

These stiff polymeric materials, used mainly for commercial carbon dioxide
removal applications (polyimides, cellulose acetate, polysulfones) have their
selectivity mainly based on molecular sieving effects. Despite their affinity for
polar compounds (those membranes are also recommended by membrane

Table 6.4 Hydrogen sulfide/methane separation performances of polymeric membranes

Membrane material	Feed composition	Operating conditions	CH_4 permeability (barrer)	H_2S permeability (barrer)	Mixture selectivity	Reference
Rubbery polymers						
Polyphosphazene	3% CH_4, 3% CO_2, 3% H_2S	2 bar, ambient T?	14	1103	79	65
PDMS	Pure compounds	1.4 bar, 23 °C	850	5100	6	61
PTMSP	Pure compounds	1.4 bar, 23 °C	—	21 400	—	61
PDMS + PEG	Pure compounds	3 bar, ambient T	—	—	140	63
PDMS + PEG + glycerol	Pure compounds	3 bar, ambient T	—	—	162	63
PDMS + PEG	Pure compounds		—	—	121/115/185	62
poly(ether urethane urea)	69.4% CH_4, 18.1% CO_2, 12.5 H_2S	13.5 bar, 35 °C	1.4	130	93	64
Block co-polymers						
PEBAX® (MX1074)	69.4% CH_4, 18.1% CO_2, 12.5% H_2S	10 bar, 35 °C	13.8	695	50	64
PEBAX 4011®	99.9% CH_4 970 ppm H_2S	67 bar, 23 °C	—	—	52	67
PEBAX 4011®	87.34% CH_4 10.8% CO_2 1.83% H_2S	67 bar, 23 °C	—	—	50	67
Glassy polymers						
Cellulose acetate	—	—	0.4	—	19	69
6FDA-HAB polyimide	55.3% CH_4 36.7% CO_2 8% H_2S	10 bar, 35 °C	0.1	—	15	9

manufacturers for gas dehydration applications) it appears that those membrane materials have rather low H_2S/CH_4 selectivities (Table 6.4).[9,69] This can be mainly attributed to the fact that the kinetic diameters of hydrogen sulfide and methane are rather close, resulting in close values of their respective diffusion coefficients through glassy polymer-based selective layers.

6.1.4 Other Membrane-based Natural Gas Treatments

6.1.4.1 Removal of Water

Water has to be removed from natural gas in order to prevent hydrate formation (which results in pressure loss and even plugging of gas conducts) and corrosion in pipelines. Generally containing around 500–1500 ppmv of water from the wellhead, natural gas has to be dehydrated up to 20–150 ppmv in order to prevent condensation or hydrate formation problems. So far, such an operation is realized by absorption on diethylene or triethylene glycol. This type of process is widely used in the gas processors community because of its very low investment and operating costs, as well as its reliability: as an illustration, 40 000 of those units are in operation nowadays on the US ground for natural gas conditioning.[70]

Though membranes are widely used for air dehydration, it seems that this technology is not yet very widely accepted by the industry for gas dehydration applications.[15] The main reason for this is that part of the methane processed through membrane modules is lost in the permeate, which does not happen with glycol-based processes.

At the beginning of the 1990s, Permea launched an ambitious developement R&D program (the KAPOF project) in order to validate the performances of the PRISM® dehydration permeators, with the guidance of four international oil companies (Phillips, Agip, Hydro and Statoil) and the Norwegian government. BP and Shell later joined the consortium and provided access to their test facilities to assess the new technology. Several operating solutions have been proposed during this project to tackle the loss of natural gas through the membrane.

6.1.4.1.1 Reduction of Water Vapor Pressure in the Permeate. In general, gas permeation modules are operated at permeate pressure higher or equal to atmospheric pressure. When operated under these conditions, at least 5% of the methane contained in the feed is lost with existing commercial membranes, which not acceptable by operators. The solution is then to maximize the driving force inducing water mass transfer through the membrane, *i.e.* the water vapor partial pressure between the feed and the permeate, by lowering the water vapor permeate. For a fixed water extraction yield, this leads to a decrease in membrane area requirement, and thus to methane overall loss. Several options have been addressed at pilot scale in order to lower the vapor pressure in the permeate compartment of permeation modules.

In option 1 permeate is operated at pressures close to atmospheric. The first pilot-scale version of dehydration separation device was tested at Shell's Research and Technology Centre in Amsterdam over a continuous period of 20 days during the summer of 1995.[71] In this test, the feed gas came from the local city supply and was compressed and humidified on purpose to feed the dehydration permeator. Typical tests were conducted at 70 bar at a feed flow equal to 300 N $m^3 h^{-1}$. It was shown that methane loss was reduced from 16% at 4 bar permeate pressure to 0.5–1% when operated the permeate compartment below 2 bar.[71]

In option 2 the permeate compartment is stripped with nitrogen-enriched air. In 1996, Agip installed a semi-commercial dehydration system on the Sinni gas field in south Italy. This system was based on Permea (nowadays Air Products) PRISM® module technology, capable of treating 100 000 N $m^3 h^{-1}$ of natural gas.[72] The water vapor pressure in the permeate was lowered by stripping with nitrogen-enriched air issued from a membrane-based air separation unit. In 1997, a similar configuration, based also on Prism dehydration membrane modules, was installed on the Brage platform, operated by Norsk Hydro and located in the Norwegian sector of the Northern Sea.[73]

In option 3 the permeate compartment is stripped with depressurized dry natural gas. This option was chosen on a membrane field testing at BP's Easington North Sea Gas Terminal which was launched in 1997 on Permea's Prism membrane modules.[74] The membranes modules have been operating successfully on a 1000 N $m^3 h^{-1}$ natural feed gas feed flow at 50 bar with a water dew point equal to 21 °C ($\sim$700 ppmv of water) when the offshore glycol operation had been progressively decommissioned in favor of 'wet' offshore pipeline operation (involving the injection of hydrate and corrosion inhibitors). The terminal operators feeling on the membrane dehydration operation was very positive, as this technology (i) was considered as very reliable, and (ii) allowed a decrease of water vapor dew point to –20 °C (with a permeate pressure equal to 0.4 bar). Operating in 1997–1998, the Nederlandse Aardolie Maatschppij B.V., a Dutch gas operator, decided to install a fully commercial demonstration unit in the Vries-1 field, which was designed to dehydrate 200 000 N $m^3 h^{-1}$ with a 70 bar feed pressure. The permeate compartment was swept this time with depressurized natural gas which, after use, was utilized for electricity production. In 1999, during a 2-month run, the dew point of the gas was lowered at values lower than –7 °C.[71]

6.1.4.1.2 Pressurizing the Water-concentrated Permeate Stream. The water-concentrated permeate stream can be pressurized in order to condensate water and recycle the resulting water-depleted permeate streams to the retentate. The cost of the compression unit is generally larger than the membrane unit itself[15] and it seems that, though this option has been considered for paper process studies,[74] it has not been tested on the field.

6.1.4.2 *Removal of Higher (C3+) Hydrocarbons*

Before treatment, natural gas contains higher hydrocarbons such as ethane, propane, butane or higher hydrocarbons at close to saturation. It is necessary to remove those compounds for the following reasons:

- To prevent condensation problems in pipelines: as an illustration, the dew point specification in US interstate pipeline is fixed at -20 °C.

- Propane, butane or higher hydrocarbons can be valued as liquids and are particularly prized as steam cracker feeds to produce short olefins (ethylene, propylene).
- If the natural gas produced is used locally as a fuel for combustion engines operating compressors or power-generating turbines, the presence of higher hydrocarbons can cause combustion problems or even accelerate aging of the engines.

The conventional technologies used in the industry to remove such 'heavies' are condensation or lean oil absorption.

Several companies have proposed vapor permeation modules to operate heavy hydrocarbon recovery from small natural gas flowstream since the 1990s: MTR Inc., Borsig and GKSS. All the membranes in operation are based on silicone selective layer which offer mixed-gas propane/methane separation selectivity[15,75] comprised between 3 and 5 and butane/methane separation selectivity[15,75] ranging from 5 to 10.

6.1.4.3 Removal of Inert Gases from Natural Gas

The US pipeline specification requires an inert content lower than 4% in natural gas. It is estimated that 14% of US present proven gas reserves contain more than 4% nitrogen.[15,76] The most trivial solution consists in diluting small flow rates of nitrogen-concentrated natural gas with main streams of natural gas containing a low concentration in inert gases. However, if this is not feasible (as, for instance, when all the surrounding gas fields contain high concentrations in nitrogen), it is necessary to install a nitrogen removal plant to comply with the regulatory natural gas composition specifications. Cryogenic distillation is the dedicated technology for large-scale nitrogen removal operations ($60\,000$–$600\,000$ N m^3 h^{-1} capacity for an operation life longer than 10 years). At the very end of the 1990s, 26 such plants were in operation on US ground.[76] Two technologies have recently been introduced in order to solve this challenge at small and medium scale: pressure-swing adsorption (PSA) with molecular sieves (Molecular Gates® by the Engelhard company)[77,78] and gas permeation (Table 6.5), which has been promoted by MTR Inc. since the end of the 1990s.

In the mid-1990s, MTR Inc. evaluated at laboratory scale (on 1–2 m^2 membrane modules) a wide array of composite membranes which selective membranes were made of various glassy or rubbery polymers (Table 6.5). Intuitively, rubbery polymers proved to be more selective toward methane (due to its higher sorption, induced by its more condensable behavior than nitrogen) while glassy polymers, especially fluorinated ones, proved to be more selective towards nitrogen (due to diffusive selectivity).

MTR Inc. evaluated the economic feasibility of a membrane-based nitrogen removal operation on a $12\,000$ N m^3 h^{-1} natural gas stream entering the process at 32 bar and containing 10% nitrogen. The target of the membrane operation was to reach the 4% nitrogen specification in the product gas while the waste gas was containing at least 50% nitrogen and the overall methane recovery

Table 6.5 Nitrogen/methane separation performances of polymeric membranes

Polymer	T (°C)	N_2 permeability (barrer)	CH_4 permeability (barrer)	CH_4/N_2 selectivity	Reference
Fluorinated glassy polymers					
Cytop (R)	−20/22	—	—	$0.3^*/0.4^*$	76
Hyflon(R) AD 60	−20/22	—	—	$0.4^*/0.5^*$	76
Hyflon AD 80	−20/22	—	—	$0.4^*/0.5^*$	76
Polyimides					
6FDA-*m*-PDA	22	—	—	0.6^*	76
6FDA-*m*-PDA	30	3.1	1.34	0.4^{**}	79
6FDA-*m*-PDA	22	—	—	0.6^*	76
6FDA-NDA	22	—	—	0.8^*	76
6FDA-BAHF	30	0.26	0.13	0.5^{**}	79
PMDA-*m*-PDA	30	0.2	0.1	0.5^{**}	80
6FDA-IPDA	30	1.34	0.7	0.5^{**}	80
Other glassy polymers					
Polycarbonate	30	0.37	0.45	1.2^{**}	81
Cellulose acetate	30	0.35	0.43	1.2^{**}	81
Polysulfone	30	0.14	0.23	1.7^{**}	81
Rubbery polymers					
Natural rubber	30	17.3	50.1	2.9^{**}	82
Poly(dimethyl siloxane)	30	230	760	3.3^{**}	83
Poly(dimethyl siloxane)	25	—	—	3.3^{**}	76
Polypropylene oxide/ epichlorohydrin/allyl glycidyl ether (Hydrin T6000)	25	—	—	3.5^{**}	76
Poly(dimethyl siloxane) with mineral oil	25	—	—	3.7^{**}	76
Poly(siloctylene siloxane)	30	91	360	4^{**}	83
Poly(*p*-siphenylene siloxane)	30	3	12	4^{**}	83
Polyether-*block*-amide (PEBAX 2533)	30	4.8	20	4.2^{**}	67

*Separation selectivity obtained at laboratory scale on composite membranes (selective layer thickness ∼0.3–0.5 µm) by MTR Inc. with the following operating conditions: feed = 60% methane/20%carbon dioxide/20% nitrogen; feed pressure = 200 psig.
**Ideal selectivity based on pure compound permeability.

yield was to be higher than 93%. Based on the performances of the various tested membrane materials, the study led to the conclusion that a single-stage membrane operation was not viable, whether the membrane material was nitrogen selective or methane selective. Therefore, two-stage recycle schemes were studied involving only nitrogen-selective membranes, or only methane-selective membranes or a combination of both. Only the use of methane-selective membranes proved to offer interesting performances,[76] though the resulting process schemes were leading to significant extra cost due to the

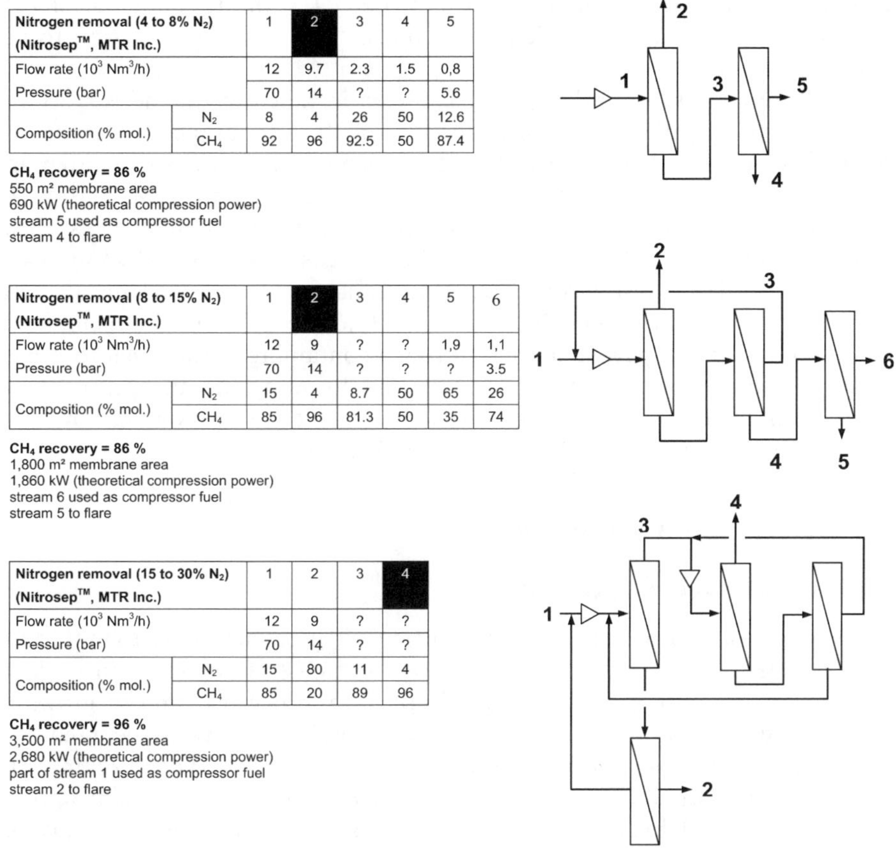

Nitrogen removal (4 to 8% N$_2$) (Nitrosep™, MTR Inc.)		1	2	3	4	5
Flow rate (10^3 Nm^3/h)		12	9.7	2.3	1.5	0,8
Pressure (bar)		70	14	?	?	5.6
Composition (% mol.)	N$_2$	8	4	26	50	12.6
	CH$_4$	92	96	92.5	50	87.4

CH$_4$ recovery = 86 %
550 m² membrane area
690 kW (theoretical compression power)
stream 5 used as compressor fuel
stream 4 to flare

Nitrogen removal (8 to 15% N$_2$) (Nitrosep™, MTR Inc.)		1	2	3	4	5	6
Flow rate (10^3 Nm^3/h)		12	9	?	?	1,9	1,1
Pressure (bar)		70	14	?	?	?	3.5
Composition (% mol.)	N$_2$	15	4	8.7	50	65	26
	CH$_4$	85	96	81.3	50	35	74

CH$_4$ recovery = 86 %
1,800 m² membrane area
1,860 kW (theoretical compression power)
stream 6 used as compressor fuel
stream 5 to flare

Nitrogen removal (15 to 30% N$_2$) (Nitrosep™, MTR Inc.)		1	2	3	4
Flow rate (10^3 Nm^3/h)		12	9	?	?
Pressure (bar)		70	14	?	?
Composition (% mol.)	N$_2$	15	80	11	4
	CH$_4$	85	20	89	96

CH$_4$ recovery = 96 %
3,500 m² membrane area
2,680 kW (theoretical compression power)
part of stream 1 used as compressor fuel
stream 2 to flare

Figure 6.5   Permeation-based nitrogen removal from natural gas.[76]

presence of two compressors: the first compression unit was used to pressurize the methane-enriched permeate produced by the first membrane stage to the pipeline pressure while the second one aimed at boosting the pressure of the permeate stream issued from the second stage in order to recycle it at the input of the first membrane stage. With this scenario, the total equipment investment cost was in the range of 4–8 million US dollars (in 2009) while the required membrane area was approximately equal to 4000 m². This appears to be acceptable by gas processors as the payback time of this type of installation was estimated to be 1 year.

The first field test of this membranes was therefore conducted with a 40-inch long module equipped with few square meters of methane-selective membranes for more than 1 year on a shut-in gas well containing 15% nitrogen operated by Butcher Energy in southern Ohio. The performances of the spiral-wound module were very similar to those observed at laboratory scale and its per-meation properties remained unchanged for the first 6 month period. After this proof of principle MTR Inc. design different membrane-based nitrogen removal process schemes, trade-named Nitrosep^{TM}, with the configurations of various modules/compressors depending on the nitrogen concentration to be addressed (Figure 6.5).

Up till now, 12 Nitrosep^{TM} units have been installed in the industry. Two examples depicting the applications addressed by this membranes process are as follows. A very small unit was installed in southern Kentucky in order to upgrade 200 N m^3 h^{-1} of natural gas containing 7% nitrogen. This two-membrane module unit was able to recover 80% of the natural gas with a nitrogen content of 3.8%. Part of the residue gas was used as compressor engine fuel while the rest was vented. A much larger unit was was installed in Rio Vista, California (14 000 N m^3 h^{-1} treatment capacity). The membrane operation aimed to upgrade a 16% nitrogen-containing natural gas to a gas stream offering 10% more heating value (thus lowering the nitrogen content to 9%). The membrane system involved three stages in series and allowed a methane recovery yield of 95% for pipeline delivery.

6.2 Petroleum Refining

6.2.1 Hydrogen Purification

The demand for hydrogen in constantly increasing in refineries due to more and more stringent sulfur content specifications for fuels (leading to an increasing hydrogen consumption in hydro-desulfurization processes) and a growing heavy crude consumption, which results in a higher worldwide demand for highly hydrogen-consuming upgrading processes, such as hydrocracking. In this changing landscape, permeation membranes constitute an elegant option for the recovery of hydrogen that is nowadays wasted in a various array of refineries off-gases, such as fuel gas, PSA tail gas, FCC gas, catalytic reformer off-gases or hydrocracker/hydrotreater off-gases.

To our knowledge, the very first hydrogen permeator was designed by Dupont and was installed in 1969 in Lousiana for synthesis gas H$_2$/CO ratio adjustment. The module was based on hollow fibers made of polyester.[18] In 1977, Monsanto gave the first demonstration of hydrogen permeator, once again for synthesis gas adjustment, and in 1979 launched the Prism$^{®}$ commercial hydrogen-selective polysulfone-based membranes. The first applications were mainly for the chemical industry, namely synthesis gas and ammonia plants. In 1986, Dupont commercialized a first generation of hydrogen selective asymmetric membranes based on polyaramide.[18]

Nowadays, three main membrane providers offer hydrogen purification permeators:

- Air Products, with the Prism$^{®}$ silicon-coated polysulfone membranes (issued from Monsanto). Air Products claims that the lifetime of the Prism$^{®}$ modules can be more than 15 years.[84]
- Ube Industries, with polyimide hollow-fibers membranes.

- Air Liquide, with the MEDAL polyimide and high selectivity polyaramide membranes.

All these membranes are based on glassy polymers and offer a diffusion-based hydrogen selectivity. Today's membranes have high H_2/CH_4 selectivities (from 35 to 200 at 80 °C).[13] For instance, Air Products claims that a single-stage array of Prism® modules is able to raise the concentration of gases from 10–30% to 70–90%,[85] while Medal membranes are able to raise the concentration of a gas at 51 bar from 86% in hydrogen to 98% with a permeate pressure at 30 bar and a residue containing 52% of hydrogen at 50 bar.[86]

In 2003, it was reported that more than 400 hydrogen permeators were installed worldwide[87] while approximately one hundred were operated in refineries.[7] As there are more than 500 refineries in the world, it is clear that the potential market for this type of membrane applications is far from saturated. Nevertheless, three main limits still hinder the wide acceptance of permeation-based hydrogen purification in the refining industry.

The first limit is that the purified hydrogen is recovered at low pressure in the permeate side and requires compression in order to feed it back to reactors. As such, PSA (pressure swing adsorption) is a more attractive process, as the produced purified hydrogen is delivered directly at high pressure.

The second limit is the sensitivity of membrane to contaminants, such as water vapor, higher hydrocarbons or acid gases. The installation of permeators generally requires at first a very detailed analysis of the contaminants present in the feed, even at traces level, in order to design pre-treatment operations. A wide array of solutions can be used in order to remove poisonous compounds: coalescing filters in order to remove aerosols, sorbent beds or even complete PSA or temperature-swing adsorption (TSA) units.[87] This can lead to a significant increase of investment and operating costs. It should noted here that recent hydrogen purification membranes are still relatively tolerant to contaminants (Table 6.6). This is particularly true if membrane modules are operated at higher temperatures (from 80 °C to 110 °C), which results in a lowering of the sorption of contaminants.

Table 6.6 Ube Industries membrane material compatibility against contaminants[88]

Contaminant	Maximum allowable content
Water vapor	Up to saturation
H_2S	3% vol.
NH_3 and amines	100 ppm vol.
Methanol	5% vol.
Methyl ether	5% vol.
Benzene	1% vol.
Toluene	2000 ppm vol.
C5+ hydrocarbons	Up to saturation

The third limit is that a membrane's mechanical integrity can be damaged in transient operating conditions, especially in the case of an emergency blowdown of the membrane-based process. In certain cases, the membrane module can be submitted to pressures differences larger than its mechanical tolerance. Solution are nowadays proposed by membrane providers in order to monitor automatically the pressure balance between the feed and the permeate compartment when operating conditions limits are reached.[87]

6.2.2 Gasoline Isomerate Fractionation

6.2.2.1 Context

For more than 20 years now, regulations have imposed increasingly tight limitations on the content in gasoline of aromatic octane number boosters produced by the reforming of straight-run gasoline (alkanes). Amongst the available alternative technologies designed to enhance the octane number of straight-run gasoline, hydro-isomerization is a catalytic technology that upgrades low-octane-number linear paraffins into higher-octane-number branched paraffins, as illustrated in Table 6.7.

Since the rate of conversion of linear paraffins in the isomerization units is limited by a thermodynamic equilibrium, an option for increasing the production yield of dibranched paraffins consists of separating the linear and monobranched paraffins from the isomerization unit effluent and recycling them in the input of the isomerization reactor. The more conventional solution consists of fractionating the output stream from the isomerization reactor through a continuous distillation column (a deisohexanizer or DIH), into three effluents. First, there is a sidestream, mainly containing unconverted normal hexane and the monobranched paraffins with six carbon atoms. This stream is recycled to the isomerization unit.

Second, the bottom stream, containing the heaviest alkanes (paraffins) with seven carbon atoms and naphthenes (cyclic paraffins) with six carbon atoms, with a Research Octane Number (RON) of 82, is sent directly to the gasoline pool.

Third, at the top of the column, there is a head stream rich in di-branched paraffins with six carbon atoms, isopentane and normal pentane. This stream, which corresponds to 90% by weight of the fresh feed, has a RON of 87 though it contains about 16% by weight of normal pentane, which has a low RON of 61.

Table 6.7 Octane numbers of main components of a light gasoline

Type of isomer	*Research octane number (RON)*
Normal pentane	61
Normal hexane	31
Isopentane	93.5
Methylpentanes	75
Dimethylbutanes	94–105
Cyclohexane	84
Benzene (reference)	120

With current distillation processes, it is not economically feasible to separate normal pentane from the other components in the top stream of the deisohexanizer since their respective boiling points are very close. This type of separation, however, can be achieved with molecular sieves, such as zeolites, implemented in cyclic adsorption processes such as a 'simulated moving bed' (UOP MOLEX process) or 'cycled pressurisation/depressurisation' (IFP IPSORB process or Exxon-Mobil ISOSIEVE process). With such processes, normal paraffins are preferentially adsorbed inside the micro-porosity of the zeolites and therefore separated from their branched isomers. Though offering excellent separation performance, this type of technology exhibits several drawbacks: high investment costs, sophisticated sequential operation (adsorption-desorption cycle), use of large quantities of solvents (desorbants) and lack of modularity.

Since the beginning of the 1990s much attention has been paid to overcoming the drawbacks of conventional zeolite adsorbents through the development of zeolite membranes that combine the technical advantages of membranes (modularity, continuous operation) with the high separation performances of zeolites (due to their sieving properties). Most of the published R&D work on that topic was carried out at laboratory scale mainly by academic laboratories, though a few companies, like Exxon-Mobil or NGK Insulators have also published results on that topic. One of the most studied topics in this research field was the separation of normal short (C4–C6) paraffins from their branched isomers through MFI-type zeolite membranes.[89–104] Indeed, MFI zeolites are crystalline aluminosilicates with a micro-porous structure that is composed of two intricate micro-pore networks: elliptical straight channels with openings of 0.51 nm × 0.55 nm and zigzag channels that are almost cylindrical with a diameter of 0.53 nm × 0.56 nm, as measured by X-ray diffraction at ambient temperature.[105] In such a confined porous system wherein the diameter of the micro-pores and the kinetic diameter of the diffusing molecules are close, the higher the kinetic diameter of a permeating molecule, the higher the friction of the molecule alongside the micro-pore wall, and therefore the lower its diffusion coefficient inside the micro-porosity of the MFI zeolite. Therefore, the diffusion coefficient of normal alkanes (which display a kinetic diameter of 0.43 nm)[101] in MFI zeolites is higher[93] than the diffusion coefficient of their monobranched isomers (with a kinetic diameter of 0.5 nm). Moreover, these materials prove to be hardly permeable to dibranched paraffins (2,2 dimethylbutane displays a kinetic diameter of 0.63 nm).[103] As an illustration, it was shown experimentally that a MFI zeolite membrane operated under close to industrial operating conditions (2–4 bar feed pressure, membrane temperature between 200 °C and 400 °C) was able to produce a permeate composed of 95% normal pentane and 5% isopentane from a vapor feed composed of 20% 2,2 dimethylbutane, 55% isopentane and only 25% normal pentane.[106]

6.2.2.2 Process Considerations

The interest of integrating a MFI zeolite membrane unit in a light gasoline isomeration unit plus a deisohexanizer was evaluated economically on a 2003 cost basis.[106] The performances of the membrane were evaluated experimentally at laboratory scale in industrial conditions with a composite membrane composed of a 20 µm-thick MFI zeolite layer deposited in a porous alumina tube. The base case for the economical evaluation was a light gasoline isomerization unit with a nominal capacity of 62 tonnes h^{-1}, which was representing an average processing capacity for this type of unit in Western refineries.

Two scenarios concerning the DIH-zeolite membrane were evaluated. The first case consists of operating the membrane separation unit at 300 °C in order to obtain high fluxes and thereby reduce the membrane area required. There was nevertheless a drawback with these operating conditions: the feed has to be vaporized with a furnace that consumed fuel gas. The second case consisted in operating the membrane separation unit at 200 °C. Although impairing the membrane permeability, vaporization of the feed can be achieved through a conventional steam-heat exchanger instead of a furnace, which led to a significant reduction of the investment and operating costs.

The financial benefit of integrating the membrane separation unit was evaluated by comparing the additional expenses (investment costs and operating costs) induced by installing and operating the membrane separator and the gain related to the increase in the isomerate octane number induced by installing the membrane separator.

On a technical basis only, inserting a MFI zeolite-based separation unit in the isomerization loop led to a significant improvement of the isomerate octane number (by more than three points). However, this substantial gain in terms of performance was offset by a significant increase in the isomerization unit operating costs. This additional cost could be mainly attributed to the energy consumption required to vaporize the feed input to the membrane separation unit and to the additional hydrogen consumption in the isomerization reactor induced by the greater flow of feed to be processed related to recycling the normal pentane extracted at the top of the deisohexanizer.

In all cases, if the cost of the octane number was low (3.1 euros/RON point/ t), addition of a MFI zeolite-based membrane separation unit was not financially justified. Under these conditions, in fact, either the energy consumption related to the vaporization of the feed weighed too heavily on the operating costs (membrane temperature 300 °C), or the margin cleared by the gain in octane number was not sufficient to make the construction of the membrane separation unit profitable, since it was too large (area > 10 000 m^2).

If the cost of the octane number was higher (5 euros/RON point/t), addition of a membrane separation unit could only be justified if the zeolite membrane permeability was high, *i.e.* elevated temperatures such at 300 °C. With this medium size installation, still the membrane area required was rather high,[106] close to 7500 m^2. In this case, the cost price of the membrane separation unit had to be less than 1000 euros per m^2 installed to justify the investment in this type of installation.

Progress is thus still needed to lower the production cost of zeolite membranes, as the present cost of zeolite membranes fabrication[107,108] is considered to be between 2000 euros per m^2 and 3000 USD per m^2. A dramatic reduction of the overall required membrane area could solve this cost issue. As demonstrated before, though zeolite membranes are designed to operate at high temperatures, increasing the operating temperature in order to increase the permeate flux is a fake 'good idea', as the energy consumption would hinder the profitability of the whole separation. The remaining option relies then in a significant increase of the membranes permeability. Great achievements concerning the laboratory scale production of sub-micron highly selective, high flux zeolite membranes have recently been reported.[104] A very attractive axis for zeolite membrane R&D in the near future is obviously the scaling up of these synthesis procedures.

6.3 Petrochemicals

Petrochemicals are base chemical intermediates that are used for the production of raw chemicals and polymers. They are divided into two main classes: olefins and aromatics.

Olefins (alkenes), principally ethylene (110 million ton y^{-1} worldwide) and propylene (65 million ton y^{-1} worldwide). In the next 10 years, the demand for propylene will grow faster than the demand for ethylene. One attractive option to boost propylene production would be to re-orient refineries fluid catalytic cracking (FCC) units, designed originally to produce gasoline, towards the production of lighter olefins, especially propylene. This choice implies that the design of the FCC gas plants (that are meant to fractionate the cracked products issued from the FCC) has to be revisited. In particular, the bottle-neck in the propylene/propane splitter column has to be removed or the system redesigned. Coupling membranes with existing columns could constitute an elegant way achieve this goal at minimal investment cost.

Within the aromatics (*e.g.* benzene, toluene and xylene) the xylenes are used as a feedstock for the production terephthalic acid and dimethyl terephthalate, both monomers that are used for the production of polyethylene terephthalate (PET), which is the main constituent of plastic bottles and polyester clothing. There are three very close boiling isomers of xylene (*ortho-*, *meta-* and *para-*) and it is mainly *p*-xylene that is used for PET production. The fractionation processes for all existing xylene isomers (crystallization or simulated moving bed) are expensive technologies and it is of interest to study the potentials of membranes for such a separation.

6.3.1 Separation of Light Olefins/Paraffins

6.3.1.1 Overview

For more than 20 years the performances of a very wide array of membrane materials for olefin/paraffin separation have been studied at laboratory scale mainly by academic teams.

Several selectivity mechanisms have been tried from molecular sieving with glassy polymers carbon membranes or zeolites to affinity with metal-ion facilitated transport membranes. We provide here a brief overview of the advantages and limits of each membrane material family. For further reading, extensive and detailed reviews concerning the use of several types of membrane materials for olefin/paraffin separation can be found in various recent papers.[109–111]

6.3.1.2 Glassy Polymer Membranes

The ethylene/ethane and propylene/propane separation performances of various glassy polymer are listed respectively in Tables 6.8 and 6.9. It appears clearly in Table 6.8 that poly(2,6-dimethyl-1,4 phenyleneoxide) offers the 'best' selectivity/permability compromise for ethylene/ethane separation.

Table 6.8 Ethylene/ethane separation performances of polymeric membranes

Polymer type		Temperature ($^{\circ}C$)	Pressure (bar)	Ethylene permeability (barrer)	P_{C2H4}/P_{C2H6} (permselectivity)	Reference
Polyimides						
Dianhydride	*Diamine*					
6FDA	mPDA	35	3.8	0.3	3.3	112
6FDA	IPDA	35	3.8	1.4	3.8	112
6FDA	6FpDA (BAAF)	35	3.8	2.1	4.4	112
6FDA	NDA	35	2	1.17	6.84	113
6FDA	NDA-Durene (75:25)	35	2	4.46	5.62	113
6FDA	NDA-Durene (50:50)	35	2	9.48	4.27	113
6FDA	NDA-Durene (25:75)	35	2	36.7	3.6	113
6FDA	Durene	35	2	76.7	2.89	113
6FDA	6FpDA (BAAF)	35	5	1.9	5.8	113
6FDA	1,5-NDA	35	5	0.87	4.2	113
6FDA	TrMPD	50	2	58	2.9	114
BPDA	TeMPD	50	2	5.8	4.3	114
Other polymers						
PPO		20		140	4.4	115
PPO		30		160	5.3	115
PPO		50		200	2	115
PPO-*co*-DPPEO		20		58	5.8	115
PPO-*co*-DPPEO		30		72	2.9	115
PPO-*co*-DPPEO		50		51	1.7	115
PPO-*co*-DPPEO		20		29	4.8	115
PPO-*co*-DPPEO		30		58	4.5	115
PPO-*co*-DPPEO		50		76	2.2	115
PETP		80		0.9	1.8	115
PTFE		20		2.6	1.4	115
PTFE		30		4.9	1.5	115
PTFE		50		7.1	1.4	115
PTFE		80		20	1.6	115

The best compromise between high propane/propylene selectivity/high propylene permeability is achieved with 6FDA-based polyimides (Table 6.9). Indeed, this trend can be achieved if two conditions are met.[121] First, the material considered should be constituted of rigid polymer chains in order to promote molecular sieving selectivity. This can be achieved if the monomers are connected to each other via multiple chemical bounds – in order to limit free rotation of monomers – and if the polymer chains are mainly constituted of aromatic groups. Second, the free volume should be high in order to favor the permeation of the fastest permeating compound.

The $C-(CF_3)_2$ group present in the dianhydride groups of 6FA-based polyimides favors these two properties, due mainly to its bulkiness induced by the presence of fluorine atoms. This explains why, for comparable selectivity values, 6FDA-based polyimides display higher permeabilities towards propylene than the other types of polymers. It should be pointed out here than the Japanese company Nitto Denko conducted an ambitious R&D program at the end of the 1990s in order to screen the olefin/paraffin (propylene/propane[116,122–126] and butadiene/butane[127,128]) separation properties of a very wide array of 6FDA-based polyimides. Among the 6FDA-containing polyimides, 6FDA-TrMP and 6FDA-TeMP offer the most interesting compromise on the technological point of view (propylene permeability higher than 10 barrer associated with an ideal propylene/propane selectivity higher than 10).

It has been shown that olefin/paraffin selectivity of glassy polymer membrane materials was mainly of diffusive nature.[110,113,114] Indeed, with the pressure and temperature conditions currently used in the literature, sorption isotherms of C2/C3 olefins and paraffins in glassy polymer are very similar.[119,110,112–114,125]

Though 6FDA-based polyimides display very appealing propylene/propane separation performances if only pure compounds permeation properties (*i.e.* ideal selectivity) or separation selectivities at low hydrocarbon partial pressures are considered, it has been shown by a significant amount of studies[112–114,118] that those materials are subject to plasticization at ambient temperature for feed propylene partial pressures higher than 4–5 bar. Plasticization results in a swelling of the polymer network, which leads to an increase of the permeability for all permeating species associating with a decrease of the propylene/propane selectivity. This phenomenon was not observed with ethylene, which is a less condensable gas than propylene.[117]

Several post-treatment of membrane materials have been considered in the literature in order to limit, or even suppress, this mechanism.

6.3.1.2.1 Thermal Treatment. A first approach consists in curing thermally the membrane material at a temperature slightly lower than its glassy transition temperature. The resulting increase in polymer mobility favors the formation of charge transfer complexes, induced by the superposition of electron acceptor groups (the carbonyl functions of the imide bound) and electron donor groups (the nitrogen atom in the aromatic diamine groups) in neighboring polymer chains. This raises the inter-chain cohesion and therefore reduced the overall polymer free volume. As a result, the permeability of

Table 6.9 Propylene/propane separation performances of polymeric membranes

Polymer type		Temperature (°C)	Pressure (bar)	Propylene permeability (barrer)	P_{C3H6}/P_{C3H8} (perm-selectivity)	Reference
Polyimides						
Dianhydride	*Diamine*					
6FDA	TrMPD	25	1.1	9.29	15.5	116
6FDA	TeMPD	25	1.1	13.8	13.5	116
6FDA	44′DPE	25	1.1	0.0823	41.6	116
6FDA	6FpDA (BAAF)	25	1.1	0.726	19.8	116
6FDA	FDA	25	1.1	0.853	24.9	116
6FDA	DABA	25	1.1	0.0475	124	116
6FDA	33′DMDB	25	1.1	0.263	39.4	116
6FDA	DSH	25	1.1	0.115	45.1	116
6FDA	BAPF	25	1.1	0.188	15.5	116
6FDA	44BAPS	25	1.1	0.0821	20.7	116
6FDA	33′DMDB	35	1.1	0.15	13.2	110
6FDA	TeMPD	50	2	37	8.6	114
6FDA	TrMPD	50	2	30	11	114
6FDA	DDBT	50	2	1.8	20	—
6FDA	DDBT	50	2	0.76	27	114
6FDA	mPDA	35	3.8	0.13	10	112
6FDA	IPDA	35	3.8	0.58	15	112
6FDA	6FpDA (BAAF)	35	3.8	0.89	16	112
6FDA	ODA	100	2	0.48	11	114
6FDA	TrMPD	50	2	27	10	117
6FDA	DDBT	100	2	2.6	15	117
6FDA	NDA	35	2	0.315	10.1	113
6FDA	NDA-Durene (75:25)	35	2	1.69	6.58	113
6FDA	NDA-Durene (50:50)	35	2	5.19	4.4	113
6FDA	NDA-Durene (25:75)	35	2	15.7	2.93	113
6FDA	Durene	35	2	30.6	2.13	113
6FDA	6pFDA (BAAF)	50	5	0.8	16.3	113
6FDA	1,5-NDA	50	5	0.24	10	113
BPDA	TeMPD	50	2	4	13	113
BPDA	TeMPD	50	2	3.2	13	114
BTDA	Matrimid	26	2	0.1	16	118
BTDA	Matrimid–Thermid 85/15	26	2	0.03	4	118
	Pyralin 2566	26	2	0.09	21	118
BTDA	Matrimid®	35	2	0.1	10	110
Other polymers						
PPO		50	2	2.9	9	119
PPO		30	3	9	4.25	120

Table 6.9 (*Continued*)

Polymer type	Temperature (°C)	Pressure (bar)	Propylene permeability (barrer)	P_{C3H6}/P_{C3H8} (perm-selectivity)	Reference
PPO	50	2	2.3	9.1	114
PPO	50	0.58[*]	1.77	17.8	115
PPO-*co*-PDMPO	50	0.58[*]	0.41	25.6	115
PPO-*co*-PDMPO	50	0.58[*]	0.22	68.8	115
EC	30	3.5	52	3.25	115
CA	30	3.5	15.2	2.6	115
PSF	30	3.5	25	1.4	115
P4MP	50	2	54	2	114
1.2PB	50	2	260	1.7	114

the membrane can dramatically reduced, though separation selectivity is improved.[37,40,118,121,129] Pushing the thermal treatment higher than the glassy transition temperature of the polymer can lead to cross-linking reactions induced by the oxidation of side chemical functions in the polymer chains.[118]

6.3.1.2.2 Chemical Cross-linking. It is also possible to limit the swelling ability of polymer by cross-linking the chains constitutive of the polymer. Chemical cross-linking has been mainly studied to limit the CO_2-induced plasticization of polymeric membrane materials applied to carbon dioxide/methane separation.[42,44]

6.3.1.3 Facilitated Transport Membranes

Initially, academic research on ionic metal olefin-selective facilitated transport membranes was focused on liquid membranes, whether immobilized liquid membranes containing silver nitrate[130–134] or solvent-swollen membranes.[135–137] The selectivity of this type of membrane materials is based on a reversible specific interaction of the pi-electrons of olefins with certain metal ions, such as ionic silver, dissolved in the polymer matrix, which results in facilitated olefin transport of the ion–olefin coupling through the membrane.

Though offering good olefin/paraffin separation performances at laboratory scale, those materials were subject to a loss of solvent by evaporation, which resulted in dramatic fall of olefin/paraffin separation performances. In order to maintain the separation performances of those membranes, the feed had thus to be saturated with vapor in order to prevent the drying of the membrane. This is a critical point as, at industrial scale, propylene/propane mixtures are dehydrated before being fractionated by distillation.

This limit was overcome by replacing water or poly(hydric) alcohol with 'heavier' polar solvents, such as glycerol.[137–140] A low vapor pressure polar 'swelling' agent, such as glycerol, was indeed not subject to vaporization when flowing a dry gaseous feed and acted as a solvent for the olefin metal carrier.

In latter years, attention has been paid to promising ionic silver-impregnated solid-state polyelectrolyte membranes based on PVP, PAZ or PEO polymer which were not suffering from desiccation phenomena when flowing dry gases.[141–149] Though the first results based on pure compounds were very promising,[142,148] for instance, an ideal separation propylene/propane factor as high as 15 000, the mixed selectivity of those types of materials was lower by orders of magnitude.[150,151] This significant discrepancy between the ideal and the mixture could be mainly attributed to a swelling-induced plasticization of the membrane materials by short olefins and paraffins.[152]

Another limit of those types of electrolyte membrane materials is their poor long-term stability. It was found in particular that the silver ions contained polymer electrolyte membrane were subject to reduction leading to the formation of silver metal nanoparticles and consequently to a loss of their olefin carrier ability.[150,153–155] This reduction phenomena occurs because polar polymers such as POZ or PVP act as reducing agents.[156,157] A first attempt to solve this problem was to use phthalates, which led to a significant improvement of the long-term stability of this type of membrane.[153] However, phthalates had technical limitations because of their hazardous behavior and their relatively low solubility in the polymer matrices. It was possible to escape from those limitations by using phthalate-containing polymers, such as poly(ethylene phthalate) which proved to offer stable olefin/paraffin performances over several days (Table 6.10).[158]

6.3.1.4 Carbon Membranes

Carbon membranes are prepared by pyrolyzing polymeric precursors. The pyrolysis aims at (i) enhancing the molecular sieving properties of the processed polymer, and (ii) rendering them less sensitive to plasticization. When they are carbonized at high temperatures, the polymer chains tend to transform to an inorganic structure and the transfer mechanism of sorbents inside the membrane materials switches from a solution-diffusion mechanism to a molecular sieving mechanism.

Therefore carbon membranes generally offer higher gas separation properties than their native polymeric precursors, as indicated in Table 6.11. It is important to mention that, if this remark is particularly true for small gas pairs (N_2/CO_2, CO_2/CH_4) separations, the carbonizing process of polymeric membranes has to be monitored with care when higher molecular weight compounds, such as propylene/propane, separation is concerned. Indeed, it has been shown experimentally[169] that a too severe pyrolysing process (induced by a too long duration of the pyrolysis or a too high temperature) would lead to a shrinking of the carbonaceous membrane network, and thus to a strong narrowing of the microporous network inside the membrane material. Though the resulting material would still prove to be very permeable for small gaseous molecules (such as nitrogen or methane) it would offer poor propylene/propane separation properties. It is probable that the main porosity of the carbonized polymer network, microporous by nature, becomes less permeable towards the olefins, which can then only diffuse through the few remaining mesoporous defaults, leading to a parallel decrease in selectivity and permeability. It seems that this trend is more pronounced if the precursor polymeric network is denser. As an illustration, it was proven that Matrimid, a commercial polyimide displaying a d-spacing (mean statistical distance between polymer chains measured by WAXD) of 5.2 Å was more susceptible to propylene/propane separation performances deterioration when 'over'-pyrolyzed than a more 'open' fluorine-containing polyimide, 6FDA/BPDA-DAM (d-spacing = 5.7 Å).[169]

The main limit of carbon membranes is their brittleness, resulting practically in a poor mechanical strength.[177,178] Several authors managed to keep the pyrolyzed membrane materials flexible by incorporating pendant sulfonic acid groups in the precursor polymer and keeping the pyrolysis temperature sufficiently low in order to allow the decomposition of the sulfonic acid groups without decomposing the skeleton of the polymer matrix.[172,179–182] This trend tends to indicate that the sulfonic acid group acts as a template that is removed from the heated pyrolyzed polymer backbone.

Several post-treatments have been screened in order to enhance furthermore the olefin/paraffin selectivity of carbon membranes, such as (i) the addition of a silicone coating over the selective layer of polyimide-derived carbon membranes seems to fill surface defects, which leads to an enhancement of the propylene/propane selectivity while lowering the overall propylene permeability of those materials;[216] and (ii) air oxidation of phenolic resin-based carbon membranes leads to a decrease of the olefin/paraffin selectivity paired with an increase of the olefin permeability, except if chemical vapor deposition with trichloroethane is applied to the membrane.[175]

6.3.1.5 Zeolite Membranes

Several studies concerning the separation of propylene from propane base on faujasite (FAU zeolite) membranes[183–186] or titanosilicate ETS[187] have been published recently. The olefin/paraffin selectivity of all those materials was comparable to glassy polymers and always below 10.

6.3.1.6 Evaluation of Membrane-distillation Couplings

The advantages of coupling a permeation membrane on a distillation-based propylene/propane fractionation were first evaluated in a study carried out at University of Colorado, Boulder, and sponsored by BP.[188] Various membrane/distillation coupling scenarios (membrane unit located on the top stream or on the bottom stream or on the side stream of the column, all cases with a recirculation of the propylene-enriched permeate inside the column) were compared with a reference case based on a propylene/propane splitter column operated with 152 theoretical stages, a reflux ratio equal to 24.1 and a feed

Table 6.10 Olefin/paraffin separation performances of facilitated transport membranes

Olefin/paraffin polymer name	Temperature	Pure gases		Mixed gases		Reference
		P_{olefin} (barrer)	$P_{olefin}/P_{paraffin}$	P_{olefin} (barrer)	$P_{olefin}/P_{paraffin}$	
Ethylene/ethane						
Nafion/poly(pyrolle)	40 °C	0.0564	2.2			159
Nafion/poly(pyrolle)	70 °C	0.2170	2.15			159
Nafion/poly(pyrolle) + AgNO$_3$	40 °C	0.2	12			159
Nafion/poly(pyrolle) + AgNO$_3$	70 °C	0.9	9			159
CA + AgBF$_4$	?	?	270			160
PA12-PTMO + AgBF$_4$	22 °C		20–25			161
Propylene/propane						
PPO	30 °C	9	2.2	6	3.6	120
PPO + RuCl$_3$	30 °C	14	7	8	5	120
PPO + PdCl$_2$	30 °C	8.4	2.9	12.5	5.2	120
PPO + IrCl$_3$	30 °C	10	5			120
CA + AgBF$_4$	?	?	200			160
POZ + AgBF$_4$	?	25–150	>10		45–240	152, 162, 163
POZ + AgBF$_4$ + asparagines					45–65	163
POZ + AgBF$_4$ + valine					45–55	163
POZ + AgBF$_4$ + glutamic acid					45–54	163
POZ + AgBF$_4$ + lysine					40–47	163
POZ + AgNO$_3$ + asparagines					1–34	163

POZ + AgNO$_3$ + valine					1–15	163
POZ + AgNO$_3$ + glutamic acid					1–13	163
POZ + AgNO$_3$ + lysine					1–4	163
POZ + AgNO$_3$					1	164
POZ + AgBF$_4$ + SiO2 (0.1 mol/mol)					90	165
POZ + AgBF$_4$					50	164
POZ + AgNO$_3$/AgBF$_4$ (0.7/0.3)					95	164
POZ + AgCF$_3$SO$_3$					15–100	152, 164
POZ + AgNO$_3$/AgCF$_3$SO$_3$ (0.3/0.7)					30	164
PEP + AgBF$_4$					55	166
PMMA + AgBF$_4$					24–42	167
PVMK + AgBF$_4$					16–48	167
PVP + AgBF$_4$					5–45	167
PVP-co-PVAc					27–55	168
Isobutene/isobutane						
LLDPE-g-AA	25 °C	6.7	2.5			120
LLDPE-g-AA-AgNO$_3$						—
PE-g-AA-AgNO$_3$	25 °C			5.25	9.01	139
SR-g-AA Ag	25 °C			2400	2.91	139
PTMSP	25 °C	28 600	3	29 600	1.8	135
PTMSP-AgClO$_4$	25 °C	3110	5.6			135, 138
PTMSP-g-AA	25 °C			1250	2.2	138
PTMSP-g-AA AgNO$_3$/glycerol	25 °C			1000	6	138

Table 6.11 Comparison of propylene/propane separation properties of pyrolyzed polymeric membranes with their non-pyrolyzed precursors (values in brackets)

Polymer name	Pyrolysis conditions	Pure gases		Mixed gases		Reference
		P_{C3H6} (barrer)	P_{C3H6}/P_{C3H8}	P_{C3H6} (barrer)	P_{C3H6}/P_{C3H8}	
Matrimid	500 °C	13 (0.1)	40 (10)			169
Kapton	1000 °C (under vac.)		>100			170
6FDA-BPDA-DAM	550 °C	196 (12)	100 (14)			169
6FDA/BPDA-DDBT	540 °C/1 h				22 (12)	171
(50:50)	550 °C/0 h				14 (12)	
	580 °C/$\frac{1}{4}$ h				23 (12)	
	650 °C/0 h				19 (12)	
NTDA/BAHFDS	450 °C	41	26	30	11	172
NTDA/BAHFDS-BAPF (80:20)	450 °C	15	21			172
NTDA/BDSA-BAPF (80:20)	450 °C	15	34	9.3	19	172
NTDA/BDSA	450 °C	6.4	29			172
BPDA/pp'ODA	700 °C in Ar		29–54		33–46	173
Novolak phenolic resin	700 °C (1 h under vac.)	20	40			174
Novolak phenolic resin	700 °C ($\frac{1}{2}$ h under vac.)	2000	15		13	174
Novolak phenolic resin	700 °C ($\frac{1}{2}$ h under vac.) + various post-treatments	120–290	31–11	70–180	4.2–54 (9.6)	175
PAEK/azide (50:50)	450 °C	17	31			176
PAEK/azide (80:20)	450 °C	46	17			176
PAEK/azide (50:50)	550 °C	16	48			176
PAEK/azide (80:20)	550 °C	48	44			176
PAEK/azide (50:50)	650 °C	6.6	24			176
PAEK/azide (80:20)	650 °C	4.4	16			176

pressure equal to 17 bar. The membrane separation unit was based on a silver/ Nafion facilitated transport membrane with a permeate pressure equal to 3.5 bar. This choice of operating conditions led to the introduction of compression stages in order to blow the permeate stream back inside the column. This study demonstrated that the addition of a permeation operation enhanced the technical performances of the overall propylene/propane fractionation operation by three methods.

First, with a fixed feed flow and propylene purity specification at the top of the distillation column, the addition of a permeation unit on an existing distillation column could lead to an increase of the capacity of the column. This conclusion though should be considered with circumspection as, in real life, the design of distillation columns generally allows only limited over-capacity. Nevertheless, when retrofitting a conventional fuel-producing FCC unit into a petrochemical FCC (*i.e.* producing more propylene), the addition of membranes could prove to be an elegant way to remove the bottle-neck in an existing propylene/propane splitter column.

Second, with a fixed feed flow and propylene purity specification at the top of the distillation column, the addition of a permeation unit on an existing distillation column could lead to a decrease of operating costs of the column induced by the reduction of the internal reflux ratio (and thus of the reboiler and condenser duties).

Third, on a new design, the addition of the membrane separation unit can lead to a reduction of the required number of theoretical stages, lowering thus the size of the distillation column. This trend was already observed in a former paper[189] which demonstrated that the addition of relatively small area of a silver-based facilitated transport membrane (several dozen square meters) to a small capacity propylene/propane distillation column was leading to a reduction of the required number of trays from 135 to less than 105 (feed flow $= 2.78$ mol s^{-1}, propylene feed content $= 0.44$ mol mol^{-1}, propylene distillate content $= 0.99$).

It should be noticed here that no real overall cost analysis of the impact of implementation of a permeation unit on a propylene/propane splitter column was carried out in this paper. Therefore, it was not demonstrated if the extra operating and capital costs due to the membrane and compressor installation and operation were over-balanced by the reduction in the operating costs of the distillation column.

A more detailed cost analysis was carried out by Kookos *et al.*[190] The base case of the standalone propylene/propane distillation column was issued from Gokkhale *et al.*:[191] the feed stream was composed of 70% mol mol^{-1} propylene and 30% 70% mol mol^{-1} propane while the top product propylene content specification was 99%. The optimal operating conditions of the base case were the following:

- The column has 248 trays
- The feed was located at tray 118 (from the bottom of the column)
- The reflux ratio was 12.84
- The column diameter was equal to 3.86 m.

It was found that the operating costs of this fractionation unit was 5.7 millions US dollars, with steam consumption alone accounting for 65% of the overall costs.

The annualized costs of the standalone column were then compared to those of an optimized distillation–permeation coupling. The membrane separation unit was located at the side-stream of the column with a recirculation of the pressurized propylene-rich permeate inside the top half of the column and a recirculation of the propylene-lean retentate in the top–bottom of the column.

The performances (selectivity/permeability) of the membrane were considered to be part of the optimisation parameters of the study and were left free to vary following the Robeson plot (permeability–selectivity upper boundary)[192] formula of the best available materials at the time, *i.e.* carbon membranes:[173,175]

$$P_{C3H6}/P_{C3H8} = 200(P_{C3H6})^{-1/2.5}$$

It was found that the optimal distillation/permeation design had the following characteristics:

- The column has 250 trays
- The feed was located at tray 120 (from the bottom of the column)
- The reflux ratio was 9.1
- The column diameter was equal to 3.3 m
- The membrane permeability and propylene/propane selectivity were respectively equal to 390 barrer and 18.4, while the permeate pressure was equal to 4.7 bar.

The annualized costs of the distillation/membrane hydrid, including compression costs, was 17.1% lower than in the case of the distillation column alone. As indicated by the authors, whereas previous studies were aiming at reducing the number of trays in the distillation unit,[188,189,193] this work proved that the main advantage offered by the coupling of a membrane unit with a distillation column was primarily a reduction of the costs of the column utilities.

Another study focused in that case on ethylene/ethane separation[194] came to the conclusion. It was shown that the coupling of a carbon-based membrane in parallel with the distillation column could lead to a little bit more than 100 000 US dollars annual savings in operating costs. As in the previous study, the membrane performances were considered as a 'floating' optimisation parameter following the Robeson plot upper limit equation. The authors chose to correlate the selectivity with the permeability after the following equation:[175]

$$Pe_{C2H4}/Pe_{C2H6} = 14(Pe_{C2H4})^{-0.2}$$

6.3.2 Separation of Xylene Isomers

In recent years, there has been significant focus on the development of MFI-type zeolite membranes for xylene separation by pervaporation. It should be mentioned that the separation of xylene isomers through those membrane materials have also been tested in vapor permeation conditions.[195–200] Most of the results were obtained at laboratory scale with very low xylene partial pressure in the feed. Though those membranes were offering rather good separation performances (*para*-xylene/*ortho*-xylene separation factor generally higher than 10 and possibly reaching several hundreds), they were tested in model conditions very different from industrial operating conditions. Moreover, it was shown that *para*-xylene/*ortho*-xylene selectivity in MFI membranes was dropping significantly when its partial pressure in the feed was raised.[201]

6.3.3 Recovery of Monomers

Rubbery polymer membrane materials have the ability to permeate preferentially condensable vapors. Rubbery membranes were found to be very profitable for the recovery of high-value monomer in petrochemical plants purges gases (Table 6.12).[202,203] In the case of large polymerization facilities, the value of purge monomers can reach amounts up to 2 million USD per year.

Table 6.12 Membrane-based olefin recovery from polymer resin degassing vent (MTR Inc. membrane modules)

Application	Feed composition	Olefin flow rate recovered (recovery yield)
LLPDE/HDPE reactor purge	Butenes, 36%	
	Ethylene, 41%	C2 =, 145 kg h^{-1} (88%)
	Ethane, 1%	*n*-C4 =, 142 kg h^{-1} (98%)
	Nitrogen, 21%	
	Hydrogen, 1%	
HDPE resin degassing	*i*-Butene, 57%	
	CO$_2$/O$_2$, 2%	*i*-C4 =, 1160 kg h^{-1} (97%)
	Nitrogen, 40%	
	Water, 1%	
Polypropylene resin degassing	Propylene, 14%	
	Propane, 0.3%	C3 =, 500 kg h^{-1} (91%)
	Nitrogen, 84.4%	
	Hydrogen, 1%	
	Water, 0.3%	

6.4 Concluding Remarks

Glassy polymer membranes have been used by gas processors for more than 30 years. Their main current application is the removal of carbon dioxide from natural gas. Despite several obvious advantages (compactness, modularity), membrane technology still occupies a minor, though continuously growing, part in the natural gas treatment market, if compared to other conventional solvent solutions. The main limit of commercial carbon dioxide removal membranes resides in their relatively moderate CO_2/CH_4 selectivity. Consistent R&D efforts are presently conducted to overcome this limitation on a wide array of membrane materials, ranging from cross-linked polymers to mineral molecular sieve membranes and very promising results have already been obtained at laboratory scale. In the refining industry, glassy polymer membranes are competing hard on hydrogen purification applications with other sophisticated technologies, such as PSA. Still in the refining industry, zeolite membranes proved to separate paraffin isomers very efficiently. The cost of these membrane materials remains still too high and R&D efforts are still needed to produce thinner zeolite selective layers and optimize the membrane synthesis scale-up. Concerning the permeation membranes applications in petrochemistry, there is a true opportunity for membrane for the purification of high-value olefins. It seems that facilitated transport membranes have achieved a real breakthrough in terms of olefin/paraffin separation performances, if compared to the first generation of glassy polymer membrane materials. As a main conclusion, it appears that membrane technology is still facing many appealing market opportunities in the fossil carbon industry, which constitutes very motivating R&D topics.

6.5 List of Abbreviations

1,2 PB	1,2-Poly(butadiene)
6-FDA	4,4′-Hexafluoroisopylidenediphtalic dianhydride
6FpDA	2,2-bis(4-Aminophenyl) hexafluoropropane
BAHF	2,2′-bis(4-Aminophenyl) bexafluoro propane
BAPF	9,9-bis(4-Aminophenyl)fluorene diamine
BAPS	bis[4-(4-Aminophenoxy)phenyl]sulfone
BTDA	3,3′,4,4′-Benzo phenonetetracarboxylic dianhydride
C4	Butanes isomers
C5	Pentane isomers
C6	Hexane isomers
CA	Cellulose acetate
DEA	Diethanolamine
DDBT	Dimethyl-3,7-diaminodiphenyl-thiophene-5,5-dioxide
DIH	Deisohexaniser
33′DMDB	3,3′-Dimethyl-4,4′-diamino biphenyl
DPPEO	Poly(2,6-diphenyl-1,4- phenyleneoxide)
EC	Ethyl cellulose
EOR	Enhanced oil recovery
HAB	3,3′-Dihydroxy-4,4′-diaminobiphenyl
HF	Hollow fibre
IPDA	3-(Aminomethyl)-3,5,5-trimethylcyclohexylamine
LLDPE	Low density poly(ethylene)
MDEA	*N*-Methyldiethanolamine
m-PDA	*meta*-Phenylene diamine
NDA	Naphthylene-1,8-diamine
P	Permeability (barrer)
P4MP	Poly(4-methylene-1-pentene)
PDMPO	Poly(2,6-dimethyl 1,4-phenylene oxide)
PDMS	Poly(dimetylsiloxane)
PE	Poly(ethylene)
PEG	Poly(ethyleneglycol)

PEP	Poly(ethylene phtalate)
PETP	Polyethylene terephthalate
PMDA	Pyromellitic Dianhydride
PMMA	Polymethyl Methacrylate
POZ	Poly(2-ethyl-2-oxaline)
PPO	Poly(2,6-dimethyl-1,4-phenyleneoxide)
PSA	Pressure swing adsorption
PSF	Polysulfone
PTFE	Poly(tetrafluoroethylene)
PTMO	Poly(tetramethylene oxide)
PTMSP	Poly[1-(trimethyl-silyl)propine]
PVMK	Poly(vinyl-1-methyl ketone)
PVP	Poly(vinylpyrolidone)
RON	Research octane number
SR	Styrene rubber
SW	Spiral wound
TEA	Triethanolylamine
TeMPD	2,3,5,6-Tetramethyl-1,4-phenylenediamine
TrMPD	2,4,6-Trimethyl-*m*-phenylenediamine
TSA	Temperature swing adsorption
VOC	Volatile organic compound

References

1. W. J. Schell and C. D. Houston, *Chem. Eng. Prog.*, 1982, **78**, 33.
2. A. L. Kohl and R. B. Nielsen, Membrane permeation processes, in *Gas Purification*, ed. R. B. Nielsen and A. L. Kohl, Gulf Publishing Co, Houston, TX, 5th edn, 1997, ch. 15, pp. 1238–1295.
3. J. J. Marquez and R. J. Hamaker, *Development of Membrane Performances During SACROC Operations*, AIChE Spring National Meeting and Petro. Expo. '86, 1986.
4. R. Prasad, F. Notaro and D. R. Thompson, *J. Membr. Sci.*, 1994, **94**, 225.
5. Air Liquide, *www.airliquide.com/ww/chapter02/medal/en/Co2_membrane/htm*, 2010.
6. P. J. Cook and M. S. Losin, *Hydrocarbon Process*, 1995, **74**, 79.
7. R. W. Baker, *Ind. Eng. Chem. Res.*, 2002, **41**, 1393.
8. S. P. Nunes and K. V. Peineman, in *Membrane Technology in the Chemical Industry*, Wiley-VCH Verlag GmbH, Weinheim, 2001.
9. J. Hao, P. A. Rice and S. A. Stern, *J. Membr. Sci.*, 2008, **320**, 108.
10. B. D. Bhide and S. A. Stern, *J. Membr. Sci.*, 1993, **81**, 209.
11. B. D. Bhide and S. A. Stern, *J. Membr. Sci.*, 1993, **81**, 239.
12. T. E. Cooley and W. L. Dethloff, *Chem. Eng. Prog.*, 1985, **81**, 45.
13. I. C. Roman, R. W. Ubersax and G. K. Fleming, *Chim. Ind.*, 2001, **83**, 1.
14. L. S. White, T. A. Blinka, H. A. Kloczewski and I. F. Wang, *J. Membr. Sci.*, 1995, **103**, 73.
15. R. W. Baker and K. Lokhandwala, *Ind. Eng. Chem. Res.*, 2008, **47**, 2109.
16. R. E. Babcock, R. W. Spillman, C. S. Godden and T. E. Cooley, *Energy Prog.*, 1988, **8**, 135.

17. R. W. Spillman, *Chem. Eng. Prog.*, 1989, **85**, 41.

18. R. Rautenbach and R. Albfecht, Gas permeation, in *Membrane Process*, ed. R. Rautenbach and R. Albrecht, J. Wiley & Sons, New York, 1989, ch.13, pp. 422–454.

19. R. S. Purgason and C. D. Houston, *Purification of carbon dioxide-containing gas streams with membranes*, AIChE National Meeting, 1983.

20. D. G. Chapel, C. L. Mariz and J. Ernest, *Recovery of CO$_2$ from flue gases: commercial trends*, Canadian Society of Chemical Engineers, Annual Meeting, 1999.

21. R. L. Mckee, M. K. Changela and G. J. Reading, *Hydrocarbon Process*, 1991, **70**, 63.

22. H. S. Meyer and J. P. Gamez, *Gas Separation Membranes: Coming of Age for Carbon Dioxide Removal from Natural Gas*, 45th Laurence Reid Gas Conditioning Conference, 1995.

23. W. H. Mazur and M. C. Chan, *Chem. Eng. Prog.*, 1982, **78**, 38.

24. W. J. Koros, in *Membrane Separation Systems. Recent Developments and Future Directions*, ed. R. W. Baker, W. J. Koros, R. L. Riley and H. Strathmann, Nayes Data Corp., Park Ridge, NJ, 1991, ch. 3, pp. 189–242.

25. C. S. Goddin, *Comparison of Processes for Treating Gases with High CO$_2$ Content*, Annual Gas Processors Association Convention, 1982.

26. K. Boustany, R. S. Narayan and D. J. Stookey, *Economics of Removal of Carbon Dioxide from Hydrocarbon Gas Mixtures*, Annual Gas Processors Association Convention, 1983.

27. R. L. Schendel and J. D. Seymour, *Integration of Membranes with Other Separation Processes*, Membrane Technology R&D Workshop, 1984.

28. R. L. Schendel and J. D. Seymour, *Oil Gas J.*, 1985, **83**, 84.

29. S. G. Kimura and G. E. Walmet, *Sep. Sci. Technol.*, 1980, **15**, 1115.

30. W. J. Schell, *J. Membr. Sci.*, 1985, **22**, 217.

31. K. Eriksen and T. Tensby, *http://chartered.net/biogas.htm*, 2000.

32. A. Ruf and S. Egli, *Gas Sep. Purif.*, 1988, **2**, 90.

33. S. W. Lloyd, in *Membrane Gas Separations*, ed. Y. Yampolskii and B. D. Freeman, J. Wiley and Sons, 2010, ch. 15, pp. 313–332.

34. A. Bos, I. G. M. Punt, M. Wessling and H. Strathmann, *J. Membr. Sci.*, 1999, **155**, 67.

35. J. H. Kim, W. J. Koros and D. R. Paul, *J. Membr. Sci.*, 2006, **282**, 32.

36. J. D. Wind, S. M. Sirard, D. R. Paul, P. F. Green, K. P. Johnston and W. J. Koros, *Macromolecules*, 2003, **36**, 6433.

37. J. D. Wind, D. R. Paul and W. J. Koros, *J. Membr. Sci.*, 2004, **228**, 227.

38. C. Zhou, T. S. Chung, R. Wang, Y. Liu and S. H. Goh, *J. Membr. Sci.*, 2003, **225**, 125.

39. C. Cao, T. S. Chung, Y. Liu, R. Wang and K. P. Pramoda, *J. Membr. Sci.*, 2003, **216**, 257.

40. T. S. Chung, J. H. Ren, R. Wang, D. F. Li, Y. Liu, K. P. Pramoda, C. Cao and W. W. Loh, *J. Membr. Sci.*, 2003, **214**, 57.

41. Y. Liu, T. S. Chung, R. Wang, D. F. Li and M. L. Chng, *Ind. Eng. Chem. Res.*, 2003, **42**, 1190.

42. J. H. Ren, R. Wang, T. S. Chung, D. F. Li and Y. Liu, *J. Membr. Sci.*, 2003, **222**, 133.
43. C. Staudt-Bickel and W. J. Koros, *J. Membr. Sci.*, 1999, **155**, 145.
44. P. S. Tin, T. S. Chung, Y. Liu, R. Wang, S. L. Liu and K. P. Pramoda, *J. Membr. Sci.*, 2003, **225**, 77.
45. J. D. Wind, C. Staudt-Bickel, D. R. Paul and W. J. Koros, *Macromolecules*, 2003, **36**, 1882.
46. J. D. Wind, C. Staudt-Bickel, D. R. Paul and W. J. Koros, *Ind. Eng. Chem. Res.*, 2002, **41**, 6139.
47. Y. C. Xiao, T. S. Chung, H. M. Guan and M. D. Guiver, *J. Membr. Sci.*, 2007, **302**, 254.
48. J. Van Den Bergh, W. D. Zhu, F. Kapteijn, J. A. Moulijn, K. Yajima, K. Nakayama, T. Tomita and S. Yoshida, *Res. Chem. Intermed.*, 2008, **34**, 467.
49. S. Himeno, T. Tomita, K. Suzuki, K. Nakayama, K. Yajima and S. Yoshida, *Ind. Eng. Chem. Res.*, 2007, **46**, 6989.
50. S. Himeno, T. Komatsu, S. Fujita, T. Tomita, K. Suzuki, K. Nakayama and S. Yoshida, *Kagaku Kogaku Ronbunshu*, 2007, **33**, 122.
51. S. Himeno, T. Tomita, K. Suzuki and S. Yoshida, *Microporous Mesoporous Mater.*, 2007, **98**, 62.
52. T. Tomita, K. Nakayama and H. Sakai, *Microporous Mesoporous Mater.*, 2004, **68**, 71.
53. S. G. Li, M. A. Carreon, Y. F. Zhang, H. H. Funke, R. D. Noble and J. L. Falconer, *J. Membr. Sci.*, 2010, **352**, 7.
54. S. G. Li, J. L. Falconer and R. D. Noble, *Microporous Mesoporous Mater.*, 2008, **110**, 310.
55. S. G. Li, J. L. Falconer, R. D. Noble and R. Krishna, *Ind. Eng. Chem. Res.*, 2007, **46**, 3904.
56. S. G. Li, J. L. Falconer and R. D. Noble, *Adv. Mater.*, 2006, **18**, 2601.
57. S. G. Li, G. Alvarado, R. D. Noble and J. L. Falconer, *J. Membr. Sci.*, 2005, **251**, 59.
58. S. Himeno, T. Tomita, K. Suzuki, K. Nakayama, K. Yajima and S. Yoshida, *Ind. Eng. Chem. Res.*, 2007, **46**, 6989.
59. S. G. Li, G. Alvarado, R. D. Noble and J. L. Falconer, *J. Membr. Sci.*, 2005, **251**, 59.
60. S. G. Li, M. A. Carreon, Y. F. Zhang, H. H. Funke, R. D. Noble and J. L. Falconer, *J. Membr. Sci.*, 2010, **352**, 7.
61. T. C. Merkel, R. P. Gupta, B. S. Turk and B. D. Freeman, *J. Membr. Sci.*, 2001, **191**, 85.
62. S. Kulprathipanja and S. S. Sudhir, *U.S. Patent* 4608060, 1985.
63. S. Kulprathipanja, *U.S. Patent* 4606740, 1985.
64. G. Chatterjee, A. A. Houde and S. A. Stern, *J. Membr. Sci.*, 1997, **135**, 99.
65. C. J. Orme and F. F. Stewart, *J. Membr. Sci.*, 2005, **253**, 243.
66. K. Amo, R. W. Baker, V. D. Helm, T. Hofman, K. Lokhandwala, I. Pinnau, M. Ringer, T. T. Su, L. Toy and J. G. Wijmans, *Low-Quality Gas Sulfur Removal/Recovery*, DE-AC21-92MC28133-01, 1998.

67. I. Blume and I. Pinnau, *U.S. Patent* 4963165, 1990.
68. K. Lokhandwala, M. Ringer, J. G. Wijmans and R. W. Baker, *Tech. Doc.* www.mtrinc.com, 1998.
69. B. D. Bhide, A. Voskericyan and S. A. Stern, *J. Membr. Sci.*, 1998, **140**, 27.
70. D. W. Choi, *Hydrocarbon Process.*, 2006, **85**, 69.
71. H. K. Rijkens and J. G. Sponselee, *Natural Gas Dehydration by Membranes*, Symposium on Membrane Technology in Petroleum, Petrochemical and Gas Processing, 221st National Meeting, American Chemical Society, 2001.
72. S. DeDonno, L. Ciccarelli and T. Johannessen, *Natural Gas Dehydration Field Testing of a Novel Membrane Separator*, Offshore Mediterranean Conference and Exhibition, 1997.
73. T. Johannessen and K. Jones, *An Advanced Membrane System for Natural Gas Dehydration Applications and Advantages*, KAPOF Conference and New Product Workshop, 1996.
74. C. Aitken, K. Jones and A. Tag, *PRISM Membrane Systems for Cost Efficient Natural Gas Dehydration*, GPA Technical Meeting, 1998.
75. A. Alpers, B. Keil, O. Ludtke and K. Ohlrogge, *Industrial & Engineering Chemistry Research*, 1999, **38**, 3754.
76. K. A. Lokhandwala, I. Pinnau, Z. J. He, K. D. Amo, A. R. DaCosta, J. G. Wijmans and R. W. Baker, *J. Membr. Sci.*, 2010, **346**, 270.
77. M. Mitariten, *Oil Gas J.*, 2001, **99**, 42.
78. M. Mitariten, *Hydrocarbon Eng.*, 2004, **9**, 53.
79. K. Tanaka, H. Kita, M. Okano and K. Okamoto, *Polymer*, 1992, **33**, 585.
80. T. H. Kim, W. J. Koros, G. R. Husk and K. C. Obrien, *J. Membr. Sci.*, 1988, **37**, 45.
81. R. W. Baker, in *Membrane Technology and Applications*, Chichester, 2004.
82. T. C. Merkel, I. Pinnau, R. Prahbhakar and B. D. Freeman, in *Materials Science of Membranes for Gas and Vapor Separation*, ed. B. D. Freeman, Y. Yampolskii and I. Pinnau, John Wiley & Sons Ltd., Chichester, 2006, pp. 251–270.
83. S. A. Stern, V. M. Shah and B. J. Hardy, *J. Polym. Sci. Part B: Polym. Phys.*, 1987, **25**, 1263.
84. Air Products website, www.airproducts.com/Products/Equipment/PRISMMembranes/page08.htm. 2010.
85. Air Products website, www.airproducts.com/Products/Equipment/PRISMMembranes/page08.htm. 2010.
86. Medal website, www.medal.airliquide.com, 2010.
87. C. Monerau, *Perméation hydrogène: de la périphérie des procédé vers le coeur des procédés?*, Intégration des membranes dans les procédés, 2, 2003.
88. Ube Industries website, http://northamerica.ube.com/page.php?pageid = 29, 2010.
89. M. Arruebo, J. L. Falconer and R. D. Noble, *J. Membr. Sci.*, 2006, **269**, 171.
90. W. J. W. Bakker, F. Kapteijn, J. Poppe and J. A. Moulijn, *J. Membr. Sci.*, 1996, **117**, 57.

91. J. Coronas, R. D. Noble and J. L. Falconer, *Ind. Eng. Chem. Res.*, 1998, **37**, 166.
92. J. Coronas, R. D. Noble and J. L. Falconer, *Ind. Eng. Chem. Res.*, 1998, **37**, 166.
93. L. Courthial, A. Baudot, M. Tayakout-Fayolle and C. Jallut, *AIChE J.*, 2008, **54**, 2527.
94. C. L. Flanders, V. A. Tuan, R. D. Noble and J. L. Falconer, *J. Membr. Sci.*, 2000, **176**, 43.
95. H. H. Funke, M. G. Kovalchick, J. L. Falconer and R. D. Noble, *Ind. Eng. Chem. Res.*, 1996, **35**, 1575.
96. H. H. Funke, A. M. Argo, J. L. Falconer and R. D. Noble, *Ind. Eng. Chem. Res.*, 1997, **36**, 137.
97. T. Q. Gardner, J. L. Falconer and R. D. Noble, *AIChE J.*, 2004, **50**, 2816.
98. C. J. Gump, X. Lin, J. L. Falconer and R. D. Noble, *J. Membr. Sci.*, 2000, **173**, 35.
99. F. Kapteijn, W. J. W. Bakker, J. Vandegraaf, G. Zheng, J. Poppe and J. A. Moulijn, *Catal. Today*, 1995, **25**, 213.
100. F. Kapteijn, W. J. W. Bakker, G. H. Zheng, J. Poppe and J. A. Moulijn, *Chem. Eng. J. Biochem. Eng. J.*, 1995, **57**, 145.
101. S. Sommer, T. Melin, J. L. Falconer and R. D. Noble, *J. Membr. Sci.*, 2003, **224**, 51.
102. J. M. van de Graaf, F. Kapteijn and J. A. Moulijn, *AIChE J.*, 1999, **45**, 497.
103. H. vanKoningsveld, J. C. Jansen and A. J. M. Deman, *Acta Crystallogr., Sect. B: Struct. Sci.*, 1996, **52**, 131.
104. J. Hedlund, J. Sterte, M. Anthonis, A. J. Bons, B. Carstensen, N. Corcoran, D. Cox, H. Deckman, W. De Gijnst, P. P. de Moor, F. Lai, J. McHenry, W. Mortier and J. Reinoso, *Microporous Mesoporous Mater.*, 2002, **52**, 179.
105. E. M. Flanigen, J. M. Bennett, R. W. Grose, J. P. Cohen, R. L. Patton, R. M. Kirchner and J. V. Smith, *Nature*, 1978, **271**, 512.
106. A. Baudot and L. Bournay, *Oil Gas Sci. Technol. – Revue de l 'Institut Francais du Petrole*, 2009, **64**, 759.
107. G. W. Meindersma and A. B. de Haan, *Desalination*, 2002, **149**, 29.
108. J. Caro, M. Noack, P. Kolsch and R. Schafer, *Microporous Mesoporous Mater.*, 2000, **38**, 3.
109. M. T. Ravanchi, T. Kaghazchi and A. Kargari, *Desalination*, 2009, **235**, 199.
110. R. L. Burns and W. J. Koros, *J. Membr. Sci.*, 2003, **211**, 299.
111. M. Azhin, T. Kaghazchi and M. Rahmani, *J. Ind. Eng. Chem.*, 2008, **14**, 622.
112. C. Staudt-Bickel and W. J. Koros, *J. Membr. Sci.*, 2000, **170**, 205.
113. S. S. Chan, R. Wang, T. S. Chung and Y. Liu, *J. Membr. Sci.*, 2002, **210**, 55.
114. K. Tanaka, A. Taguchi, J. Q. Hao, H. Kita and K. Okamoto, *J. Membr. Sci.*, 1996, **121**, 197.

115. O. M. Ilinitch, G. L. Semin, M. V. Chertova and K. I. Zamaraev, *J. Membr. Sci.*, 1992, **66**, 1.
116. A. Shimazu, T. Miyazaki, M. Maeda and K. Ikeda, *J. Polym. Sci. Part B: Polym. Phys.*, 2000, **38**, 2525.
117. S. I. Semenova, *J. Membr. Sci.*, 2004, **231**, 189.
118. J. J. Krol, M. Boerrigter and G. H. Koops, *J. Membr. Sci.*, 2001, **184**, 275.
119. A. Shimazu, T. Miyazaki, M. Maeda and K. Ikeda, *J. Polym. Sci. Part B: Polym. Phys.*, 2000, **38**, 2525.
120. S. Bai, S. Sridhar and A. A. Khan, *J. Membr. Sci.*, 1998, **147**, 131.
121. H. H. Hoehn, Treatment of membranes of selected polyimides. *U.S. Patent* 3822202, 1974.
122. T. Miyazaki, A. Shimazu and K. Ikeda, *Polymer*, 2000, **41**, 8167.
123. T. Miyazaki, A. Shimazu, K. Ikeda and T. Kanaya, *Polymer*, 2003, **44**, 1553.
124. A. Shimazu, T. Miyazaki and K. Ikeda, *J. Membr. Sci.*, 2000, **166**, 113.
125. A. Shimazu, T. Miyazaki, S. Katayama and Y. Ito, *J. Polym. Sci. Part B: Polym. Phys.*, 2003, **41**, 308.
126. A. Shimazu, T. Miyazaki, M. Maeda, O. Tozawa and K. Ikeda, *Sen'i Gakkaishi*, 2000, **56**, 85.
127. T. Miyazaki, A. Shimazu, T. Matsushita and K. Ikeda, *J. Appl. Polym. Sci.*, 2000, **78**, 1818.
128. A. Shimazu, T. Miyazaki, T. Matsushita, M. Maeda and K. Ikeda, *J. Polym. Sci. Part B: Polym. Phys.*, 1999, **37**, 2941.
129. H. Kawakami, M. Mikawa and S. Nagaoka, *J. Membr. Sci.*, 1996, **118**, 223.
130. M. Teramoto, H. Matsuyama, T. Yamashiro and S. Okamoto, *J. Membr. Sci.*, 1989, **45**, 115.
131. M. T. Ravanchi, T. Kaghazchi and A. Kargari, *Sep. Sci. Technol.*, 2009, **44**, 1198.
132. M. T. Ravanchi, T. Kaghazchi and A. Kargari, *Desalination*, 2010, **250**, 130.
133. M. T. Ravanchi, T. Kaghazchi, A. Kargari and M. Soleimani, *J. Taiwan Inst. Chem. Eng.*, 2009, **40**, 511.
134. M. T. Ravanchi, T. Kaghazchi and A. Kargaric, *Chem. Eng. Process.*, 2010, **49**, 235.
135. J. S. Yang and G. H. Hsiue, *J. Membr. Sci.*, 1996, **120**, 69.
136. G. H. Hsiue and J. S. Yang, *Macromol. Symp.*, 1996, **105**, 51.
137. G. H. Hsiue and J. S. Yang, *J. Membr. Sci.*, 1993, **82**, 117.
138. J. S. Yang and G. H. Hsiue, *J. Membr. Sci.*, 1996, **111**, 27.
139. J. S. Yang and G. H. Hsiue, *J. Membr. Sci.*, 1997, **126**, 139.
140. J. S. Yang and G. H. Hsiue, *J. Polym. Sc. Part B: Poly.r Phys.*, 1997, **35**, 909.
141. J. H. Kim, J. Jang, D. Y. Lee and W. C. Zin, *Macromolecules*, 2002, **35**, 311.
142. J. H. Kim, B. R. Min, C. K. Kim, J. Won and Y. S. Kang, *Macromolecules*, 2002, **35**, 5250.

143. J. H. Kim, B. R. Min, J. Won and Y. S. Kang, *Macromolecules*, 2003, **36**, 4577.
144. J. H. Kim and S. C. Kim, *Macromolecules*, 2003, **36**, 2867.
145. J. H. Kim, B. R. Min, J. Won, S. H. Joo, H. S. Kim and Y. S. Kang, *Macromolecules*, 2003, **36**, 6183.
146. S. Sunderrajan, B. D. Freeman, C. K. Hall and I. Pinnau, *J. Membr. Sci.*, 2001, **182**, 1.
147. I. Pinnau, L. G. Toy, S. Sunderrajan and B. D. Freeman, *Abstr. Papers Am. Chem. Soc.*, 1997, **214**, 63.
148. I. Pinnau and L. G. Toy, *J. Membr. Sci.*, 2001, **184**, 39.
149. T. C. Merkel, Z. He, A. Morisato and I. Pinnau, *Chem. Commun.*, 2003, 1596.
150. J. H. Kim, B. R. Min, H. S. Kim, J. Won and Y. S. Kang, *J. Membr. Sci.*, 2003, **212**, 283.
151. J. H. Kim, J. Won and Y. S. Kang, *J. Membr. Sci.*, 2004, **237**, 199.
152. D. Ko, J. H. Kim, S. W. Kang, D. H. Lee, J. Won and Y. S. Kang, *J. Ind. Eng. Chem.*, 2009, **15**, 8.
153. B. Jose, J. H. Ryu, B. G. Lee, H. Lee, Y. S. Kang and H. S. Kim, *Chem. Commun.*, 2001, 2046.
154. H. S. Kim, J. H. Ryu, B. Jose, B. G. Lee, B. S. Ahn and Y. S. Kang, *Langmuir*, 2001, **17**, 5817.
155. J. H. Kim, B. R. Min, J. Won and Y. S. Kang, *J. Membr. Sci.*, 2003, **227**, 197.
156. H. H. Huang, X. P. Ni, G. L. Loy, C. H. Chew, K. L. Tan, F. C. Loh, J. F. Deng and G. Q. Xu, *Langmuir*, 1996, **12**, 909.
157. Z. T. Zhang, B. Zhao and L. M. Hu, *J. Solid State Chem.*, 1996, **121**, 105.
158. S. W. Kang, J. H. Kim, K. S. Oh, J. Won, K. Char, H. S. Kim and Y. S. Kang, *J. Membr. Sci.*, 2004, **236**, 163.
159. A. Sungpet, J. D. Way, P. M. Thoen and J. R. Dorgan, *J. Membr. Sci.*, 1997, **136**, 111.
160. J. H. Ryu, H. Lee, Y. J. Kim, Y. S. Kang and H. S. Kim, *Chem.-A Eur. J.*, 2001, **7**, 1525.
161. A. Morisato, Z. J. He, I. Pinnau and T. C. Merkel, *Desalination*, 2002, **145**, 347.
162. J. H. Kim, S. M. Park, J. Won and Y. S. Kang, *J. Membr. Sci.*, 2004, **236**, 209.
163. S. W. Kang, J. H. Kim, J. Won, K. Char and Y. S. Kang, *J. Membr. Sci.*, 2005, **248**, 201.
164. J. H. Kim, S. M. Park, J. Won and Y. S. Kang, *J. Membr. Sci.*, 2005, **248**, 171.
165. S. W. Kang, J. H. Kim, K. Char, J. Won and Y. S. Kang, *J. Membr. Sci.*, 2006, **285**, 102.
166. S. W. Kang, J. H. Kim, K. S. Oh, J. Won, K. Char, H. S. Kim and Y. S. Kang, *J. Membr. Sci.*, 2004, **236**, 163.
167. J. H. Kim, B. R. Min, J. Won and Y. S. Kang, *J. Polym. Sci. Part B: Polym. Phys.*, 2006, **44**, 1168.

168. S. H. Joo, J. H. Kim, S. W. Kang, J. Jang and Y. S. Kang, *J. Polym. Sci. Part B: Polym. Phys.*, 2007, **45**, 2263.
169. K. M. Steel and W. J. Koros, *Carbon*, 2005, **43**, 1843.
170. H. Suda and K. Haraya, *Chem. Commun.*, 1997, 93.
171. M. Yoshino, S. Nakamura, H. Kita, K. Okamoto, N. Tanihara and Y. Kusuki, *J. Membr. Sci.*, 2003, **215**, 169.
172. M. N. Islam, W. L. Zhou, T. Honda, K. Tanaka, H. Kita and K. Okamoto, *J. Membr. Sci.*, 2005, **261**, 17.
173. J. Hayashi, H. Mizuta, M. Yamamoto, K. Kusakabe, S. Morooka and S. H. Suh, *Ind. Eng. Chem. Res.*, 1996, **35**, 4176.
174. T. A. Centeno, J. L. Vilas and A. B. Fuertes, *J. Membr. Sci.*, 2004, **228**, 45.
175. A. B. Fuertes and I. Menendez, *Sep. Purif. Technol.*, 2002, **28**, 29.
176. M. L. Chng, Y. C. Xiao, T. S. Chung, M. Toriida and S. Tamai, *Carbon*, 2009, **47**, 1857.
177. Y. Kusuki, H. Shimazaki, N. Tanihara, S. Nakanishi and T. Yoshinaga, *J. Membr. Sci.*, 1997, **134**, 245.
178. N. Tanihara, H. Shimazaki, Y. Hirayama, S. Nakanishi, T. Yoshinaga and Y. Kusuki, *J. Membr. Sci.*, 1999, **160**, 179.
179. Y. K. Kim, H. B. Park and Y. M. Lee, *J. Membr. Sci.*, 2003, **226**, 145.
180. W. L. Zhou, M. Yoshino, H. Kita and K. Okamoto, *Ind. Eng. Chem. Res.*, 2001, **40**, 4801.
181. W. L. Zhou, T. Watari, H. Kita and K. Okamoto, *Chem. Lett.*, 2002, 534.
182. W. L. Zhou, M. Yoshino, H. Kita and K. Okamoto, *J. Membr. Sci.*, 2003, **217**, 55.
183. V. Nikolakis, G. Xomeritakis, A. Abibi, M. Dickson, M. Tsapatsis and D. G. Vlachos, *J. Membr. Sci.*, 2001, **184**, 209.
184. I. G. Giannakopoulos and V. Nikolakis, *J. Membr. Sci.*, 2007, **305**, 332.
185. I. G. Giannakopoulos, K. Kalambaliki, V. Dracopoulos and V. Nikolakis, *Studies Surf. Sci. Catal.*, 2005, **158**, 137.
186. I. G. Giannakopoulos and V. Nikolakis, *Ind. Eng. Chem. Res.*, 2005, **44**, 226.
187. I. Tiscornia, S. Irusta, P. Pradanos, C. Tellez, J. Coronas and J. Santamaria, *J. Phy. Chem. C*, 2007, **111**, 4702.
188. T. Pettersen, A. Argo, R. D. Noble and C. A. Koval, *Sep. Technol.*, 1996, **6**, 175.
189. S. Moganti, R. D. Noble and C. A. Koval, *J. Membr. Sci.*, 1994, **93**, 31.
190. I. K. Kookos, *Ind. Eng. Chem. Res.*, 2003, **42**, 1731.
191. V. Gokhale, S. Hurowitz and J. B. Riggs, *Ind. Eng. Chem. Res.*, 1995, **34**, 4413.
192. L. M. Robeson, *J. Membr. Sci.*, 1991, **62**, 165.
193. W. Stephan, R. D. Noble and C. A. Koval, *J. Membr. Sci.*, 1995, **99**, 259.
194. J. A. Caballero, I. E. Grossmann, M. Keyvani and E. S. Lenz, *Ind. Eng. Chem. Res.*, 2009, **48**, 9151.
195. M. O. Daramola, A. J. Burger, M. Pera-Titus, A. Giroir-Fendler, L. Lorenzen and J. A. Dalmon, *Sep. Sci. Technol.*, 2010, **45**, 21.

196. Y. F. Yeong, A. Z. Abdullah, A. L. Ahmad and S. Bhatia, *J. Chem. Technol. Biotechnol.*, 2010, **85**, 216.
197. Y. F. Yeong, A. Z. Abdullah, A. L. Ahmad and S. Bhatia, *J. Membr. Sci.*, 2009, **341**, 96.
198. X. H. Gu, J. H. Dong, T. M. Nenoff and D. E. Ozokwelu, *J. Membr. Sci.*, 2006, **280**, 624.
199. H. Sakai, T. Tomita and T. Takahashi, *Sep. Purif. Technol.*, 2001, **25**, 297.
200. G. Xomeritakis, Z. P. Lai and M. Tsapatsis, *Ind. Eng. Chem. Res.*, 2001, **40**, 544.
201. G. Xomeritakis and M. Tsapatsis, *Chem. Mater.*, 1999, **11**, 875.
202. R. W. Baker, K. A. Lokhandwala, M. L. Jacobs and D. E. Gottschlich, *Chem. Eng. Prog.*, 2000, **96**, 51.
203. R. W. Baker and M. Jacobs, *Hydrocarbon Process.*, 1996, **75**, 49.

Membrane Gas Separation Processes for Post-combustion CO_2 Capture

PETER MICHAEL FOLLMANN*, CHRISTOPH BAYER,
MATTHIAS WESSLING AND THOMAS MELIN

RWTH Aachen University, AVT Chemical Process Engineering,
Turmstraße 46, 52056 Aachen, Germany

7.1   Introduction

CO_2 capture and storage is considered an important option to control and reduce greenhouse gas emissions. Among currently investigated capture technologies only post-combustion capture enables retrofitting of existing power plants and usage in existing industrial units like cement plants or steel mills.

Besides a number of technical challenges, amine absorption is presently considered the best available technology for CO_2 capture from power plant flue gases. Recent publications, however, indicate that the energy requirement for solvent recovery might cause overall power plant efficiency losses up to 16%-points.[1] Furthermore, chemical degradation requires continuous solvent replacement leading to additional material and disposal costs.

In contrast, membrane systems do not require any solvent or additional chemical for proper operation. Moreover, they potentially offer higher energy efficiencies, greater operational flexibility as well as simplicity of operation and maintenance. The impressive number of large scale research projects like

Membrane Engineering for the Treatment of Gases, Volume 1:
Gas-separation Problems with Membranes
Edited by Enrico Drioli and Giuseppe Barbieri
© Royal Society of Chemistry 2011
Published by the Royal Society of Chemistry, www.rsc.org

'NanoGLOWA' or 'Membrain' mirrors the increasing interest in membrane technology for post-combustion CO_2 capture.

This chapter presents an integral view on membrane-based post-combusion CO_2 capture from coal fired power plants. First, the process boundary conditions will be defined and the different types of driving force generation for mass transfer across gas permeation membranes will be studied. A final techno-economic analysis provides insight into process economics detailing energy consumption, CO_2 recovery and CO_2 avoidance costs.

7.2　Boundary Conditions

The design, optimization and integration of any separation process generally requires the definition of process boundary conditions, which typically include flow rates, compositions, temperatures and pressures of all entering and leaving streams. The flue gas stream establishes the inlet conditions of any CO_2 capture facility. CO_2 transport and CO_2 storage, in contrast, set constraints on the composition of the CO_2-rich product stream (Figure 7.1). Restrictions with respect to the maximum concentration limit of species other than CO_2 may be subject to technical, safety, environmental or legal issues. The CO_2 recovery of the capture process depends mostly on economic considerations.

7.2.1　Upstream Boundary Conditions: the Power Plant

To avoid major modifications of the power plant, the optimal position of a CO_2 capture process seems to be in between the flue gas desulfurization unit (FGD) and the stack. Particulate matter as well as SO_x and NO_x concentrations are reduced to a large extent, which lowers the requirements regarding chemical resistance and reduces the risk of plugging. To quantify the flue gas condition subsequent to the FGD unit, a model of a state of the art hard coal fired power plant was developed and implemented in Aspen Plus®. The considered power plant features a thermal duty of 1210 MW and a net thermal efficiency of 45% (based on lower heating value of the feedstock).

The calculated composition given in Table 7.1 is in good agreement with literature data.[2] For simplicity this work focuses only on the major flue gas constituents, N_2, CO_2, O_2 and H_2O. The flue gas stream exits the

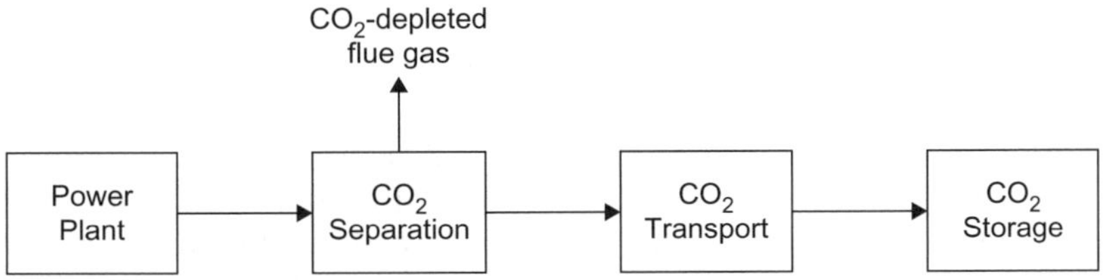

Figure 7.1　Schematic overview of post-combustion CO_2 capture.

Table 7.1 Flue gas conditions at the outlet of the FGD unit of a coal fired power plant

N_2 vol.%	CO_2 vol.%	H_2O vol.%	O_2 vol.%	Flow rate $kg\ s^{-1}$	Temperature $^\circ C$	Pressure bar
71.2	13.6	12	3.2	560	50	1.013

desulfurization unit saturated in water at 50 °C and 1.013 bar. The volume flow accounts for 1.36 million $m_N^3\ h^{-1}$ (at 0 °C and 1.013 bar) on a dry basis.

7.2.2 Downstream Boundary Conditions: CO_2 Transport

Captured CO_2 will most likely be transported to an appropriate sink by pipelines, a state of the art technology.[3] Today's commercial CO_2 pipelines are operated in the dense phase, *i.e.* at subcritical temperature and supercritical pressure.[4] However, several literature sources indicate that pipeline operating conditions are very case specific.[4–6] Since it is beyond the scope of this work to design a specific CO_2 pipeline, the authors assume a pipeline injection pressure of 130 bar. This assumption is in line with the majority of engineering studies regarding CO_2 sequestration. de Visser *et al.*[7] and Svensson *et al.*[8] recommend that the injection temperature should not exceed 30 °C.

Small amounts of non-condensable gases such as N_2, O_2 or Ar have an adverse effect on pipeline flow properties, operating pressures and compression requirements.[4,9] Therefore, the concentration of non-condensable gases should be as low as possible. Since neither general quality standards for CO_2 transmission exist nor a conclusive reflection is possible at this moment, it is necessary to rely on published data to quantify concentration limits of non-condensable gases.[7] Table 7.2 summarizes CO_2 specifications for pipeline transport available in literature.

According to Table 7.2, the concentration of all impurities must not exceed 5%, which implies a minimum CO_2 concentration of at least 95%. Air based gases (N_2, O_2, Ar) are restricted to 4% in total.

Due to the considerable vapor concentration in power plant flue gases (compare Table 7.1), it is very likely that some water is captured in conjunction with CO_2. Because carbon and low alloy steals are typical construction materials for CO_2 pipelines, corrosion is a major concern if moisture is present.[5,6] Mohitpour *et al.* state that industry generally accepts up to 500 ppm residual water content for carbon steel pipelines.[6] This limitation, however, seems to be valid only for CO_2 mixtures with low O_2 quantities (compare Table 7.2). Membrane-based CO_2 separation may yield elevated O_2 levels in the captured CO_2, which may require lower residual water concentrations than 500 ppm for pipeline transmission. Since a dehydration process is necessary in any case subsequent to the capture unit, the allowed water content for pipeline transmission has no direct impact on the membrane gas separation process. Therefore, the authors leave out a quantification of the residual water content.

Table 7.2　Review of CO_2 quality requirements for pipeline transport

Gas	Reference			
	3	7	12	13
CO_2 (mole%)	>95	>95.5	97–99	96
Total HC (mole%)	<5	—	—	—
N_2 (mole%)	<4	<4	<3	0.6
Ar (mole%)	—	<4	—	—
O_2 (ppm)	<10	0.1–4	—	100
Total air gases (mole%)	—	<4	—	—
H_2O (ppm)	<600	<500	<600	20

7.2.3　Downstream Boundary Conditions: CO_2 Storage

In view of CO_2 storage, there is strong evidence that sub-surface sinks offer the least disadvantages. Among potential subsurface sinks, depleted oil and gas reservoirs as well as deep aquifers have received the highest attention. To guarantee the effective use of the available storage volume, Bachu recommends geological CO_2 storage only at supercritical conditions.[10] In that case, reservoir and thus CO_2 injection pressures correspond well with the pressure requirements for CO_2 pipeline transmission.

CO_2 injection into oil reservoirs is a proven technology to enhance oil recovery (EOR). The injected CO_2 usually mixes with the oil and reduces its viscosity. At this state, the capillary forces holding the oil in the reservoir pores are suppressed and more oil can be recovered.[11] Since the recovered oil offsets CCS costs to some extent, enhanced oil recovery may improve CCS economics significantly.

Lighter impurities such as O_2 or N_2 make the injected CO_2 a poorer oil extractor. Most publications focusing on EOR suggest a total content of 4% air based gases or even less.[3,12,13]

Apart from the reduction of extraction properties, O_2 supposedly accelerates the corrosion of oil processing equipment. Moreover, O_2 may react with the reservoir oil making its extraction even more difficult. Thus, strict limits with respect to the residual O_2 content have been defined for EOR applications. Metz *et al.*,[3] for instance, mention a residual O_2 content of 10 ppm or lower, while Liljedahl *et al.*[13] allow 100 ppm O_2. After all, CO_2 purity requirements for EOR applications are highly site specific. Thus, it is only possible to quantify them in a genertic way, however they are not applicable to all reservoirs.

In contrast to the CO_2 purity requirements for enhanced oil recovery, only little data about the purity requirements for CO_2 storage in gas reservoirs are available in literature. Thus, it is currently not possible to quantify the allowed concentration of O_2 and other impurities accompanying the CO_2. However, since natural gas reservoirs and oil reservoirs may be considered comparable by a first approximation, the authors assume that purity requirements derived for EOR are applicable to gas reservoirs as well.

CO_2 injection in deep saline aquifers is considered an important option for CO_2 storage, particularly in Germany. After injection, CO_2 remains first in a separate phase until it dissolves in the formation water. Geochemical reactions

Table 7.3 CO_2 composition for transport and storage

	Air gases vol.%	O_2 vol.%	H_2O ppm
Scenario 1	<4	<4	n. a.
Scenario 2	<4	<0.01	n. a.

finally convert some of the dissolved CO_2 to stable minerals.[14–16] Light components (N_2, O_2, Ar) co-injected with CO_2, however, do not dissolve in the formation water as well as the CO_2 does.[17] They remain mostly in a separate phase occupying pore volume that is not available for CO_2 storage.[17,18] Thus, it appears that considerable quantities of light components reduce the storage capacity. Besides, O_2 may change the redox conditions in an aquifer. So far, only very few publications address the impact of O_2 on aquifer storage and give no specific indications on acceptable concentration levels. Low O_2 concentrations may be neglected, but if membrane processes are applied for CO_2 capture, O_2 concentrations up to some mole percent are likely. Therefore, this work will take into account two different scenarios. In the first scenario it is assumed that O_2 has a severe effect on the reservoir and needs to be restricted to very low concentrations. For simplicity the same limit as for EOR applications is assumed. In the second scenario O_2 is only regarded as an impurity, which reduces the available storage volume for CO_2. To meet the requirements for pipeline transmission, the maximum allowed O_2 concentration shall not exceed 4%.

7.2.4 Summary of Boundary Conditions

Table 7.3 summarizes the downstream boundary conditions, apart from CO_2 recovery. In view of CCS, the optimal CO_2 recovery depends predominantly on the associated CO_2 avoidance costs, which will be evaluated as part of the techno-economic analysis.

7.3   Membranes and Membrane Model

Since temperatures are moderate at the outlet of the FGD unit (compare Table 7.1), polymeric membranes appear to be the most suitable ones for post-combusion CO_2 capture. In the recent past, particularly, PEO-based block copolymers received considerable attention. The high polarity of PEO leads to a high CO_2/N_2 gas separation factor in combination with a high CO_2 permeance. Due to the rubbery nature of PEO-based block copolymers, the presence of water hardly influences the transport of CO_2 and other gases.[19] The chemical and mechanical resistance of the polymer to flue gas conditions are currently evaluated in the NanoGLOWA project. Table 7.4 summarizes the rather favorable separation characteristics of PEO, which will be further used to investigate the economics of membrane gas separation processes of post-combusion CO_2 capture.

Table 7.4 Separation performance of the PEO membrane; permeance in m_N^3 $(m^2\ h\ bar)^{-1}$

	CO_2 permeance	H_2O permeance	CO_2/N_2 selectivity	CO_2/O_2 selectivity
PEO	1.25	$\gg CO_2$	45	15

To perform the techno-economic analysis, a permeation model implemented in FORTRAN was interfaced with the process design software Aspen Plus. A summary of the respective model equations is available elsewhere.[20] The model has been validated with experimental data.[21–23] Most engineering studies on post-combusion CO_2 capture are based on model assumptions that are very similar or even identical to those in this work.[24–33]

7.4   Driving Force

The driving force for mass transfer in gas permeation processes relies on the partial pressure difference of the respective component across the membrane.[20] Three options generally exist for its generation:

1. To increase the partial pressure of the respective component on the upstream side by compression of the feed stream
2. To decrease the partial pressure on the downstream side (permeate) by applying vacuum and/or sweeping with an inert gas
3. To combine options 1 and 2.

In their recent paper Merkel *et al.* pointed out that the pressure ratio is the limiting factor for post-combusion CO_2 capture, even for average performing membranes.[31] Consequently, not only the development and improvement of CO_2–N_2 selective membranes but also the generation of the required driving force is crucial, particularly with respect to the technical implementation in power plants. In the subsequent paragraphs the authors will therefore examine different options to generate the driving force for CO_2 capture in more detail. Given the flue gas conditions in Table 7.1, the equipment used needs to be able to handle large volume flows efficiently.

7.4.1   Feed Compression

Follmann *et al.* reviewed compression machines with a particular focus on the requirements for post-combustion CO_2 capture.[34] Dynamic machines, *e.g.* axial and centrifugal compressors, appear to be most suitable to pressurize large volume flows. Among dynamic machines, axial compressors feature the highest volume capacity and the highest isentropic efficiency. Pressure ratios up to 4 may be realized by a multi-stage axial compressor without inter-stage cooling. Higher pressure ratios are usually generated by a combination of a multi-stage axial compressor and several subsequent radial stages. Inter-stage cooling is performed prior to each radial stage. Such mixed compression concepts are common practice in large scale air separation units.

The expansion of the CO_2 depleted residual stream allows the recovery of mechanical energy, which reduces the total energy requirement for flue gas compression. The extent of the recovered energy depends on the pressure level and on the temperature of the residual stream.

7.4.2   Suction at the Permeate Side

The driving force for mass transfer across the membrane may also be realized by suction at the permeate side, while the flue gas passes the membrane unit at ambient conditions. In contrast to compression, suction features significantly lower energy consumption, but the low flux across the membrane leads to much larger membrane areas. To provide sufficient driving force, most engineering studies came to the conclusion that very low suction pressures are necessary. Nevertheless, the majority of these studies did not examine the large scale implementation of such low pressures and did not consider the tremendous volume flows. If, for instance, 0.075 bar are applied at the permeate side, a PEO membrane recovers about 90% CO_2 from the flue gas given in Table 7.1 and enriches it into a gas stream of approximately 55% CO_2. The permeate stream accounts for 0.34 million m_N^3 h^{-1} (dry basis), which corresponds to approx. 4.5 million m^3 h^{-1} (dry basis) at the respective pressure. Water vapor is not considered here and it further increases the volume flow. Thus, in view of the capacity of machines for vacuum generation, the huge volume flows must be taken into account. A review concerning vacuum machines came to the following conclusions:[34]

- Due to their low energetic efficiency, steam ejectors are unattractive for CO_2 capture applications.
- Liquid ring pumps, rotary lobe blowers and root pumps presently allow only small suction volumes, which lead to a large number of machines. At a suction pressure of 100 mbar the compression of 0.34 million m_N^3 would require, for example, about 60 liquid ring pumps.

Besides state of the art machines for vacuum generation, the authors have been in contact with a supplier of compression equipment to discuss the application of axial and radial compressors. According to the supplier, the use of axial/radial machines is possible, but it would require some significant modifications of the machine. The available suction pressure is limited to approximately 200 mbar. Axial compressors may provide a suction capacity of about 0.5 million m^3 h^{-1} at this pressure. The isentropic efficiency is approximately 75%, which is significantly higher in comparison to liquid ring pumps or jet pumps.

7.4.3   Feed Compression and Suction at the Permeate Side

The discussion of feed compression and suction reveals that each concept encounters specific advantages and drawbacks. While compression allows the realization of high pressure ratios at the expense of energy consumption, the more energy efficient suction concept has a limit at approximately 200 mbar suction pressure. Therefore, a combination of both concepts may overcome the pressure limitation of suction as well as limit the energy consumption of feed compression.

7.4.4 Sweep Operation

The dilution of the permeated CO_2 stream by sweep gas represents another option to induce CO_2 transfer across the membrane. Sweeping as a standalone concept is not able to provide a sufficiently large driving force. Thus, it is always applied in conjunction with feed gas compression, suction or a combination of both. Since the sweep gas usually has to be separated easily from the permeate stream, condensable vapors are most suitable. In the context of power plants, only steam represents a suitable sweep gas. Regardless of temperature and pressure, the steam cannot be extracted from the steam cycle of the power plant. The very strict purity standards for boiler feed water (impurities on ppb level allowed only) require extensive purification of the condensate prior to the recycling into the steam cycle. As a result, a second steam cycle including a complex steam piping and -distribution systems has to be installed for that purpose. Thus, sweeping with vapor at the low pressure side appears unattractive for driving force generation.

7.5 Techno-economic Analysis

7.5.1 Process Configurations

Single stage membrane processes are presently not able to achieve an appropriate degree of CO_2 separation and enrichment at the same time. Therefore, the use of multi-stage membrane processes is necessary. According to Carapellucci and Milazzo, the total number of membrane stages usually does not exceed two, as additional stages do not significantly improve efficiency but increase system complexity.[35] Therefore, engineering studies about membrane cascades for CO_2 capture mostly focus on two stage processes, which represent the simplest alternative to a single membrane stage.[25,28,29,32,33]

   In such cascades, two categories of gas recycling exist: product/permeate recycling and retentate recycling. Rautenbach and Albrecht observed that permeate recycling increases the concentration of the target component in the permeate stream, but it would be expensive with respect to capital investment and energy consumption.[36] Furthermore, from a theoretical point of view, product recycle causes substantial mixing losses and consequently leads to very low thermodynamic efficiencies.[37] In contrast, retentate recycling to the feed of the previous stage offers a simple mean to enhance the overall recovery of more permeable species.[38] Rautenbach and Albrecht reported that two stage membrane processes are optimal for O_2 enrichment in air.[36] In terms of

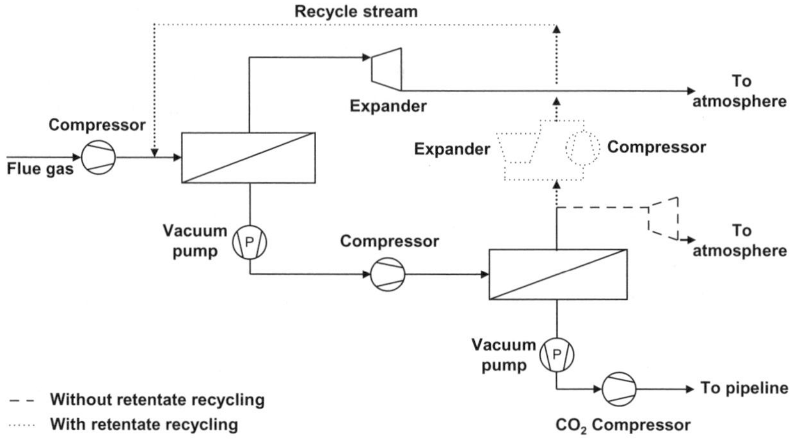

Figure 7.2 Simplified flow sheet of a two-stage membrane process with and without retentate recycling.

production costs, cascades with and without residue recycling showed only minor differences with higher possible recovery rates of the former concept.

In short, product recycling appears to be very unattractive, while residue recycling might lead to economic benefits. The increase in recovery may result at the expense of more membrane area as well as enhanced energy consumption. To evaluate that trade-off and to determine the impact of recycling on energetic and economic performance, cascades with and without recycling of the residual stream of the second stage will be further explored.

To be in line with Section 7.4, driving force generation in both stages relies on a combination of compression on the feed and suction at the permeate side. If recycling is neglected, the retentate streams of both stages are expanded prior to their release into the atmosphere. If recycling is considered, the residual stream of the second stage is mixed with the flue gas stream. The pressure of the flue gas stream determines whether the recycled gas stream is either compressed or expanded prior to the mixing point. A multi-stage compressor finally boosts the recovered CO_2 product up to pipeline pressure (Figure 7.2).

7.5.2 Key Performance Indicators and Economics

To evaluate the technical and economic performance of membrane based CO_2 capture, the authors have chosen as key performance indicators (KPIs) the specific energy consumption (eqn (7.1)), the total membrane area (eqn (7.2)) and the specific CO_2 avoidance costs (eqn (7.3)).

$$\text{specific energy consumption} = \frac{\text{total energy cons. for CO}_2 \text{ capture}}{\text{mass flow CO}_2 \text{ captured}} \left[\frac{\text{kJ}_{\text{electric}}}{\text{kg}_{\text{CO}_2,\text{captured}}} \right]$$

$$(7.1)$$

$$\text{total membrane area} = \text{area 1}^{\text{st}}\text{ stage} + \text{area 2}^{\text{nd}}\text{ stage } [\text{m}^2] \qquad (7.2)$$

$$\text{specific CO}_2 \text{ avoidance costs} = \frac{\text{COE}_{\text{capture}} - \text{COE}_{\text{ref}}}{\text{CO}_2 \text{ emission}_{\text{ref}} - \text{CO}_2 \text{ emission}_{\text{capture}}} \left[\frac{\text{Euro}}{\text{ton}_{\text{avoided}}} \right]$$

$$(7.3)$$

The costs of CCS are the additional costs associated with the installation and operation of a CO$_2$ capture facility. The amount of CO$_2$ avoided represents the net CO$_2$ reduction taking into account energy consumption losses. Thus, the specific avoidance costs correspond to the specific carbon credit that would be required to match the present value of the respective capital and operating expenses. Rao and Rubin introduced a commonly used method for the calculation of the avoidance costs.[39] As eqn (7.3) indicates, it relates the costs of electricity (COE) generation with and without CO$_2$ capture to the respective emission reduction.

Table 7.5 summarizes investment and operation costs as well as the costs for electricity generation of the coal fired reference power plant given in Table 7.1. The costs were defined and agreed in a joint effort by power plant and utility providers within the NanoGLOWA project.

Table 7.6 provides the technical and economic assumptions used for the design and costing of the membrane gas separation process.

The applied cost estimates are in line with the recent work by Merkel.[31] A replacement interval of 5 years is considered for membranes. The book life of components other than membranes, such as compressors, vacuum equipment,

Table 7.5 Economics of the reference power plant without CO$_2$ capture

Total plant cost	*euro/kW$_{el}$*	*1500*
Cost of electricity (COE)	euro/MWh$_{el}$	40.30
Capital	euro/MWh$_{el}$	20.50
Fuel	euro/MWh$_{el}$	15.20
O&M	euro/MWh$_{el}$	4.60
Book life	years	20
Interest rate	%	8
Operation factor	—	0.85

Table 7.6 Economics of membrane gas separation process

	Isentropic efficiency	*Mechanical efficiency*	*Specific costs*
Compressors	0.90	0.96	500 euro/kW
Vacuum equipment	0.75	0.96	1000 euro/kW
Expander	0.75	0.96	500 euro/kW
Membrane	—	—	50 euro/m^2
Membrane piping etc.	—	—	30 euro/m^2

expanders, membrane frames and piping, is assumed to be equal to the book life of the reference power plant. The annual operation and maintenance costs (O&M) of the capture facility account for 6% of the respective capital investment excluding membranes.[28] The interest rate and operation factor correspond to those of the reference power plant.

7.5.3 Process without Retentate Recycling

Figures 7.3 to 7.6 represent the KPIs as function of the overall CO_2 recovery for different feed pressures in both stages. The permeate pressure in both stages was fixed to 0.2 bar. The stream at the low pressure side of the second stage exhibits the targeted CO_2 concentration.

Figures 7.3 to 7.6 indicate:

- The feed pressure of the first stage affects energy consumption, CO_2 recovery and membrane area considerably (Figures 7.3 and 7.4). This fact points out a dominating role of the first membrane stage within the two stage process.
- Higher feed pressures in both stages enhance the overall CO_2 recovery. This effect on the recovery and the membrane area diminishes with increasing pressure level (Figures 7.3 and 7.4).

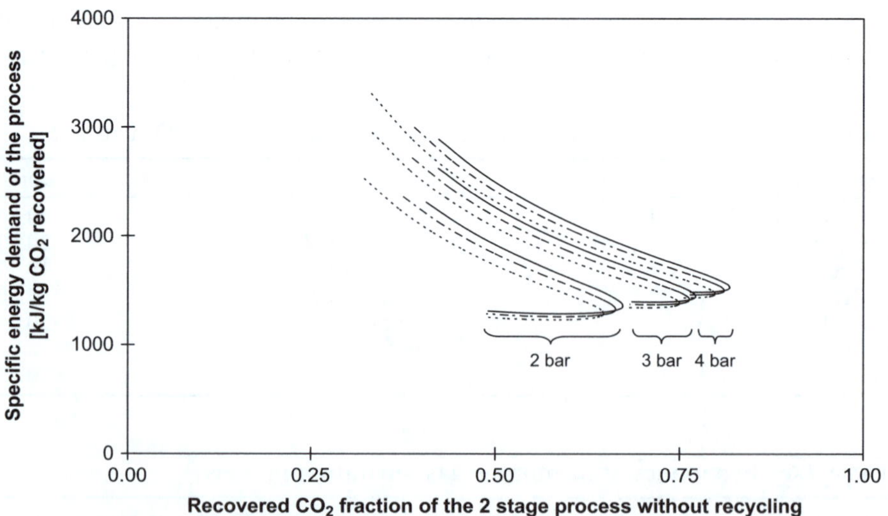

Figure 7.3 Specific energy demand of a two-stage membrane process without retentate recycling as a function of the overall CO_2 recovery for different flue gas pressures (curly brackets) and different feed pressures of the second stage (dotted line: 3 bar, dashed line: 4 bar, solid line: 5 bar); 0.2 bar permeate pressure in both stages.

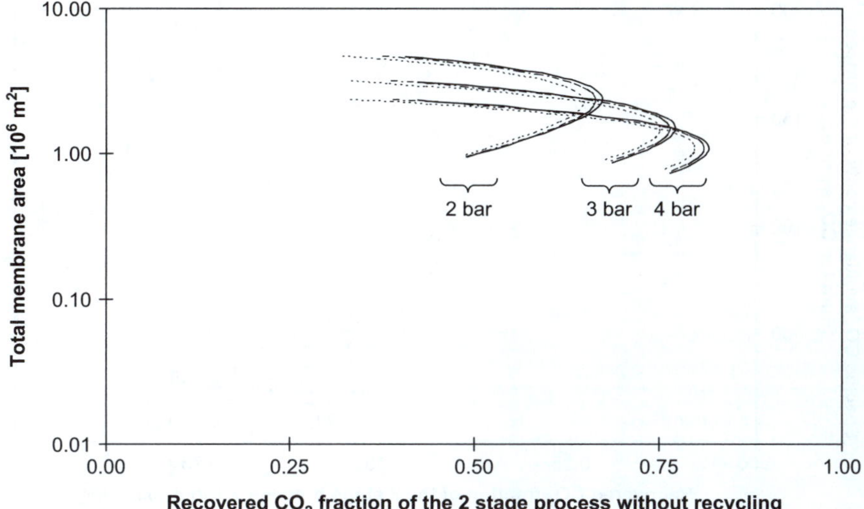

Figure 7.4 Total membrane area of a two-stage membrane process without retentate recycling as a function of the overall CO_2 recovery for different flue gas pressures (curly brackets) and different feed pressures of the second stage (dotted line: 3 bar, dashed line: 4 bar, solid line: 5 bar); 0.2 bar permeate pressure in both stages.

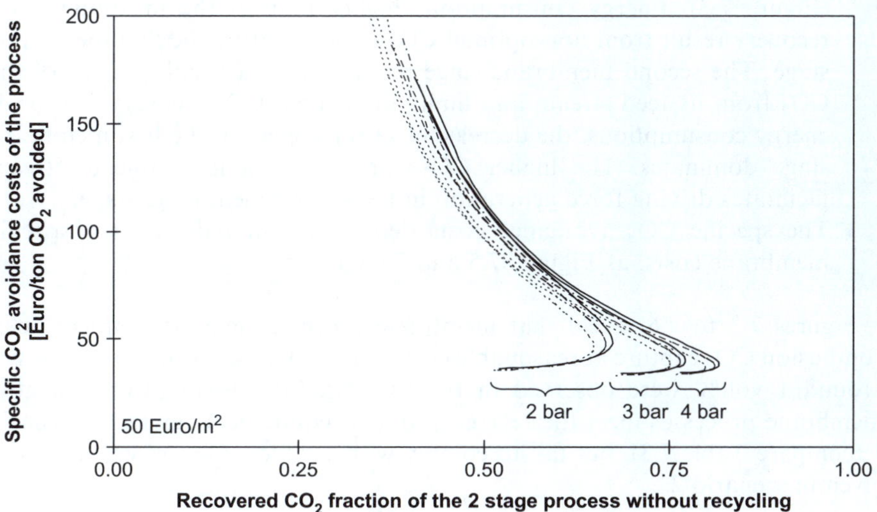

Figure 7.5 Specific CO_2 avoidance costs of a two-stage membrane process without retentate recycling as a function of the overall CO_2 recovery for different flue gas pressures (curly brackets) and different feed pressures of the second stage (dotted line: 3 bar, dashed line: 4 bar, solid line: 5 bar); 0.2 bar permeate pressure in both stages; 50 euros per m^2 membrane.

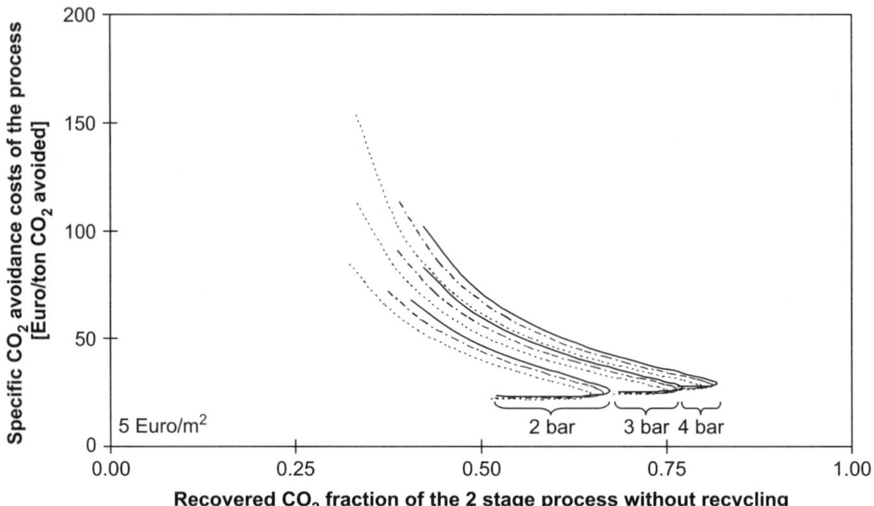

Figure 7.6   Specific CO_2 avoidance costs of a two-stage membrane process without retentate recycling as a function of the overall CO_2 recovery for different flue gas pressures (curly brackets) and different feed pressures of the second stage (dotted line: 3 bar, dashed line: 4 bar, solid line: 5 bar); 0.2 bar permeate pressure in both stages; 5 euros per m^2 membrane.

- A maximum CO_2 recovery exists for each combination of feed pressures (Figure 7.3). Energy consumptions higher than at the maximum CO_2 recovery result from non-optimal CO_2 enrichment by the first membrane stage. The second membrane stage recovers only a small amount of the CO_2 from its feed stream and limits the overall CO_2 recovery. At lower energy consumptions, the decreasing CO_2 recovery of the first membrane stage dominates. The higher CO_2 enrichment in its permeate stream facilitates driving force generation in the second membrane stage.
- The specific CO_2 avoidance costs depend substantially on the specific membrane costs, as Figures 7.5 and 7.6 indicate.

Figures 7.3 to 7.6 reveal that membrane processes may accomplish post-combusion CO_2 capture at reasonable costs. Nevertheless, O_2 concentrations of around 1 vol.% were observed in the CO_2 product stream. Consequently, membrane processes meet the relaxed product requirements given in scenario 1 (compare Table 7.3), but fail to comply with very low O_2 concentrations as given in scenario 2.

7.5.4   Process with Retentate Recycling

Figures 7.7 to 7.10 present the technical and economic KPIs as function of the overall CO_2 recovery for a two stage membrane process with retentate recycling.

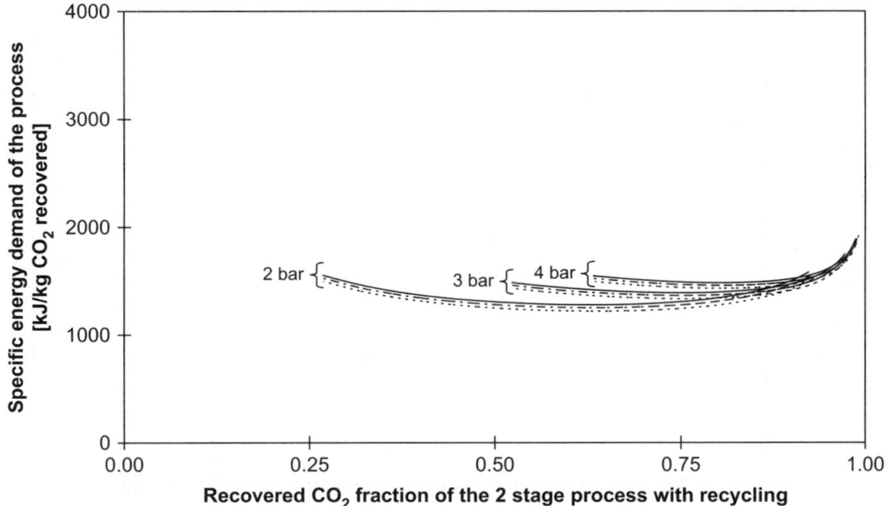

Figure 7.7 Specific energy demand of a two stage membrane process with retentate recycling as a function of the overall CO_2 recovery for different flue gas pressures (curly brackets) and different feed pressures of the second stage (dotted line: 3 bar, dashed line: 4 bar, solid line: 5 bar); 0.2 bar permeate pressure in both stages.

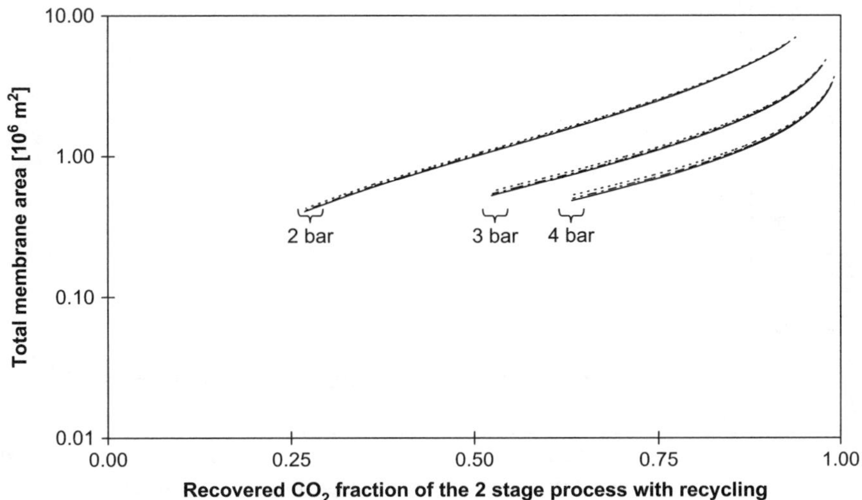

Figure 7.8 Total membrane area of a two-stage membrane process with retentate recycling as a function of the overall CO_2 recovery for different flue gas pressures (curly brackets) and different feed pressures of the second stage (dotted line: 3 bar, dashed line: 4 bar, solid line: 5 bar); 0.2 bar permeate pressure in both stages.

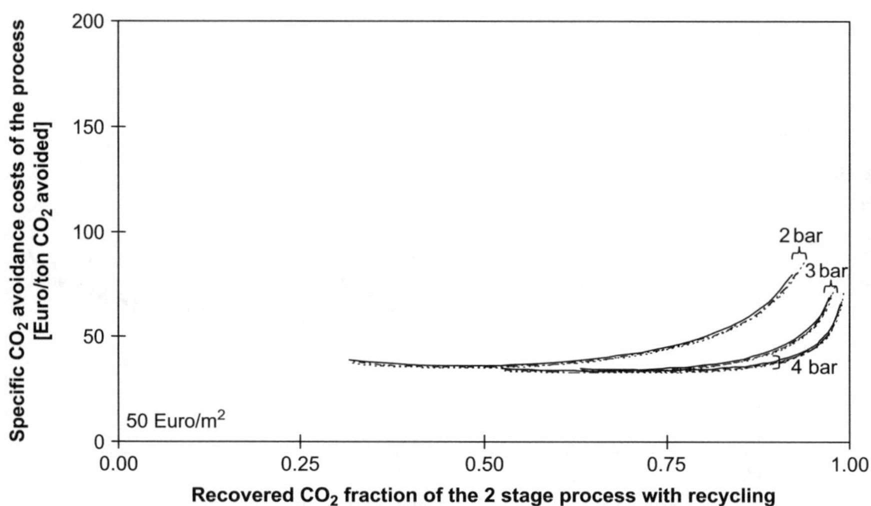

Figure 7.9 Specific CO_2 avoidance costs of a two stage membrane process with retentate recycling as a function of the overall CO_2 recovery for different flue gas pressures (curly brackets) and different feed pressures of the second stage (dotted line: 3 bar, dashed line: 4 bar, solid line: 5 bar); 0.2 bar permeate pressure in both stages; 50 euros per m^2 membrane.

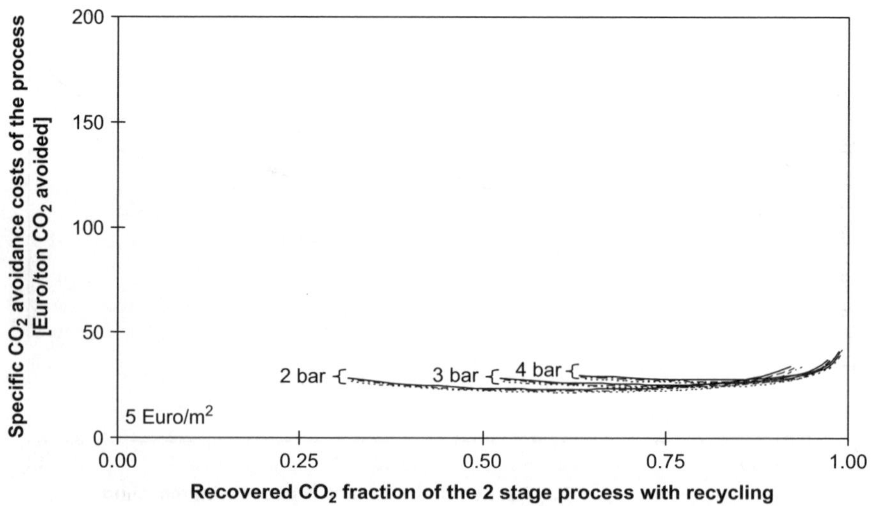

Figure 7.10 Specific CO_2 avoidance costs of a two-stage membrane process with retentate recycling as a function of the overall CO_2 recovery for different flue gas pressures (curly brackets) and different feed pressures of the second stage (dotted line: 3 bar, dashed line: 4 bar, solid line: 5 bar); 0.2 bar permeate pressure in both stages; 5 euros per m^2 membrane.

The analysis of Figures 7.7 to 7.10 leads to the following conclusions:

- Residual recycling improves the overall CO_2 recovery. CO_2 capture up to 97% is possible, but at the expense of specific energy consumption (Figure 7.7). Furthermore, recycling increases the volume flows in the process, which results particularly at high CO_2 recoveries in larger membrane areas (Figure 7.8).
- The flue gas pressure has a significant impact on all KPIs, while the feed pressure of the second membrane stage is less important.
- In analogy to non-recycling systems, the specific membrane costs have a considerable impact on the process economics (Figures 7.9 and 7.10). At high membrane costs (50 euro per m^2) more energy intensive configurations (4 bar flue gas pressure) yield the lowest costs at high CO_2 recoveries. At low membrane costs, however, there are only slight differences among the different configurations. Since the application of residual recycling leads to a flat minimum of the avoidance costs, capture levels between 80 and 95% seem attractive.

Nevertheless, O_2 concentrations of 1.1 to 1.5 vol.% were observed in the CO_2 product stream. Thus, in analogy to membrane processes without retentate recycling, membrane systems with retentate recycling fail to meet strict O_2 concentration limits as required for enhanced oil recovery.

7.6 Competing Technologies

Since amine absorption is presently considered the best available technology for post-combusion CO_2 capture, it sets the benchmark for membrane processes. In relation to amine absorption (about 1000 kJ kg^{-1} CO_2), membrane cascades feature in general a higher specific energy demand (Figures 7.3 and 7.7).[40] Moreover, amine processes comply with strict purity requirements regarding co-captured species and they yield a CO_2 product concentration, which is higher than with membrane processes. Nevertheless, while the steam supply for solvent recovery in MEA processes requires modifications of the steam cycle or even the replacement of the low pressure steam turbines, the integration of membrane processes is much easier.[41,42]

7.7 Concluding Remarks

This chapter provided an integral view on membrane based post-combusion CO_2 capture from coal fired power plants. Since membranes generally do not feature the same selectivity towards CO_2 as chemical absorption processes, a precise specification of the boundary conditions up- and downstream of the membrane unit was performed first. While the flue gas condition have been derived on the basis of a simulation, literature presently provides only little tangible information regarding CO_2 purity requirements for pipelining and storage. The ongoing uncertainty with respect to technical and legal issues resulted in the derivation of different CO_2 purity scenarios.

Because flue gas temperatures are very moderate at the outlet of the FGD unit, polymeric membranes appear most suitable for post-combusion CO_2 capture. A PEO membrane, which is currently investigated in the Nano-GLOWA project, was chosen for the techno-economic analysis of membrane based capture processes. In addition to the membrane material itself the driving force for the CO_2 transfer across the membrane constitutes the second critical issue. Particularly the ambient pressure of the flue gas stream in conjunction with the rather low CO_2 concentrations make efficient driving force generation a tricky task. Flue gas compression appears a feasible technology for driving force generation, regardless of the volume stream and the required end pressure. The large scale generation of low pressures at the permeate side exhibits a limit at about 200 mbar. The combination of flue gas compression and suction at the permeate side appears presently the most practical option. Sweeping with steam originating from the power plant encounters substantial problems in terms of practical implementation.

The techno-economic analysis of two stage membrane processes for post-combusion CO_2 capture revealed that CO_2 capture by membranes appears to be feasible. Nevertheless, the CO_2 product stream contains at least 1 vol.% O_2, thus violating strict purity requirements, *e.g.* for enhanced oil recovery. Membrane costs substantially affect the optimal operation mode of the capture process. Residual recycling is particularly attractive to enhance overall CO_2 recovery. Since the application of residual recycling leads to a flat minimum of the avoidance costs, capture levels between 80% and 95% seem to be most attractive.

Acknowledgements

This study is part of the NanoGLOWA project (NMP3-CT-2007-026735), which has been financially supported by the EU Commission within the thematic priority NMP of the Sixth Framework Program.

References

1. P. G. Cifre, K. Brechtel, S. Hoch, H. García, N. Asprion, H. Hasse and G. Scheffknecht, *Fuel*, 2009, **88**, 2481.
2. M. R. M. Abu-Zahra, L. H. J. Schneiders, J. P. M. Niederer, P. H. M. Feron and G. F. Versteeg, *Int. J. Greenhouse Gas Contr.*, 2007, **1**, 37.
3. B. Metz, O. Davidson, H. de Coninck M. Loos and L. Meyer, *IPCC Special Report on Carbon Dioxide Capture and Storage*, Cambridge University Press, Cambridge, 2005.
4. C. B. Farris, *Energy Prog.*, 1983, **3**, 150.
5. C. B. Wallace, *Oil Gas J.*, 1985, **24 June**, 98.

6. M. Mohitpour, H. Golshan and A. Murray, *Pipeline Design & Construction: A Practical Approach*, The American Society of Mechanical Engineers, New York, NY, 2nd edn, 2003.

7. E. de Visser, C. Hendriks, M. Barrio, M. J. Molnvik, G. de Koeijer, S. Liljemark and Y. Le Gallo, *Int. J. Greenhouse Gas Contr.*, 2008, **2**, 478.

8. R. Svensson, M. Odenberger, F. Johnsson and L. Strömberg, in *Greenhouse Gas Control Technology, Proceedings of the 7th International Conference*, ed. M. Wilson, T. Morris, J. Gale and K. Thambimuthu, Elsevier, Amsterdam, 2004, p. 2531.

9. H. Li and J. Yan, ASME Turbo Expo: Power Land, Sea and Air, 2006, GT2006-90954.

10. S. Bachu, *Prog. Energy Combust. Sci.*, 2008, **34**, 254.

11. S. Bryant and L. Lake, in *Carbon Dioxide Capture for Storage in Deep Geologic Formations*, ed. D. C. Thomas and S. M. Benson, Elsevier, Amsterdam, 2005, vol. 2, p. 983.

12. E. Johnson and F. B. Walter, *Hydrocarbon Process*, 1985, **October**, 62.

13. G. Liljedahl, D. Turek, N. Nsakala, C. Mohn and T. Fout, *31st International Technical Conference on Coal Utilization & Fuel Systems*, 21–25, May 2006, Clearwater, Florida.

14. S. Holloway, *Energy*, 2005, **30**, 2318.

15. D. A. Voormeij and G. J. Simandl, *Geosci. Can.*, 2004, **31**, 11.

16. B. Z. Saylor and B. Zerai, in *Energy, Waste, and the Environment: A Geochemical Perspective*, ed. R. Gieré and P. Stille, Geological Society, London, Special Publications, 2004, **236**, 285.

17. S. Bachu and D. Bennion, *Int. J. Greenhouse Gas Contr.*, 2009, **3**, 458.

18. IEA GHG, *Impact of Impurities on CO$_2$ Capture, Transport and Storage*, Report Number PH4/32, August 2004.

19. T. Melin and R. Rautenbach, *Membranverfahren: Grundlagen der Modul- und Anlagenauslegung*, 3, aktualisierte u. erw. Aufl., Springer Verlag, Berlin, 2007.

20. S. R. Reijerkerk, R. Jordana, K. Nijmeijer and M. Wessling, *Int. J. Greenhouse Gas Contr.*, 2010, doi:10.1016/j.ijggc.2010.06.014

21. R. Rautenbach and A. Struck, *Chem. Ing. Technol.*, 1996, **68**, 290.

22. B. Klinkhammer, C. Matthias and T. Melin, *Chem. Ing. Technol.*, 2003, **75**, 1598.

23. M. Ajhar, M. Follmann, C. Matthias and T. Melin, *J. Membr. Sci.*, 2008, **323**, 105.

24. L. Zhao, E. Riensche, R. Menzer, L. Blum and D. Stolten, *J. Membr. Sci.*, 2008, **325**, 284.

25. L. Zhao, R. Menzer, E. Riensche, L. Blum and D. Stolten, *Energy Proc.*, 2009, **1**, 269.

26. R. Bounaceur, N. Lape, D. Roizard, C. Vallieres and E. Favre, *Energy*, 2006, **31**, 2556.

27. E. Favre, *J. Membr. Sci.*, 2007, **294**, 50.

28. M. T. Ho, G. W. Allinson and D. E. Wiley, *Ind. Eng. Chem. Res.*, 2008, **47**, 1562.

29. D. Yang, Z. Wang, J. Wang and S. Wang, *Energy Fuels*, 2009, **23**, 4755.
30. J. Kotowicz, T. Chmielniak and K. Janusz-Szymañska, *Energy*, 2010, **35**, 841.
31. T. C. Merkel, H. Lin, X. Wei and R. Baker, *J. Membr. Sci.*, 2010, **359**, 126.
32. L. Zhao, E. Riensche, L. Blum and D. Stolten, *J. Membr. Sci.*, 2010, **359**, 160.
33. J. P. Van Der Sluijs, C. A. Hendriks and K. Blok, *Energy Convers. Mgmt.*, 1992, **33**, 429.
34. P. M. Follmann, C. Bayer, M. Wessling and T. Melin, *Driving force generation for membrane based post combustion CO_2 capture from coal fired power plants*, in preparation.
35. R. Carapellucci and A. Milazzo, *Proc. Inst. Mech. Eng., Part A*, 2003, **217**, 505.
36. R. Rautenbach and R. Albrecht, *Membrane Processes*, John Wiley and Sons, Chichester, 1989.
37. R. Agrawal and J. Xu, *J. Membr. Sci.*, 1996, **112**, 129.
38. C. Pan and H. Habgood, *Can. J. Chem. Eng.*, 1978, **56**, 197.
39. A. B. Rao and E. S. Rubin, *Environ. Sci. Technol.*, 2002, **36**, 4467.
40. P. M. Follmann, C. Bayer, R. File, L. Peters and T. Melin, *Hybrid systems for CO_2 capture from coal fired power plants*, presented at the International Conference on Applied Energy (ICAE 2010), 21–23 April 2010, Singapore, Conference proceedings, pp. 469–478.
41. M. Lucquiaud, H. Chalmers and J. Gibbins, *Energy Proc.*, 2009, **1**, 1411.
42. Ö. Korkmaz, G. Oeljeklaus and K. Görner, *Energy Proc.*, 2009, **1**, 1289.

Commercial Applications of Membranes in Gas Separations

PUSHPINDER S. PURI

PuriMem, LLC, Emmaus, Pennsylvania, USA

8.1 Gas Separation Membrane Systems

Commercial scale gas separations[1] using membrane systems were applied for the first time in late 1970s to early 1980s. Their applications were limited to separation of hydrogen from petroleum refineries and in the H_2/CO ratio adjustments for the synthesis gas. Since then, membrane-based gas separation systems have made tremendous progress and have gained wider acceptance in a variety of applications. The sustainability and energy area has also attracted attention to gas separation membranes. Apart from the traditional energy area applications such as natural gas processing, membranes are gaining interest in CO_2 recovery and sequestration, and energy generation areas. Numerous membrane gas separation systems are in operation today for a wide variety of separation applications. Although much research is under way in developing novel membrane materials, only polymeric membranes are commercially used in large scale. Therefore, the discussion in this chapter is limited to the application of polymeric membranes only. There are three generic membrane gas separation applications: gas generation, gas separation and gas purification. A list of these generic membrane gas separation applications is given in Table 8.1. Most of the current commercial applications are summarized in Table 8.2 and are described in details in this chapter.

Membrane Engineering for the Treatment of Gases, Volume 1:
Gas-separation Problems with Membranes
Edited by Enrico Drioli and Giuseppe Barbieri
© Royal Society of Chemistry 2011
Published by the Royal Society of Chemistry, www.rsc.org

Table 8.1   Generic membrane separation systems

Application	Examples
Gas generation	Nitrogen production
	Helium recovery
	Hydrogen recovery
	Monomer recovery
Gas separation	Syngas (H_2/CO) ratio adjustment
	Natural gas dehydration
	Removal of acid gases from natural gas
	Air drying
	Enriched oxygen air
Gas purification	Hydrogen purification

Table 8.2   Commercial applications of gas separation membranes

Separation area	Product	Application
Air	Nitrogen	Fruit and vegetable preservation
		Grain storage
		Inerting in chemical and petroleum processes
		Purging of pipes and vessels
		Enhanced oil recovery
Air	Dry air	Instrument air
		Spray paint shops
Air	Oxygen	Enriched oxygen air for combustion
SynGas	CO/H_2 ratio	Methanol synthesis
	Hydrogen	Hydrogen production
	Carbon monoxide	Carbon monoxide production
Ammonias syngas	Hydrogen	Hydrogen recovery and recycle
Petroleum refinery	Hydrogen	Hydrogen recovery in petrochemical processes
Olefin polymerization	Olefins	Recovery and recycle of monomers
Natural gas	Pipeline Grade Gas	Acid gas removal
		Natural gas dehydration
Greenhouse gases	Clean Air	CO_2 removal from stack gases
		CO_2 removal from pre-combustion gases
		Removal of chlorofluoro hydrocarbons
		SO_x and NO_x removal from exhaust gases
Carbon dioxide/ hydrocarbons	Carbon dioxide	Enhanced oil recovery
Helium	Helium	Helium recovery and recycle
Hydrogen	Hydrogen	Ultrapure H_2 by metallic (Pd-alloys) membranes

8.1.1   Gas Separation Processes

Separation of gases at commercial scale is done by the technologies like cryogenic distillation, absorption, adsorption and membranes. Of these, cryogenic separations are universally used for the large scale separation of atmospheric gases and absorption technologies are well established for the scrubbing of carbon dioxide and removal of water from natural gases. In the cryogenic distillation of a gas mixture, the gases are liquefied and then distilled at cryogenic temperatures to separate it into its components. In gas absorption process one or more components of the gas mixture is preferably absorbed in a chemical or physical solvent in a gas–liquid contacting device. The separation of the non-absorbing/non-reacting gases from the gas mixture is thus achieved. Both these methods are complex and capital intensive but cost competitive.

Adsorption and membrane technologies for gas separation have emerged over the last three decades and are now widely accepted in a few commercial applications. These two technologies have competed neck to neck and have their own pros and cons. The adsorption process, similar to its counterpart absorption process, uses high surface area solids to contact gas mixture where one component of the gas mixture adsorbs leaving others behind. In this gas–solid contact process there are several material and process challenges which need to be overcome. Whereas development of adsorbents and the design of adsorption cycles are within the main stream of chemical engineering, adsorption separations have an inherent disadvantage of being a cyclic process therefore for continuous separation of gases, one need to use more than one adsorbent bed. This is because an adsorbent bed needs to be regenerated after saturation (or breakthrough) has occurred. Adsorbent processes are best suited to produce high purity gases, especially for the removal of trace gas impurities from a gas mixture.

Membrane gas separation processes are simple and operate at a steady state requiring little attention under smooth operations. Gas mixture is fed on one side of the membrane, one or more components of the gas mixture preferably permeates through the membrane, leaving the residual impermeable gases as retentate. The concentration driving force for the permeating components is achieved either by compressing the feed gas or using a vacuum on the permeate side. Because it is not possible to have membrane materials which will be totally impermeable to certain gases, membrane gas separation process is not best suited to produce high purity gases. Multi-stage gas separation and membrane–hybrid processes have been developed to produce high purity gases using membrane systems. Membranes are very attractive in applications where product demand is not constant. Membrane separation processes offer easier scale-up and scale-down of the production rates. This is because membrane systems are made up of several banks of membrane modules which can be easily added or taken off the operation without causing efficiency losses or process upsets. Membrane systems also offer operational flexibility with respect to the product purification and production rates.

8.1.2 Polymeric Gas Separation Membrane Systems

Gas separation membrane systems offer several advantages, these are: reliability, flexibility, high efficiency, low cost, quiet, reliable and simple operation over broad operating conditions (temperature up to 90 °C and pressure up to 8000 kPa).

8.1.2.1   Membrane Materials

A wide variety of materials (polymers, zeolites, ceramics and metals) have been reported for various gas separation applications. Polymeric membranes are the only ones which are extensively used. Inorganic and metallic membranes have limited applications in the gas separations. Palladium alloy membranes have been applied for the purification of hydrogen and mixed metal oxide membranes have been developed for high temperature air separations by ion transport mechanism. The state-of-the-art of gas separation membrane materials is presented in a recent review article.[2]

In spite of this vast wealth of the membrane materials, only a handful of materials are used to make commercial membranes. This is because, while permeability and selectivity of the membrane material is important, it is not the only requisite for the commercial membranes. For a material to be commercial, in addition to its separation characteristics, it must be amenable to make into a commercial membrane in an inexpensive and reproducible process. The membrane material must have certain chemical, physical, mechanical and thermodynamic properties to qualify. It must be able to do a wide variety of separations and stable at the use conditions.

Much of the effort of the membrane and polymer scientists has been in the development of the new membrane materials of higher selectivity and permeability; sufficient attention has not been given to the prerequisites of the materials to meet the fabrication process needs. In addition to the polymer structure properties, orientation and packing of the polymer molecules achieved by the use of unique chemistry and/or processing conditions during the membrane fabrication can led to membranes of superior separation properties. The evidence of this is available in the published work.[3–5] Much of the practical information in this area of membrane fabrication is kept proprietary by the commercial membrane manufacturers. Understanding of the impact of polymer scale-up and membrane fabrication processes for the new materials is also important.

Generalization of the material issues which need to be overcome for all gas separation applications is difficult; however there are a few specifics which can be pointed out. These are the ability to: (i) withstand high pressure and pressure cycling, (ii) operate over a wide range of temperatures, (iii) resist chemical reactions with the trace impurities, and (iv) resist plasticization. Membrane and membrane material compatibility generally is satisfactorily matched with the pressure, temperature and chemical resistance. However, the plasticization of the membrane by the gases to be separated or by the contaminants present in the gas mixture still remains to be solved. Plasticization of the membrane over the course of its use decreases their separation ability and may also adversely affect the gas flux due the compaction of the membrane sub-layer over which a thin gas separation film may be supported. The plasticization of polymeric membranes by high pressure CO_2 is well documented.[6] Cross-linking highly branched polymers is one of the methods practiced by the membrane scientists to produce membranes which are resistant to plasticization by CO_2.[7]

In one recent report, aromatic polyimide polymer membranes cross-linked at elevated temperatures (up to 450 °C) showed resistance to plasticization to at least up to 3000 kPa (450 psia) CO_2.[8] Achieving uniform high temperature cross-linking of polymers in commercial membrane manufacturing has its own challenges. Solution-based chemistry to cross- link membrane may be more commercially amenable than the high temperature thermal rearrangements.

Zeolite membranes have also been proposed for CO_2/CH_4 separation as an alternative to solving plasticization problems associated with polymeric membranes. These membranes offer superior thermal, mechanical and chemical stability and good erosion resistance and stability at high CO_2 pressures.[9] However, economic production of defect-free zeolite membranes at commercial scale still remains a challenge.

Ion transport membranes made using mixed metal oxides have been developed to produce pure oxygen at elevated temperatures. These membranes are now at tonnage quantity demonstration stage. These membranes are targeted to compete with adsorption and cryogenic methods of large scale production of oxygen.

8.1.2.2 Membrane Fabrication

The majority of gas separation applications employ glassy polymers because of their high gas selectivity and good mechanical properties. As pointed out in the previous section, in spite of a very large volume of work done on the development of specialty polymers for gas separation, only a handful of commercially available polymers are used to make membranes. This is because of two factors: first, scaling up of laboratory scale synthesis to polymer production at commercial levels is expensive; and, second, specialty polymers produced in small quantities are much more costly than the large scale commercially produced polymers. Further, the polymer variability in small batch productions is significant enough that it may hinder production of uniform and reproducible membranes. These factors need to be recognized by polymer researchers and developers.

A successful commercial membrane need to have a thin defect free and stable skin of a thickness less than 1000 Å in order to provide high gas flux. The ability to form large quantity of high quality membrane at reasonable costs is a prerequisite of a successful gas separation membrane commercialization. The manufacturing process must preferably use environmentally friendly solvents and coagulants for a safe and economical commercial production. The post-modification process of membrane to achieve selective properties (such as chemical or physical surface modification or cross-linking), should be simple and must yield uniform membrane properties all through the membrane surface. Any non-uniformity in the membrane surface is equivalent to defects that contributes to non-uniform gas flux through the membrane resulting in membrane selectivity drop. Fortunately, the surface skin of majority of the glassy polymer membranes can be repaired by coating a high permeability polymer (generally a rubbery polymer) on the thin gas separation skin of the membrane.[10,11]

The majority of the commercial gas separation membranes are made by wet phase inversion method which results in an integrally skinned asymmetric membrane. This method was first used by Loeb and Sourirajan[12] to produce cellulose acetate membranes for desalination of sea water. An alternative method for making gas separation membranes uses an ultra-porous skinned asymmetric membrane over which a thin polymer film is deposited by either coating or by interfacial polymerization. This method was developed by Cadotte[13] for the creation of *in situ* dense skin thin film composite membranes for water desalination. These membrane fabrication techniques were made commercially successful for gas separation membranes by a brilliant empirical discovery for *in situ* sealing of the tiny pinhole defects on the skin of the membrane.[10,11]

8.1.2.3 *Membrane Modules*

Membranes are packaged in membrane modules and provided channels to separate the feed gas from the permeate gas. A module also acts as a pressure vessel for the high pressure applications. In some module designs additional ports are provided for the use of a sweep gas on the shell side. One of the key factors in the membrane module design is to provide maximum possible area density in the module, barring the pressure drop considerations. For the flat sheet membranes the preferred module design is a spiral wound module, where several membrane leaves are wound across a central mandrill in which the permeate is collected. The retentate flows cross-flow to permeating gas. Spiral wound modules are made by semi-automatic processes, have lower area per unit volume of the module and are expensive to make. The hollow fiber membranes, the fibers are gathered as a bundle either by winding the fibers on a mandrill or are gathered as a jelly roll and their ends are potted to open the lumen side of the fiber, thus separating the inside volume of the membrane from the outside. Hollow fiber membrane modules are relatively easy to make and the manufacturing process can be easily automated. The surface area per unit volume for the hollow finer can be very high (500–1000 m^2 m^{-3}) and depends on the fiber dimensions. Membrane mechanical properties especially the tensile strength and abrasion resistance are important to prevent any damage to the membrane during the fabrication process.

Marriott and Sorensen[14] have developed models for the design of hollow fiber and spiral wound modules that are applicable to generic gas separations. Module design and process design are important for a membrane material to operate at its intrinsic value. Membrane module design field has marginally advanced over the last three decades. Better membrane module configurations, especially those which can operate in multi-stage mode by internal manipulations of gas streams to minimize external connections, fittings to lower fabrication cost and make operation safer, are desired.

8.1.2.4 Membrane Systems

Design of membrane separation system is a little more complex and to some extent limited in comparison of the other separation processes. In addition to the selection of a given membrane material and module design, all membrane gas separation systems require some type of pre-treatment to get rid of all the particulates, aerosols, oil and organic condensable vapors. The feed gas may be heated or cooled depending on the feed stock and the nature of separation. Theoretically, a membrane will operate at its best if the feed stream consists nothing but the gases which need to be separated.

One key challenge in the gas separation membrane system design is how to produce desired quality gases economically by using membranes which do not have very high selectivity. Another challenge, especially for the high flux gases and vapors, is to design a membrane module and system which will permit membrane to function at or close to the intrinsic properties of the membrane material itself. Further, all other factors, such as gas mixing, mass transfer and heat transfer which alter effective membrane selectivity in a gas separation module need to be minimized.

Although there are no universal design rules which can be adopted for all membrane gas separation applications, the essential components which need to be selected in most of the membrane gas separation systems are as follows:

1. The membrane material and all other parts of the membrane module which come in contact with the gases should be stable and chemically compatible not only with the gas mixture to be separated but also with the trace impurities present therein. All materials used in the membrane module should also have sufficient mechanical properties to resist any creep at operating temperature and pressure. The membrane should offer high gas flux and high selectivity for compact design, higher gas purity and lower gas losses.

2. A pre-treatment process for cleaning the feed gas is needed for stable and reliable operation of the membrane module. Ideally, a membrane system operates best when the gas feed is free of all entrained liquids and particulate matter.

3. The membrane geometry and module design selection should be based on feed composition and separation conditions (feed pressure and temperature) for optimum use of the membrane module. A flat sheet spiral wound module may be preferred in certain applications over a hollow fiber module and *vice versa*.

4. The membrane system configuration to achieve a given separation needs to be carefully designed. A membrane system may contain more than a single stage operation. In most of the applications, a multi-stage operation with intermediate compression, permeate sweep and feed or product proportion amongst different stages offers a better separation system. The genius of membrane system cascade design is one of the keys to the successful deployment of the membrane modules.

5. A membrane separation system may sometimes be coupled with another gas separation unit operation (such as adsorption, absorption or cryogenic distillation) to obtain an economically optimum hybrid process.

There are several membrane system design and simulation tools used to determine the membrane area needed for a given separation. These tools can also be used for the design of the multi-stage membrane systems. These tools are based on the equations developed to estimate the membrane area for a given gas separation. The development of these equations and the assumptions used are well documented in the literature.[15] The earliest tool was an Excel®-based spreadsheet using basic design equations for counter current and cross-current flows.[16] Presently several proprietary HYSIS® and Aspen Plus® modules are available for the membrane system design which are used by the membrane system designers. Since the major membrane module vendors are unwilling to share details of their membranes, it is essential for the process engineers to work closely with the membrane module manufacturers to develop membrane systems for their applications.

Membrane gas separations systems can also be purchased that are supplied by the vendors as fully instrumented skids and require only tie-in with the existing processes. Examples of commercial scale skids for three applications are shown in Figure 8.1.

8.1.2.5 *Staged and Cascaded Systems*

We pointed out earlier that gas separation membranes are not most suited for producing high purity gases. This is because selectivity of the various membranes for a given gas pair is generally limited. For a single stage system

Figure 8.1 Commercial gas separation membrane systems.

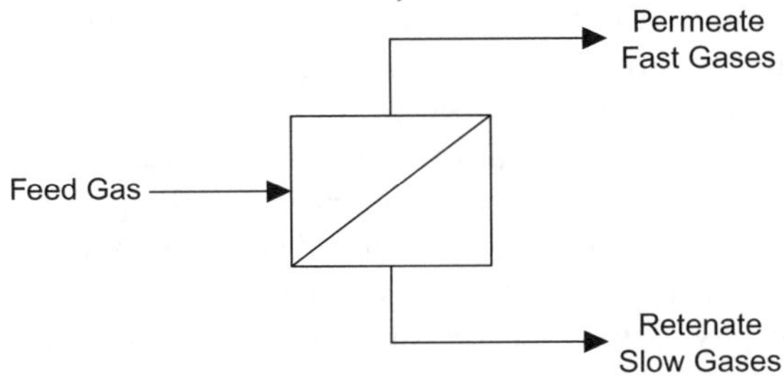

Figure 8.2　Basic membrane system.

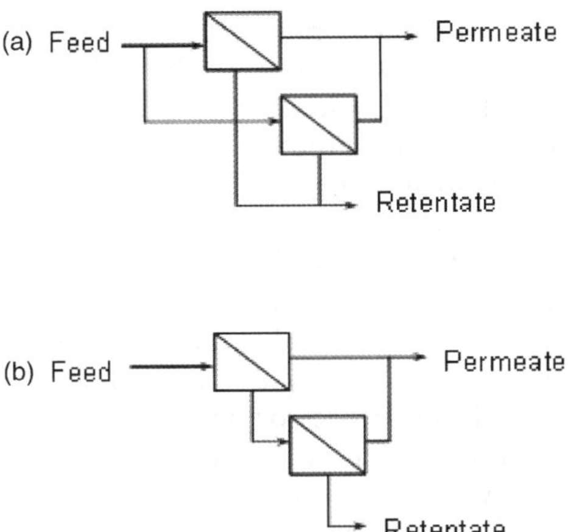

Figure 8.3　(a) Parallel feed, (b) staged feed.

schematics of which is shown in Figure 8.2, the pressurized feed gas is fed on one side of the membrane and due to pressure driving force more permeable gases permeate through the membrane leaving retentate of less permeable gases.

The quality of the permeate and retentate gases depends on the selectivity of the membrane. In a commercial system several modules are stacked either in parallel to increase the production capacity or cascaded in a certain fashion to get higher purity gases. An example for two membrane module system stacked in parallel and cascaded is shown in Figure 8.3. In the design of the multi-stage operations, each stage must be operated to obtain the highest possible efficiency. This can be done by the manipulation of the membrane selectivity and productivity, pressure driving force and the stage cut.

Different schemes therefore can be derived by arranging the modules in series and parallel and by directing product or part of the product from one stage to sweep some other stage. Likewise, the permeates from some modules may be compressed and mixed with a feed to some other stage or separated in a subsequent stage. In fact the concept of manipulating various gas streams in multi-stage operations to overcome the limitation of low selectivity for the membrane systems is widely used.[17–20] The disadvantage of the multi-stage processes is that the capital cost may be high due to the use of a number of modules and the interstage compressors.

Most commercial gas separation membrane cascades use one or two compressors. The schemes are generally drawn by the genius and intuition of the developers. A systematic method for designing multi-stage cascade for membrane gas separation applications was proposed by Agrawal and Xu[21] to find an economically viable cascade for a given separation application.

8.1.2.6   *Hybrid Separation Processes using Membranes*

An alternative way to use membranes efficiently to produce high purity products is to use membranes as a part of the hybrid separation processes. We pointed out earlier that various separation processes can be employed to achieve a given separation. For example, nitrogen from air can be obtained by cryogenic, adsorption or membrane separations; acid gases can be removed from natural gas streams via scrubbing methods (such as using amines) or by the use adsorbents or membranes. Various separation processes can be combined to develop a better separation process where each of the separation process is operating at its optimum domain. The choice of the separation technology or combination technologies depends on the gas composition, amount of gas processed, gas processor and economic realities. The combination of various separation processes to achieve a given goal is referred as 'hybrid processes'. The basic advantages and disadvantages of a single separation technology *versus* the hybrid process vary from the separation to separation and depend on the quality and quantity of the gas processed.

The most common gas separation hybrid process is the membrane/ adsorption hybrid system which can be very cost effective if each of the separation unit is operated for what it is most suited. For example, for air separation, membranes are best suited for bulk separation and adsorption is preferred for the removal of low concentration of the contaminats, an adsorption unit followed by a membrane can offer a economical solution to produce high putrity nitrogen and oxygen. The hybrid processes are generally preferred for large size plants because of the capital and operation expense of two separate processes.

Hybrid processes are gaining momentum in the natural gas deacidification[19] and in the recovery and sequestration of CO_2 from stack gases. A hybrid process consisting of a single stage membrane system upstream of an amine absorber, schematic of which is shown in Figure 8.4, may be advantageous in some cases.

Because a membrane system operates best with higher CO_2 concentration feed and an amine system is best suited to produce pipeline spec gases.[22] For high volume units where two or more trains may be required, hybrid process can reduce the system to a single, less expensive train. Membranes can also be very useful in debottlenecking the existing units. There are many hybrid systems operating in the world today.

8.2   Major Gas Separation Membrane Producers

A list of the major membrane module and membrane systems manufacturers is given in Table 8.3. The name of the companies, markets they serve and type of polymer and membrane they use are listed.

This is a snap in time as the landscape of the membrane companies has been constantly changing through acquisitions and mergers. There are several other equipment manufacturers as well who supply small membrane systems for applications like nitrogen generators and air drying systems.

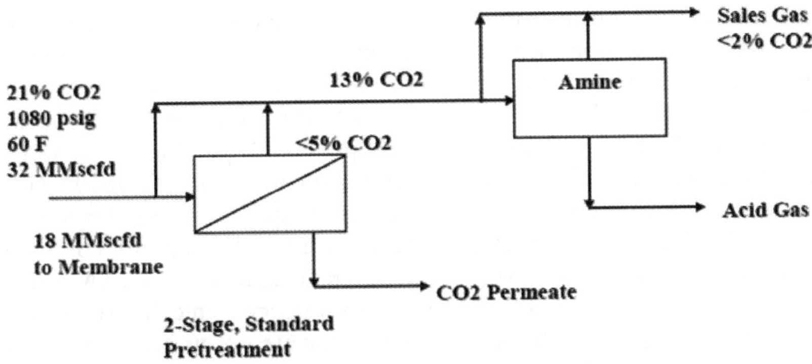

Figure 8.4   Membrane/absorption hybrid process (Courtesy: UOP, LLC a Honeywell Company).

Table 8.3   Major membrane manufactures

Company	Market	Polymer	Module type
Air Products (Prism)	H_2, N_2, H_2O, CO_2	Polysulfone and others	Hollow fibers
Medal (Air Liquide)	H_2, N_2, CO_2	Polymide, polyaramide	Hollow fibers
Generon (IGS)	N_2	Tetrabromo polycarbonate, TPX	Hollow fibers
Ube	H_2, N_2, H_2O	Polyimide	Hollow fibers
Parker (Aquilo)	O_2/N_2	Polyphenylene oxide	Hollow fibers
Cynara (Natco)	CO_2	Cellulose acetate	Hollow fibers
UOP (Separex + GMS)	CO_2	Polymide, cellulose acetate	Spiral wound
MTR	CO_2, olefins	Perfluoro polymers silicon rubber, polyimide	Spiral wound
Kvaerner	Gas/liquid contactors for CO_2	Polysulfone, PVDF, PTFF	Hollow fibers

8.3   Gas Separation Membrane Applications

8.3.1   Air Separation Membranes

The majority of the air separation is done by cryogenic distillation. Cryogenic processes are well established and are extremely energy efficient in large scale plants. Alternatively, adsorption processes, like pressure swing operation (PSA) using oxygen selective activated carbons for the production of nitrogen and pressure vacuum swing operations (PVSA) or vacuum swing operations (VSA) using nitrogen selective zeolite materials are used for the production of oxygen. Polymeric air separation membranes are by far the most accepted commercial membrane gas separation application and are used exclusively to generate nitrogen gas of low purities in small scale and remote locations. Polymeric membranes for nitrogen separation primarily compete against the PSA processes. Very few polymeric membrane systems are used for the production of oxygen because industrial oxygen applications require 90% O_2, which cannot be produced economically by commercial polymeric membranes. Inorganic membranes and specifically, ion transport membranes made with the compositions of various mixed metal oxide materials are being developed for oxygen production from air at elevated temperatures. Though not commercial, ion transport membranes are expected to compete against both VSA and cryogenic processes. Membrane/PSA and membrane/cryogenic hybrid processes have been proposed in the literature, though they are not used commercially as the excessive capital needs negate the energy savings.

In membrane air separation systems, the compressed air is fed on one side of the membrane; oxygen, argon, carbon dioxide and water which have higher permeation than nitrogen, permeate through the membrane leaving mostly nitrogen behind. The membrane selection and operating conditions used depend on the product gas quality requirements.

8.3.1.1   Nitrogen Generation System

All nitrogen generation systems, irrespective of the technology used, require a source of clean air which is fed to the separation unit to separate nitrogen from oxygen and other minority components in air. Membrane nitrogen generation systems are relatively simpler than the cryogenic and pressure swing adsorption (PSA) processes because membrane systems do not require drying of the feed air, have no moving parts and operate at a steady state. A typical nitrogen generation membrane skid consists of three main parts; air compressor, air purification and air separation membrane modules. A schematic of a membrane nitrogen production system is shown in Figure 8.5 and briefly described here.

Air compression: In air compression section, the atmospheric air is sucked through a depth filter to capture any particular matter in the incoming air by a single stage lubricated air compressor which is generally used to generate compressed air. The compressed air is laced with oil particulates and aerosol and is at a temperature about 80–100 °C and a pressure of about 1000 kPa

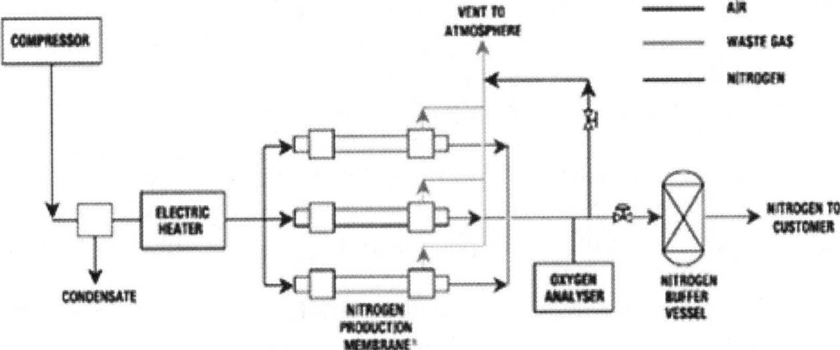

Figure 8.5 Schematic of nitrogen generation system (Reproduced with permission of Air Products).

(135–150 psi). The oil particulates and aerosols are removed using an oil filter. Oil-free compressors can also be used, but are not preferred. The lubricated oil compressors are preferred because they run at lower temperature and higher pressure than the oil free compressors. Also, oil lubricated compressors use less energy and are more reliable. The compressed air is cooled to bring down the air temperature to the membrane separators inlet temperature (20–40 °C).

Air purification: In the air purification section, any residual oils carried by the cold compressed air are condensed and removed in the condensation traps. A coalescing filter may be used to remove residual oil and water aerosol droplets. Sometimes, a carbon filter, consisting of activated carbon bed, is also added in line to capture any further oil vapor or other chemical pollutants which may exist in the environment and may harm the membrane.

Membrane modules: In the membrane module section a battery of membrane modules is used and arranged depending on the purity and productivity needs of the system. The clean compressed air is fed to one side of each of the membrane module through a manifold. Since several membrane modules are used in a given system, and all modules used may not have same performance, the air fed to each of the module is balanced depending on the performance of the module. A product pressure valve is used to regulate the amount of the feed to each of the module. An oxygen analyzer may also be used for the feed air proportion to the different modules. The product flow is controlled by PLC either by purity regulation or by the product flow rate control. The compressed air may be reheated before feeding to the membrane module depending on the optimal temperature of membrane operation. Modern nitrogen membrane system can run remotely using telemetry. A nitrogen product tank is sometimes installed to dampen any cyclic effects due to disruptions.

Membrane systems produce nitrogen at concentration ranging from 95% to all the way 99.99%. However, due to the modest selectivity (O_2/N_2 selectivity

<8) of air separation membranes, production of high purity nitrogen is not economically favorable. Membranes compete best with other nitrogen generation technologies in the 95–99% purity range.

Key producers for the nitrogen separation membranes are Air Products, Air Liquide and Ube. They all use either polysulfone or polyimide polymer membranes in hollow fiber configuration. Other nitrogen membrane module producers (IGS Generon systems, Parker Heniffin, Aquillo, *etc.*) use poly(4-methyl-1-pentene), tetrabromobisphenol polycarbonate and polyphenelyne oxide membranes.

Today, several thousand membrane-based nitrogen generation systems in sizes up to 50 ton N_2 day^{-1} are operating in various parts of the world.

8.3.1.2 *Specific Nitrogen Applications*

Membrane-based nitrogen systems can be used in almost every application in which low purity nitrogen can be used to replace high purity nitrogen produced by cryogenic distillation. However, one has to address the economics of the membrane systems. Membrane nitrogen systems have greater suitability for niche applications requiring low purity nitrogen, portability, ease of operation. They are highly desired in the remote areas where nitrogen transport by tank cars is not economical. A few of these applications are described here.

8.3.1.2.1 Food Storage and Preservation. For food transportation via road, rail and sea, portable on site inert atmosphere generators are needed to increase the shelf life of the food. Especially designed membrane-based nitrogen generation systems are supplied by various vendors which are custom designed for the trucks, rail cars and sea containers and fit in the desired area along with the refrigeration systems. The portable units are self contained and require only electric power from the carrier to produce nitrogen and refrigeration.

8.3.1.2.2 Grain Storage. Portable membrane-based nitrogen systems are also used to provide inert atmosphere in the grain storage silos and other storage facilities. Here the inert environment offers a safe method to prevent infestation of the grains and reduce the consumption of grains by rodents, *etc.*

8.3.1.2.3 Dry Nitrogen Seal. The dry gas seal for the offshore platforms and LNG terminals use portable inert gas generators to reduce the emission of the greenhouse gases like methane. Compact membrane systems have been designed by various vendors for use in these hazardous environments.

8.3.1.2.4 Inflation of Tires. Portable membrane systems to produce high pressure nitrogen for tire inflation to increase tire life and prevent a high pressure air source in case of a fire is commonly used for the aircrafts.

Claims have also been made on the use of nitrogen for inflation of auto tires due to its lower permeability and its use has been justified on the basis of gasoline savings.

8.3.1.2.5 On-board Inert Gas Generator System. Creating an on board inert environment in the aircraft fueling tank ($>88\%$ inert gas) via membrane-based OBBIGS systems to prevent ignition of fuel tank vapors is an emerging application. The nitrogen enriched air is generated by a membrane system on board the aircraft using engine bleed air to displace the flammable fuel/air mixture in the fuel tanks. This protects the aircraft from vulnerability to lighting, static discharge, fire, *etc.* A special system design is needed due to the variable amounts of nitrogen needed at the same purity depending on the flying conditions (descent, cruising, *etc.*). Because the membrane generator is on board the aircraft, its size and weight are very important. OBBIGS are currently used on USAF F-22 aircraft and it is expected that this application may soon expand to civil aviation as well.

8.3.1.2.6 Nitrogen Enhanced Oil Recovery. For enhanced or tertiary oil recovery from the old oil reservoir, a variety of chemicals and fluids are injected to extract oil. Membrane nitrogen injection in the old mines offers one of the safest and effective fluids for the EOR. Nitrogen provides a low down-hole hydrostatic pressure distributed throughout the reservoir. The nitrogen used is low purity ($>95\%$ N_2) at 12 bar pressure and can increase oil production by two to three times over the existing production rate. The nitrogen requirements for this application can vary between 500 to over 5000 N m^3 h^{-1}.

8.3.1.2.7 Mining Applications. The use of nitrogen gas as an alternative to carbon dioxide for fighting underground fires in mines has been successfully practiced in many countries. For decades, nitrogen has been used widely in the modern coal mining industry all over the world. A membrane system for nitrogen production is ideally suited at the remote mining locations.

8.3.1.3 Oxygen Generation Systems

Membrane gas separation systems are used in some rare instances for the production oxygen enriched air (OEA) with oxygen contents $<30\%$ O_2. For this membranes of modest selectivity (3–5) and higher oxygen flux are preferred. The pre-treatment and membrane skid is generally identical to the nitrogen membrane system when the system is operated under pressurized feed. Sometimes, in OEA generation systems, a vacuum on the permeate side is preferred over a compressor in the feed side to facilitate oxygen transport across the membrane. The system therefore may use a fan to feed air to the system as opposed to a compressor. For the production of $>90\%$ O_2, which is rarely done with polymeric membranes, a two stage operation may be employed. In this system, OEA permeate is compressed and fed to a second stage, preferably made of a membrane of higher selectivity (6–8), to recover high purity oxygen. For making OEA and oxygen of high purity, care must be taken to comply with all the safety regulations associated with oxygen generation and transport.

Vacuum swing adsorption (VSA) is commonly used to make 90% O_2 at small scale. An efficient hybrid process for oxygen production can be designed by combining a single stage membrane system with a VSA, where the membrane permeate after drying is fed to VSA using zeolite adsorbent.

OEA is mostly used to enhance combustion in furnaces and kilns. The use of OEA is gaining use in water and waste water treatments, especially in conjunction with the membrane bio-reactors. In some instances, OEA has been considered for use in fish and shrimp ponds. However, vacuum swing adsorption continues to be a leader in this application. In a recent application, OEA generated by membrane systems is reported to have been used to for the passengers onboard world's highest elevation railway (Qinghai–Tibet).[2]

8.3.1.4 Simultaneous Production of Oxygen and Nitrogen

The simultaneous production of low purity nitrogen and OEA has not been exploited commercially. However, in the developing countries, this concept is gaining popularity, especially for small remote locations. The system looks similar to the nitrogen separation process described above. Nitrogen produced is generally low purity (95–98%) and is used for inerting applications. The OEA (23–27% O_2) is fed to the boilers or power generation systems. In some instances membrane/PSA (pressure swing adsorption) hybrid process, using activated carbon in the PSA is used to remove small quantities of oxygen to produce high purity nitrogen while producing OEA.

8.3.2 Air Drying

Dry air is needed in several consumer and industrial applications. The removal of water vapors from atmospheric air is commercially practiced by the use of refrigeration or adsorption methods. In the refrigeration method, the wet air is brought to the refrigerator coils by the use of an air fan. Air along with water vapors is cooled to temperature below condensation to recover water from air. This process is not energy efficient as a substantial part of energy is consumed in transforming water vapors to liquid water. Refrigeration method is not very effective for producing very dry air (low dew point). Alternatively, adsorbents are used to capture water from air and are very effective, especially for the production of very dry air. Adsorbent-based processes are cyclic and need regeneration step (by heat, vacuum or sweep) to remove adsorbed water.

Membrane systems are very attractive for air drying applications because they offer a quiet, vibration free, continuous steady state supply of dry air.

Further, they are lightweight and explosion proof, requiring no desiccants or refrigerants. Almost all the polymers have higher water permeability than air permeability; however, in order to make a membrane air drying system with low air loss, the selectivity of the membrane in the module should be >1000. Highly selective water vapor permeation membranes can produce dry air dew points as low as –40 °C (–40 °F). Various commercial products use either a glassy polymer like polysulfone, polyimide, *etc.*, or ionomers which are fluorine-containing polymer membrane.

In air drying system, wet air is fed on one side of the membrane; water vapors permeate to the other side of the membrane leaving dry air in the retentate. The membranes used for air drying generally have a very high water permeability and selectivity. Due to the high water vapor flux, the water vapor build-up occurs in the close vicinity of the membrane on the permeate side. This causes serious mass transfer resistance to the transport of further water through the membrane. To alleviate this, water vapor on the permeate side need to be removed away from the membrane surface. One way to move water vapors away from the membrane surface is by sweeping the membrane surface with dry air. This is done by using a part of the dry air produced to sweep the permeate side as illustrated as a schematic in Figure 8.6.

In some module designs, sweep gas is internally generated by modifying the membranes in the module.[23] The amount of sweep gas required varies with the saturation temperature of the feed gas, pressure of operation and moisture contents desired in the dry product gas. The sweep air used in the membrane systems to achieve a –40 °C dew point air can be as high as 15–20% of the feed air. Lower amounts of purge may be needed for producing less dry air. Another way to move water vapor away from the permeate surface is the use of vacuum on the permeate side. Very moderate vacuum is needed and generally can be provided with water seal vacuum pumps or steam ejectors.

Air drying membrane modules are produced by the major membrane manufacturing companies such as Air Products, Air Liquide, Ube and Asahi Glass. However, most of the air dryer business is handled by other equipment manufacturers. Commercial membrane modules can handle flows from 1 L(STP) min^{-1} to 2.3 Nm^3 h^{-1} (100 scfm). The membrane systems are capable of supplying oil and particulate free, dry compressed air to dew point as low as –40 °C (–40 °F) at air pressure of 275–2050 kPa (40–300 psig).

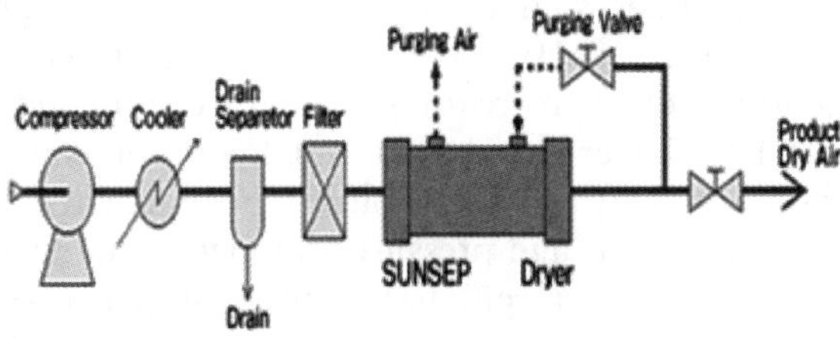

Figure 8.6 Typical air drying system (Courtesy: AGC Chemicals Americas).

8.3.2.1   Membrane Air Drying System

The front end of the membrane system used for air drying is very similar to that of the nitrogen generation system. As discussed in the nitrogen generation system, the quality of the air being fed to the membrane module is very important. The compressed air feed has to be free of particulate matter, oil and other contaminants. In case of product sweep design, the purge valve is adjusted to produce dry air of the desired dew point. The product valve is adjusted adjust the product rate, pressure and dew point.

8.3.2.2   Applications of Membrane Air Dryers

The membrane air dryers can be used in virtually almost any application where dry air is needed. However, they are ideally suited for the smaller volume remote applications. The other applications for which membrane dryers are preferred over the desiccant and refrigeration dryers are: low dew point instrument air, purging electronic cabinets and environmental chambers, power air bearings, machine tool accessories, for spray paint applications, for medical and dental applications, dry powder coating, air logic circuits, ozone generators, dry glove boxes, specialty chemical manufacturing, *etc.*

8.3.2.3   High Pressure Air Drying

For high pressure air drying specially designed four port gas separation modules were used.[24] High pressure wet feed air enters the shell side of the module and the sweep gas is used on the bore side which is vented in the atmosphere. When feed gas is not available at high pressure, a dehydration process with permeate compression and recycle is used.[25] In this process all the moisture leaves the system as condensate on the compressor/cooler system, thus conserving air loss.

8.3.3   Hydrogen Separation Membrane Systems

Over 50 million tonnes of hydrogen is produced worldwide annually for use in synthesis gas, ammonia and methanol production. The most common method of making large scale hydrogen is steam methane reforming, where H_2 is separated from CO_2. Other sources of low cost hydrogen are refinery and chemical industry waste gases. Hydrogen demands are growing steadily partly due to the increased use of hydrotreating (to remove sulfur from gasoline and diesel fuels to the new lower standards) and hydrocracking (conversion of heavy hydrocarbons to light hydrocarbons). The unreacted hydrogen from these processes is generally at high pressure and membrane processes are ideal for their recovery. Also, the recent focus on hydrogen economy has also highlighted additional hydrogen recovery opportunities from the gases which were earlier being flared.

A recent review on the hydrogen separation membranes describes the current state of the art.[26] Polymeric membrane material used for hydrogen recovery applications offer high selectivity and permeability but need to be chemically robust, resistant to ammonia, hydrochloric acid, benzene, toluene and xylene (BTX) and operate at high temperature. All membrane systems used for hydrogen separation or recovery is custom designed with unique pre-treatment suitable for the feed gas. Typical polymers used for hydrogen separation membranes are polysulfone, polyimides and polyaramids. Key suppliers of hydrogen membrane systems are Air Products, Ube and Air Liquide and UOP.

8.3.3.1 Application of Hydrogen Separation Membranes

8.3.3.1.1 Hydrogen Recovery in Ammonia Loop. The first commercial application of hollow fiber hydrogen separation membranes was in the ammonia synthesis loop. Today, almost all ammonia synthesis loops utilize membrane systems for hydrogen recovery and recycle. In this process hydrogen is selectively separated from nitrogen in ammonia purge gases by permeation of hydrogen from the membrane. The membrane system operating pressure can be as high as 100 bar. A typical high pressure (>50 bar) purge gases containing $>80\%$ hydrogen along with contaminants like nitrogen, methane, hydrogen sulfide, ammonia and water are fed on one side of the membrane (preferably, bore side of the membrane module) and high purity ($>98\%$) hydrogen is recovered at a lower pressure (30 bar). The partial pressure of hydrogen in the purge gas offers the driving force for hydrogen permeation. Since the membranes used are highly selective to hydrogen over nitrogen, the residual gases are lean in hydrogen ($<50\%$) and are at pressure (50 bar).[27] The membranes used in ammonia loop must be tolerant to very low levels of ammonia otherwise; feed gas must be treated to remove trace amounts of ammonia from it. The membrane systems are generally skid mounted with custom designed pre-treatment skid in sizes over a wide range, up to hundreds of $Nm^3 h^{-1}$ hydrogen production rate.

8.3.3.1.2 Refinery Hydrogen Recovery. Hydrogen membrane systems offer an economic method of recovery and recycle of hydrogen from various refinery operations and refinery off-gases. Hundreds of commercial plants for hydrogen recovery in the refineries are operational. Hydrogen is recovered and recycled in the refinery operations such as hydrotreating, hydrocracking and hydrodesulfurization. With the increasing focus on the reduction of sulfur in gasoline and diesel, the demand for the recovery of hydrogen within a refinery is increasing and membranes are finding newer applications. The membrane systems used in the refineries are large ($>300\,000$ $Nm^3 h^{-1}$) and operate at high pressure of 120 bar. High purity hydrogen (99.9%) can be produced with 98% of hydrogen recovery.

8.3.3.1.3 Natural Gas Reforming. Almost all large scale commercial hydrogen production is done by natural gas reforming. In this process, natural gas is reacted with steam over a catalyst at high temperatures to produce hydrogen and CO. When the intent of the process is to produce maximum hydrogen, the CO is further reacted with water in the shift reaction to produce CO_2 and H_2. Because at high CO_2 concentrations in H_2–CO_2 mixture produced, membranes are not very economical for the H_2/CO_2 separation because most of the commercial membranes are hydrogen selective. The preferred process for hydrogen separation from CO_2 is adsorption where hydrogen is produced at pressure. However, it is desirable to use rubbery polymer membranes which are selective for CO_2 over hydrogen. When the intent of the reforming process is to produce synthesis gas (a mixture of CO and H_2), membrane systems are ideally suited to adjust the CO to H_2 ratio for the given feed stock intended for a specific product for further synthesis.

8.3.4 Natural Gas Upgrading Systems

Natural gas (NG) exists in nature with several impurities in it like CO_2, N_2, H_2S, H_2O, He, C_2–C_4 hydrocarbons, *etc.* The gas is generally at 20–80 bar pressure and 35–75 °C temperature. The composition of these gases varies from well to well. All gas fields require some treatment before the gas is usable. The minimum required is to remove C_3^+ hydrocarbons and water. After removal of hydrocarbons and water, sometimes gases from different wells are combined together to meet the pipeline specifications. In about 20% of the wells, in addition to water and C_3^+ hydrocarbons, gases must be upgraded for the required fuel value $10\,000$ kcal m^{-3} (1000 BTU/cu ft gas). This requires removal of the excessive CO_2 and inert gases (N_2 and He). In the high sulfur fields, additional need is also to remove H_2S and SO_2. Membrane systems are mostly deployed for the removal of acid gases from the NG. However, applications of membrane systems for NG dehydration and removal of inert gases are being developed and tested at the demonstration stages. An excellent overview of natural gas processing with membranes is given by Baker and Lokhandwala.[19]

8.3.4.1 Removal of Acid Gases from Natural Gas

The acid gases associated with the natural gases are CO_2, SO_2 and H_2S. Of these CO_2 is in highest concentration. Removal of CO_2 from natural gas streams is essential because CO_2 reduces the energy contents of the gas and is corrosive in the presence of water. The pipeline specifications for the CO_2 contents in the natural gas are <2%. The sulfur-based acid gases are undesirable due to their toxicity and corrosive nature. They must be removed from the natural gas to the parts per million levels. The most widely practiced method for the removal of acid gases from the natural gas is scrubbing with chemical or physical solvents; amines are the most popular amongst them. A typical absorption process consists of packed towers where the acid gases (CO_2, SO_2 and H_2S) are chemically reacted with amines and the absorbed gases are recovered in a stripping column at a higher temperature to decompose the amine salt. The absorption-based processes are well established and have been in use for several decades. Absorption-stripping units are complex and require constant monitoring and routine maintenance. They are not most desirable in remote locations and on offshore platforms. The CO_2 stripping processes lose 2–5% CH_4; a part of CH_4 is lost with the flue gases in the CO_2 absorber (1%) and a large portion is consumed as heat in the stripper (1–4%).[19]

Membrane systems to remove acid gas from the natural gas emerged more than three decades ago. Today about half a dozen venders provide membrane systems for natural gas upgrading. Amongst them are UOP (Separex and Grace), who uses cellulose acetate spiral wound module; and Natco (Cynara) who uses cellulose acetate hollow fiber membrane modules. Ube, Air Liquide and Air Products offer polyimide hollow fiber membrane modules, and MTR produces perfluoropolymer membranes in spiral wound modules.

Large membrane systems to process several million cubic meters of gas per day are currently used (1 MMSCFD to over 250 MMSCFD range). Natco recently reported[28] supplying an 830 000 Nm^3 h^{-1} system for natural gas sweetening in an offshore platform in the Gulf of Thailand. This plant is designed for reducing the CO_2 concentration in the natural gas from 36% to 16% for combustion in a gas power plant.

8.3.4.1.1 Membrane System. A typical membrane system for acid gas removal from natural gases consists of a pre-treatment skid and a series of membrane modules. A schematic of which is shown in Figure 8.7.

The specifically designed skid can accommodate a wide variety of feed stocks. The heavy hydrocarbons associated with the natural gases is first removed in a dew point unit which is generally available at the gas field. The dew point unit consists of a chiller to condense any of the higher hydrocarbons and water coming with the gas from the field. The feed gas then passes through a series of coalescing filters and carbon filters to trap any hydrocarbon residues to

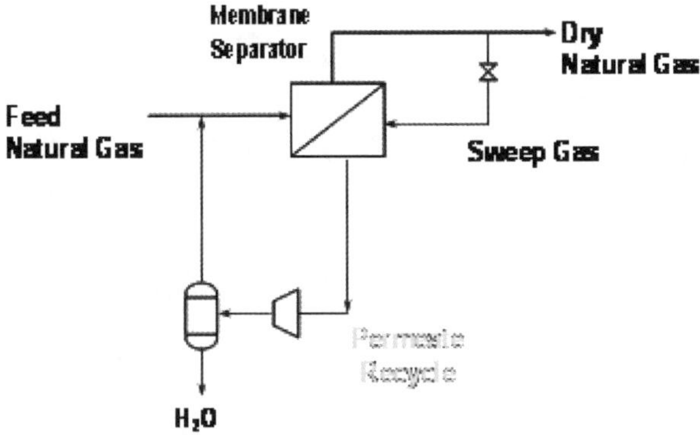

Figure 8.7 Schematics of natural gas dehydration process.

produce as clean gas as possible. Because the pressure drop across the membrane causes cooling of the gas due to the Joule–Thompson effect, a pre-heater is used for the feed gas to assure that no condensation occurs on the membrane surfaces. The pre-treatment process increases the reliability and performance of the membrane modules.

The membrane system operation is relatively simple and the system design is modular permitting easier turn-up and turn-down. The hydrocarbon losses and the system costs are dependent on the performance of the membrane. Current polymeric membranes are not selective enough to permit separation of acid gases from the natural gas with acceptable methane losses in a single stage operation. Therefore, a two stage process is commonly used as shown in Figure 8.8.

The gas permeate from the first stage, which is moderately rich in hydrocarbons, is compressed, cooled and sent to a second stage of pre-treatment to remove any entrained oil and maintain at the feed temperature before it is fed to second stage membrane modules. A bulk of CO_2 is removed from the gas and the residue gas is sent to recycle to the first stage. The capital and operating cost of a two stage membrane system is higher than a single stage process because of the recompression and pre-treatment of the permeate gas from the first stage. Therefore, one has to determine a trade-off between the loss of hydrocarbons in a single stage operation *vs.* the additional capital and operating costs of the two stage operation. The total system cost will depend on the NG flow rate, operation temperature and pressure and CO_2 contents of the feed gas.

Depending on the membrane used, the hydrocarbon losses can be 2 to 10% for a single stage membrane system and 2 to 5% of the feed gas hydrocarbons for a two stage membrane system. The higher hydrocarbon permeates can be used as a fuel, whereas the lower hydrocarbon CO_2 is vented or flared off. Hydrogen sulfide generally has the same permeation rate as CO_2; therefore the ratio of H_2S over CO_2 is generally same in the permeating gases as in the feed gases. For low H_2S gases, one may meet the pipeline specifications of <4 ppm. However, for higher H_2S gases, H_2S concentration may be as high as 100 ppm.

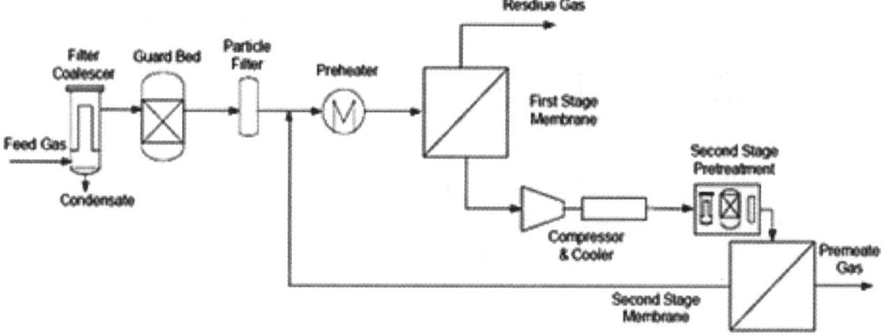

Figure 8.8 A two-stage membrane system for removal of acid gases from natural gas (Courtsy UOP LLC, A Honeywell Company).

For those feed streams, an amine scrubbing is preferred to meet the pipeline specifications.

A typical operation in which the feed gas is at 60 bar pressure, permeate will be at a lower pressure like 3 bar, with residual gases at slightly lower pressure than the feed pressure, meeting pipeline specifications of 2% CO_2 (4 ppm H_2S). For high selectivity membranes (CO_2/CH_4 >25), 95^+% methane can be recovered from the feed gas.[27]

Membrane systems are very desirable for remote gas fields where utilities are scarce. The membrane units have small footprint and low in weight. They do not generate any waste or contaminated chemicals to be disposed off. Therefore, they are also preferred on the offshore platforms.

8.3.4.2 Dehydration of Natural Gas

Natural gas occurs in nature generally fully saturated with water. The presence of water under pressure causes the formation of natural gas dehydrates which can plug valves and pipelines and causes corrosion. Therefore, water must be removed from natural gas before it is further processed and transported. The water contents of water saturated high pressure gas are around 1000 ppm which needs to be reduced to about 100 ppm to meet pipeline specifications. Conventional method of drying of natural gas is scrubbing with triethylene glycol where natural gas flows in counter current flow to triethylene glycol which absorbs water from it. However, this method is energy intensive and causes environmental hazards (by releasing hydrocarbons and volatile organic compounds) during the regeneration of triethylene glycol.

Membrane technology is an attractive alternative because membrane materials are readily available with high water/methane selectivity (>1000). Ideally, these polymers should yield membrane systems for dehydration which use less energy, generate no wastes, and offer a simple low maintenance process. The equipment footprint should be smaller and of light weight. Therefore, membrane dehydration systems are highly suited for both offshore platforms and land locations. Unfortunately, membrane dehydration systems have excessive methane loss. The reason for the loss of methane is due to the pressure ratio limits on membrane performance. Reducing the water partial pressure on the permeate side by the use of vacuum[29] improves the NG dehydration process. Alternative methods for moving water vapor away from the permeate side of the membranes have been proposed. In one method,[30] the membrane system is operated at permeate pressure close to atmospheric pressure; permeate is recompressed to condense water out and methane containing gases are sent back to the high pressure feed side. In an alternative process,[31] a part of the dry methane is used to sweep the permeate side to increase separation effectiveness. The sweep gases from the permeate side are then dried in a secondary drying loop.

More efficient membrane, membrane module and system design are all needed to achieve the benefits of the membrane dehydration system for natural gas.

8.3.4.3 Removal of Nitrogen from Natural Gas

About 14% of the natural gas in the United States contains excess nitrogen which needs to be separated from the natural gas to meet the pipeline specifications ($<4\%$ N_2).[30] Separation of nitrogen from CH_4 is difficult; only cryogenic liquefaction is commercially viable at large scale. Pressure swing adsorption technology using nitrogen selective zeolites has been also developed and is at the development stage for the small to medium size gas fields. For small gas fields the current practice of meeting nitrogen pipeline specifications is by blending the low and high nitrogen gases to achieve the target values. When such low nitrogen gases are not available, the high nitrogen containing gas wells are generally closed. Membranes have been proposed[31] for removal of nitrogen from high nitrogen concentration natural gases. Two families of membrane materials have been proposed, one, rubbery polymers, which are selective to methane (methane/nitrogen = 3–4) and other, glassy polymers, which are selective to nitrogen (nitrogen/methane = 2.5). Unfortunately, the selectivity of either of these membranes is low and a multi-stage, multi-step process is needed to remove nitrogen without excessive methane loss. Methane selective membranes offer a better separation; however, they are at disadvantage because methane permeates at low pressure and has to be recompressed to the pipeline pressure. A higher selectivity membrane and well designed membrane system may become cost competitive in the future.

8.3.4.4 Other CO_2/CH_4 Separation Applications

8.3.4.4.1 Substitute Natural Gas. City gas or substitute natural gas is produced by methanation and reforming of naptha or LPGs in which process CO_2 is generated. Membrane systems are economically used to remove CO_2 from the product gases to make them to the pipeline spec gases. A key advantage of the membrane systems is that a modular system can be designed to match the volume and the quality of the gases produced by a specific plant.

8.3.4.4.2 Coal Bed Methane. Methane and CO_2 mixtures (typically 88% CH_4 and 12% CO_2) are generally present in the coal mines and deposits. Membrane systems can be effectively used to recover methane and separate CO_2 for subsequent sequestration.

8.3.4.4.3 Biogas Systems. Methane is routinely generated at the garbage landfill sights and can be effectively removed from CO_2 and other contaminated gases by the use of a membrane system. Since the landfill gases are very dirty, a more rigorous pre-treatment is needed for the use of membranes.

8.3.4.4.3 CO_2 Enhanced Oil Recovery. Similar to the nitrogen enhanced oil recovery, in CO_2 enhanced oil recovery processes CO_2 is injected in the oil reservoirs to improve oil recovery. A typical EOR process starts with $>50\%$ CO_2 and high pressure (up to 140 bar); the CO_2 contents increase significantly over time. The CO_2 from the CO_2 enriched natural gas is recovered using membrane systems and pressurized and re-injected in the EOR wells.[32] The first membrane plant for this application was installed by Cynara (now Natco). The plant was designed for reducing CO_2 concentration from 45% to 28% CO_2 for processing 60 000 Nm^3 h^{-1} gas. This plant was later expanded[32] and is processing 120 000 Nm^3 h^{-1} gas, decreasing CO_2 concentration from 80% to less than 10%).

8.3.5 Carbon Dioxide Separation Membrane Systems: CO_2 Capture from Flue Gases

The CO_2 separations by membranes from the natural gases and enhanced oil recovery were described earlier. In contrast, CO_2 separation from stack gases is a very different application because the stack gases are modest in CO_2 concentration (generally between 6 and 16%), feed gas is at low pressure (almost atmospheric), flue gas are at high temperature (gases need to be cooled to below 100 °C for polymeric membranes) and require different pre-treatment to remove any harsh chemicals from flue gases. At present there are no commercial operations for the capture of CO_2 from flue gases. In order to use existing CO_2 gas separation membrane modules in this application one may have to either compress the flue gases or use vacuum on the permeate side to provide driving force across the membranes. However, due to the enormous size of the flue gas stream from the power plants, a standalone membrane system may not be economical; a membrane/absorption hybrid process may be commercially viable in the future. Powell and Qiao[33] have presented an excellent review of the polymeric membranes for the separation of CO_2 and nitrogen mixtures. Economic viability of the membrane-based CO_2 capture and sequestration systems have been presented in a few studies. None of these show a rosy picture of the polymeric membranes for this application.[34–37]

Alternatively, membrane gas–liquid contactors have been proposed for CO_2 capture from the stack gases. The overall absorption process for a membrane gas–liquid contactor is similar to the conventional absorbers with the exception that the gas–liquid contact area is being provided by a membrane instead of column packing. A membrane gas–liquid contactor consists of microporous, hydrophobic membranes which contact flue gases on one side of the membranes to the scrubbing liquid at the pore opening on the other side. The membrane gas–liquid contactors are operated at atmospheric pressure therefore feed gas pressurization is not needed. Membrane gas–liquid contactors for CO_2 absorption have been developed using, PVDF, PP and PTFE microporous hollow fiber membranes which can tolerate corrosive solvents. Carbon dioxide from the gas feed side diffuses through the pores of the membrane and travels to the liquid side where it absorbed/reacted with the absorbing liquid. The absorbing liquid flows away from the membrane surface by circulation.

Membrane gas–liquid contactors offer several advantages over the conventional packed and plate columns which have severe operational issues due to flooding, weeping, channeling, foaming *etc.*, all of which reduce the operational efficiency of these absorbers. The hollow fiber membrane modules can provide a very large surface area per unit volume of the module; however, due to the other design considerations such as pressure drops modest area density (500–1000 m^2 m^{-3}) is more practical. Because of the large organized contact area offered by a membrane contactor than the packed beds, gas liquid contactors offer a better contacting efficiency. The overall mass transfer of CO_2 achieved in a gas–liquid contactor is two to four times of that obtained in the same volume conventional scrubber. This means that the foot print needed for a gas–liquid contactor can be two to four times smaller than the conventional scrubber thus making them very attractive for large volume applications. Unfortunately, the membrane gas–liquid scrubbers currently available lose their effectiveness overtime due to surface wetting and penetration of the liquids in the membrane pores. Advances in materials and processes are needed in this application.

The Institute of Environmental and Energy Technology (TNO)[38] in Netherlands have developed macroporous polypropylene membrane contactors and used it with alkali aminoacid salt as absorbant. No wetting or degradation of polypropylene membrane surface was observed and hence stable membrane performance was reported. The Kvaerner Process is used for Teflon™ membranes[39] with aqueous amine solutions of MEA, DEA, MDEA and DIPA. The membrane system was used to recover acid gases from natural gas for offshore platform applications. No performance data are available for the flue gas treatment.

8.3.6   Organic Vapor Separation Systems

Separation of organic vapors from various gas streams is an emerging application. Membranes which permit the transport of condensable vapor such as propane, butane, higher hydrocarbons, alcohols and water are useful in separating condensable vapors from non-condensing gases like methane, ethane, nitrogen, hydrogen, *etc.* The largest of these opportunities, which is yet to be exploited, is in the recovery of natural gas liquids (NGL) from natural gas to reduce the dew point of the condensable to –20 °C. Engine gas conditioning for heavy hydrocarbon removal is widely used.[40] Others opportunities are in the recovery of monomers in the polymerization processes such as: vinyl chloride monomer recovery in the polyvinyl chloride polymer synthesis, propylene and ethylene recovery in their respective polymerization processes, *etc.* Examples of chemical manufacturing plants include ethylene recovery in ethylene oxide and vinyl acetate synthesis.

Membrane Technology & Research (MTR) has taken the lead in the development of organic vapor selective rubbery polymer membranes. These membranes offer mixed gas propane to methane selectivity of 3–5 and butane to methane selectivity of 5–10. The actual selectivity, however, also depends on the pressure and composition of the gas being processed.[18] In their system, the feed gas is first compressed and sent to a condenser where gases are cooled and condensate is removed. The cooled, compressed gases which still contain condensable are fed to a membrane separator. The hydrocarbon-rich permeate is either compressed or chilled to recover hydrocarbons and non-condensable feed is sent to further processing, *etc.* These systems are capable of separating 99% of the organic vapors from the permanent gases.[40]

Recovery of NGLs offers a sizable opportunity for membrane systems. Traditionally, this is done by complex process using refrigeration and turbo-expanders or scrubbing with lean oil. These processes use high energy and therefore alternatives are needed. Rubbery polymer membranes developed by MTR permit the transport of heavier hydrocarbons and water vapor selectively over methane at ambient temperature. Permeate is compressed and condensed to recover NGLs from the water hydrocarbon condensate. This thus offers a simple economic alternative process to recover NGLs. An additional benefit of membrane process used for the NGL recovery is that the degree of water removal from the natural gas is much higher than that for the refrigeration method, thus resulting in partially dry natural gas.

MTR has also brilliantly used membranes for the recovery of olefins from the polyolefin polymerization processes. In this process, schematic of which is shown in Figure 8.9, the vent gases from the polymerization process which contains hydrocarbons and nitrogen compressed, condensed to recover liquid hydrocarbons and then fed to a membrane separator to recover olefins which are recompressed and condensed. The commercial offering of this process is done under the trade name VaporSep® and is used by polyolefin producers like ExxonMobil, Formusa Plastics, SABIC and Sinopec. MTR has similar processes for propylene, vinyl chloride and ethylene oxide.

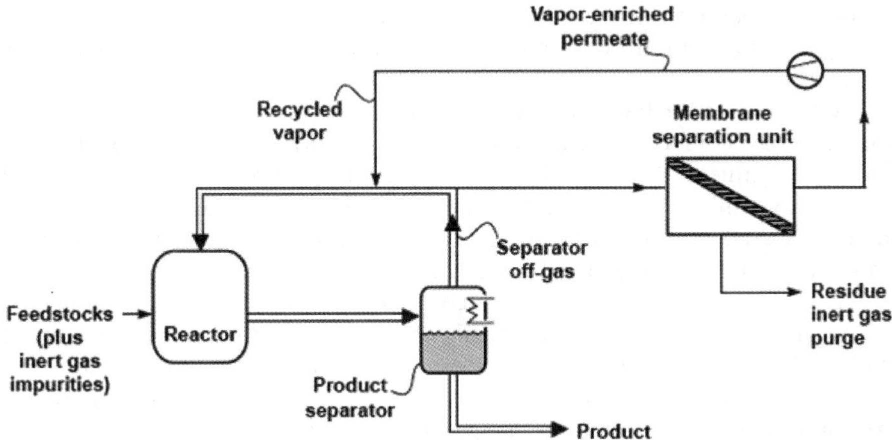

Figure 8.9 Vapor recovery and recycle system (Courtesy: Membrane Technology Research, Inc.).

8.4 Concluding Remarks

A key driver behind the adoption of membrane technology has been the energy saving it offers over competing separation technologies. It was demonstrated many decades ago that for processes like desalination, the energy requirements for a reverse osmosis process is less than half those needed for thermal desalination processes. Since then there has been great motivation to find other separation applications where one can have substantial energy saving by virtue of avoiding phase changes in the separation processes. While energy saving motivation has been behind most of the membrane gas separation applications, the commercial drivers have been different. More notably, has been the ease of operation, operational flexibility, remote operations, compact weight and volume, *etc.* For gas recovery and re-use applications, such as hydrogen recovery in petroleum processes and monomer recovery in polymerization processes, membranes have been the process enablers. The challenge for the process engineers and chemists is to find more such applications and make membranes as integral part of the chemical processes.

There are several challenges which remain for the membrane scientists and membrane manufacturers. The material related technical challenges of the membrane separation systems have been widely reported in the literature. The gas separation membranes need to maintain a stable performance over a long period of time, generally 5 years or longer. The life and replacement frequency of the membrane modules influence the overall capital cost of the equipment. The membranes must be able to withstand the process upsets, operate over a broad range of temperature and, most importantly, not have catastrophic failure. In spite of great advancement of materials in the field of gas separation membranes, we still lack commercially viable materials which are more solvent/ chemical resistant and can operate at elevated temperatures.

The challenge for the process engineers is to develop simple and robust pre-treatment processes for the various gas separation applications. This will make membrane operations more stable, reliable and increase membrane life. All this has a great impact on the economic viability of several of the existing and future applications.

The challenges for the membrane manufacturers are to produce membranes and membrane modules with reproducible performance. The performance variability is due to the variability associated with the polymer, membrane and the flow dynamics of the membrane module. Therefore, when more than one module is used in a gas separation system, each module feed flow and back pressure need to be adjusted to get the best performance of the membrane system. This makes module replacement and substitution difficult. Standardization and compatibility of the membrane modules with various membrane skids will make membrane gas separation applications more acceptable.

Another major deterrent for the advancement of the gas separation membranes is that it requires a special art to make efficient membranes and membrane modules from a given polymer. A better understanding of the role of membrane fabrication processes and module and system design on achieving gas separation effectiveness in the various applications and under various operating conditions is needed.

Finally, broader availability of the software for the design of the membrane modules and systems and other tools will popularize membrane systems amongst the process and design engineers. The current state of understanding makes gas separation membrane and membrane module and systems design and manufacturing a specialized field, knowledge of which resides with a few.

References

1. D. Maclean, W. A. Bollinger, D. E. King and R. Narayan, in *Recent Development in Separation Science*, ed. N. N. Li and J. M. Calo, CRC Press, Boca Raton, Fl, 1986.
2. P. Bernardo, E. Drioli and G. Golemme, *Ind. Eng. Chem. Res.*, 2009, **48**, 4638–4663.
3. T. D. Kusworo, A. F. Ismail, A. Mustafa and T. Matsura, *Sep. Purif. Technol.*, 2008, **61**, 249–257.
4. S. J. Shilton, A. F. Ismail, C. R. Dunkin and S. L. Gallivan, *Polymer*, 1997, **38**, 2215–2220.
5. T. S. Chung, S. K. Tooh, W. Y. W. Lau and M. P. Srinivasan, *Ind. Eng. Chem. Res.*, 1998, **37**, 3930–3938.
6. W. J. Koros and R. Mahajan, *J. Membr. Sci.*, 2000, **175**, 181.
7. W. J. Koros, D. Wallace, J. D. Wind, S. J. Millerand and C. Staudt-Bickel, *Crosslinked and Crosslinkable Hollow Fiber Membranes and Method of Making Same*, U.S. Patent 2003/0221559A1, 2003.
8. H. B. Park, C. H. Jung, Y. M. Lee, A. J. Hill, S. J. Pas, S. T. Mudie, E. V. Wagner, B. D. Freeman and D. Cookson, *Science*, 2007, **318**, 254–257.
9. S. Li, J. L. Falconer and R. D. Noble, *Adv. Mater.*, 2006, **18**, 2601–2603.
10. J. M. S. Henis and M. K. Tripodi, *J. Membr. Sci.*, 1981, **8**, 233.
11. J. M. S. Henis and M. K. Tripodi, *Multicomponent Membranes for Gas Separations*, U.S. Patent 4 467 001, 1984.
12. S. Loeb and S. Sourirajan, *ACS Adv. Chem. Ser.*, 1963, **38**, 117.
13. J. E. Cadotte, *Interfacially Synthesized Reverse Osmosis Membrane*, U.S. Patent 4 227 344, 1981.
14. J. Marriott and E. Sorensen, *Chem. Eng. Sci.*, 2003, **58**, 4975–4990.
15. M. Mulder, *Basic Principles of Membrane Technology*, Kluwer Academic Publishers, 1998, ch. 8.
16. D. T. Coker, B. D. Freeman and G. K. Fleming, *AIChE J.*, 1998, **4**, 1289.
17. R. W. Spillman, *Chem. Eng. Prog.*, 1989, **85**, 41.
18. R. W. Baker, *Ind. Eng. Chem. Res.*, 2002, **41**, 1393–1411.
19. R. W. Baker and K. Lokhandwala, *Ind. Eng. Chem. Res.*, 2008, **47**, 2109–2121.

20. G. Clarizia and E. Drioli, in ed. E. G. Derouane, V. Permon, F. Lemos and F. R. Ribeiro, *Sustainable strategies for the upgrading of natural gas: Fundamentals, Challenges and Opportunities, Physics and Chemistry*, Springer, Dordrecht, 2005, vol. 191, ch. 14, pp. 287–292.
21. R. Agrawal and J. Xu, *AIChE J.*, 1996, **42**, 2141.
22. R. L. Mckee, *Hydrocarbon Process.*, 1991, April, 63.
23. W. H. Morgan, L. K. Bleikamp and D. G. Kalthod, *Hollow Fiber Membrane Dryers with Internal sweep*, U.S. Patent 5 525 143, 1996.
24. D. J. Stookey, K. Jones, D. G. Kalthod and T. Johannessen, *Membrane dehydrators: A new alternative to drying high pressure gases*. The 1996 Membrane Technology/Planning Conference, Newton, MA, 29 October 1996.
25. S. R. Auvil, J. S. Choe and L. J. Kellogg, Jr., *Use of membrane separation to dry gas streams containing water vapor*, U.S. Patent 5 259 869, 1993.
26. N. W. Oekwig and T. M. Nenoff, *Chem. Rev.*, 2007, **107**, 4078.
27. www.medal.airliquide.com
28. A. Callison and G. Davidson, *Oil Gas J.*, 2007, 105.
29. K. Ohlrogge and T. Birkman, *Ann. N. Y. Acad. Sci.*, 2003, **984**, 306–317.
30. R. W. Baker, K. A. Lokhlandwala, I. Pinnau and S. Segelke, *Methane/ Nitrogen Separation Process*, U.S. Patent 5 669 958, 1997.
31. R. W. Baker, K. A. Lokhandwala, J. G. Wijmans and A. R. Decosta, *Nitrogen Removal from Natural Gas using two types of membranes*, U.S. Patent 6 630 011, 2003.
32. www.natcogroup.com/content
33. C. E. Powell and G. G. Qiao, *J. Membr. Sci.*, 2006, **279**, 1–49.
34. M. T. Ho, G. Allinson and D. E. Wiley, *Desalination*, 2006, **192**, 288–295.
35. M. T. Ho, G. Leamon, G. W. Allinson and D. E. Wiley, *Ind. Eng. Chem. Res.*, 2006, **45**, 2546–2552.
36. D. Shekhawat, D. R. Luebke and H. W. Pennline, DOE/NETL-2003/1200, National Energy Laboratory, Department of Energy, 2003.
37. R. Carapellucci and A. Milazzo, *J. Power Energy*, 2003, **217**, 505.
38. P. Feron and A. Jansen, *Energy Covers Mgmt.*, 1996, **36**, 411–414.
39. O. Falk-Pederson, M. S. Gronvoid, P. Nokleby, F. Bjerve and H. Svendsen, *Int. J. Green Energy*, 2005, **2**, 157–165.
40. A. Jariwala, K. Lokhandwala and R. W. Baker, *Only Raw Sour Gas for Engine fuel? Proven Membrane Process Cleans Gas for Engine*, Presented at Laurence Reed Gas Conditioning Conference, University of Oklahoma, Norman, OK 2006.

Novel Hybrid Membrane/ Pressure Swing Adsorption Processes for Gas Separation Applications

ISABEL A.A.C. ESTEVES AND JOSÉ P.B. MOTA*

Requimte/CQFB, Departamento de Química, Faculdade de Ciências
e Tecnologia, Universidade Nova de Lisboa, 2829-516 Caparica, Portugal

9.1 Gas Separation Technologies

9.1.1 Introduction

Membrane permeation has a single profile of strengths and weaknesses. Apart from being a mature technology that is applied at industrial scale to separate gaseous mixtures, the magnitude of the benefit expected from gas separation membranes is highly reliant on the exact nature of the separation problem.

Gas separations currently comprise a membrane market of two hundred million euros per year. Most of this business involves a variety of applications, such as separation of H_2 from gases like N_2, CH_4, and CO; separation of CO_2 or O_2 from N_2; H_2 recovery in oil refinery processes; CH_4 separation from biogas; removal of water vapor; CO_2 and H_2S from natural gas (NG); and removal of volatile organic compounds (VOCs) from air of exhaust streams.[1-4] A much larger potential market lies in the separation of condensable gases, such as C_{3+} from H_2 or CH_4, and C_3H_6 from C_3H_8.

Membrane Engineering for the Treatment of Gases, Volume 1:
Gas-separation Problems with Membranes
Edited by Enrico Drioli and Giuseppe Barbieri
© Royal Society of Chemistry 2011
Published by the Royal Society of Chemistry, www.rsc.org

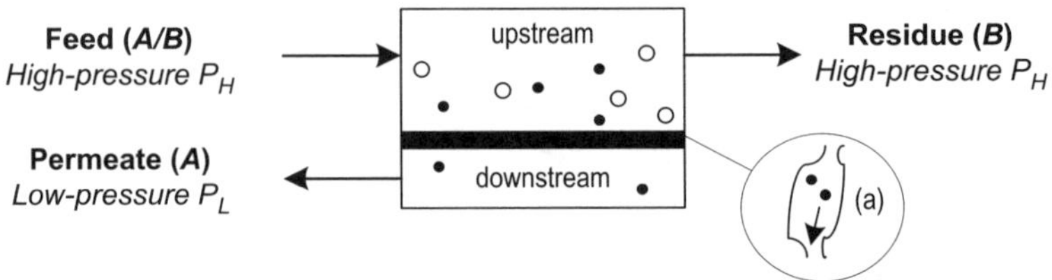

Figure 9.1 General schematic of the membrane separation process. (a) Details the solution-diffusion mechanism used in gas separations.

Non-porous polymeric membranes are usually employed for gas separation, although porous ones can also be used. Composite polymeric membranes developed in the 1970s made the separation of gas streams commercially feasible. The first large-scale gas separation modules were developed by DuPont in early 1970s, but the first successful commercial membrane gas separation processes (PRISM) were announced by Monsanto in late 1970s.

In non-porous membranes or membranes with pore size in the 1–10 Å range, vapors and gases are separated due to their different solubility and diffusivity in the polymeric material. In other words, gas separation occurs via pressure-driven solution-diffusion. The permeating molecules diffuse and move among polymer chains according to the formation of local gaps by thermal motion of polymer segments.[5] Thus, the polymer free volume, its distribution, and local changes of distribution are of the greatest importance. Figure 9.1 shows a general schematic of a membrane separation process.

The components that dissolve and then diffuse more rapidly through the membrane become enriched in the low-pressure permeate stream, while the slower ones are concentrated in the high-pressure residue stream. The diffusion process takes place when the change in chemical potential of the diffusing species is negative due to the pressure gradient that acts as driving force.

The permeability provides a quantitative measure of the easiness with which a certain species penetrates the membrane. The permeability depends on the nature of the gas, on the polymeric matrix, on temperature, and on the pressure values, P_L and P_H, on both sides of the membrane. For simple gases the solubility increases as the gas diameter increases, since the gas becomes easier to condense. The gas diffusivity through a membrane is inversely proportional to its molecular size, because large molecules interact more with the polymer chains.

In addition to the more conventional membranes, such as those of polysulfone and silicone rubber, there has been an exponential growth in the R&D of other membrane materials, such as polyimides and polyaramides.[6–9] Inorganic or carbon sieving materials can also separate many gases, although their fragility and cost have made them less attractive.[10–12] Robeson[13] has noted a trade-off between the permeability (productivity) and selectivity (efficiency) of typical polymeric membranes. To transcend this well-known upper-bound trade-off curve and maximize both membrane permeability and selectivity, new types of materials are needed, which lie in the so-called region of industrial applicability. Mixed matrix materials, composed by zeolites or carbon molecular sieves dispersed in a continuous polymer matrix, are excellent candidates to overcome the upper-bound polymer constraint. Nonetheless, while this upper-bound curve has been surpassed using mixed matrix membranes in some cases,[14–17] there are still many aspects of this new technology that are not well understood. Controlling matrix phase swelling, for example, is a major requirement for heavily interacting feeds containing strongly adsorbing components, such as CO_2. This gas has a swelling tendency for polymeric membranes, facilitating local segmental motions and thereby undermining the polymer ability to discriminate between penetrants of similar size and shape. Currently, polymer cross-linking suppresses swelling and provides increased stability and efficiency in the presence of such aggressive agent in separations like CO_2/CH_4 of natural gas streams, when compared to the more traditional polymeric membranes.[18–21]

Membrane separations are usually favorable for low-to-moderate processing rates but less advantageous when a high-purity product is required, and are thus typically considered to be more suitable for bulk separation than for high-purity purification or polishing. Unless permeate is recycled, usually high product purity is accompanied by lower product recovery. Often, the separation may be inexpensively upgraded by a subsequent process.[21] Another disadvantage of membranes is fouling that decreases the durability and robustness of the material.

On the other hand, one advantage of membranes is that they are easy to use and to integrate into an existing process flow sheet. In fact, the separation performance of many established non-permeating technologies could often be improved through their synergistic coupling with membranes under the form of hybrid schemes. This fact has motivated active research on the integration of membranes with other separation processes.[3]

This chapter examines newly developed hybrid membrane processes for gas separation, which integrate membrane permeation with pressure swing adsorption (PSA) technology. A brief review of the theory underlying membrane and PSA processes is provided. The discussion will focus on the evaluation of the present state of the art of both processes and on the practical application of both technologies into a hybrid membrane/PSA concept.

9.1.2 Pressure Swing Adsorption

In gas separation by adsorption, the gaseous feed is put into contact, at adsorption conditions, with an adsorbent that selectively adsorbs some of the components. The feed is then removed from the adsorbent, and the adsorbed components are recovered by desorption at desorption conditions. The adsorbent is usually packed into one or more fixed-bed columns. The preferentially adsorbed species are then removed and the beds are regenerated by either increasing the temperature, as in temperature swing adsorption (TSA), or by purging with a displacement agent, as in displacement purge adsorption

(DPA), or by reducing the total pressure, as in pressure swing adsorption (PSA). Industrially, PSA is the predominant type of adsorption cycle used for gas separation, although TSA is also quite common.[22] These processes are alternatives to other traditional and more energy-intensive cryogenic and distillation processes.

In a typical adsorption process, a multitude of adsorbent beds are operated in a cyclic manner, with each bed repeatedly undergoing the same steps but in a staggered sequence so that a continuous flow can be maintained. Adsorption is an inherently dynamic process and suffers from the disadvantages associated with cyclic operation.

The separation can be achieved by three different mechanisms: selective binding of the adsorbate to the solid surface; adsorbate exclusion based on molecular sieving; and kinetic separation based on significantly different intra-particle diffusion rates of the adsorbates. The first two mechanisms are mostly employed, while the third one is only applied industrially in air separation to produce pure nitrogen.

The standard adsorbent contactor is a randomly packed bed of pelletized or particulate adsorbents. The commonly used adsorbents include activated carbons, which separate mostly based on dispersive interactions; zeolites that separate based on polarity and size; carbon molecular sieves, which use the relative differences in intra-particle diffusion rates; or silica gel and alumina, generally hydrophilic. The particle size and shape should provide a suitable compromise between pressure drop (ΔP) and mass transfer resistance. Note that ΔP is rarely a dominant economic problem, except for the largest systems. In order to minimize it, the cross-sectional area must be increased, leading to a small length-to-diameter ratio (L/D). In small systems, pressure drop is relatively less important than performance in terms of separation efficiency.

The ideal adsorbent for a PSA system should exhibit a high adsorption capacity and selectivity, *i.e.* a high microporous volume with pore size and surface chemistry tailored for the particular separation. However, the solid must also provide easy desorption to prevent retention of the more strongly adsorbed species on the bed, which would increase the heat requirements for its regeneration and shorten its operating life. Lastly, the adsorbent should possess high thermal conductivity, in order to dissipate and subsequently recover the heats of adsorption and desorption, which would allow an effective restoring of its original capacity using heat management techniques. These temperature changes have design implications and, thus minimizing them is particularly important.

The amount of adsorbed gas decreases as the temperature increases. Hence, during adsorption or bed charge, the adsorbent capacity is diminished by the rise in its temperature, although part of this heat is evacuated from the bed with the exit effluent. As a rule of thumb, if the adsorbate concentration in the feed is higher than about 10% (bulk mixture), then a significant amount of heat will remain trapped and increase the bed temperature. Otherwise, in diluted mixtures, the heat of adsorption has little influence and the bed remains essentially isothermal.[1] In contrast, when the adsorbent cools during desorption or bed regeneration, the gas pressure drops and the adsorbed gas is not released.

Depending on the complexity of the PSA cycle, it is possible to obtain useful products either from the adsorption or desorption stages, or from both. The raffinate, enriched in the less strongly adsorbed species, is obtained during adsorption, at high pressure, while the desorbate or extract is the product recovered from the bed regeneration, at low pressure.

The use of multiple adsorbent beds allows near-continuous production of the target gas. The cycle also allows the so-called pressure equalization, where the gas leaving a bed being depressurized is used to partially pressurize another bed. This common industrial practice results in significant energy savings due to minimization of the recompression work. The main weaknesses associated with PSA is that the typical trade-off between purity and recovery persists, and it cannot be applied to too strongly adsorbed species unless high vacuum is employed in the desorption step, which makes the process more expensive.

The PSA concept has been in the public domain since the early 1930s. The first patents about PSA appeared at that time,[23–27] but PSA processes are commonly attributed to Skarstrom[28] and Guerin de Montgareuil and Domine[29] in late 1950s, who introduced the basic principles of the PSA steps that are known today.

The Skarstrom process comprises two steps, adsorption and depressurization, that are carried out in two adsorbent beds operated in phase, thus enabling the processing of a continuous feed. Although the operation of each bed is batchwise, the system, as a whole, is a continuous one that is operated in a cyclic steady state (CSS), in which the conditions at the end of each cycle are identical to those at its start. At CSS, the profile varies in a mean position inside the bed. Generally, the feed step is terminated before the breakthrough of the strongly adsorbed species, while the regeneration is stopped before the bed is fully desorbed. In both cases, the temperature fronts should remain inside the beds. The increase of the pressure swing frequency results in short cycles and, hence, high feed throughputs, but can only be increased up to the limit imposed by the intraparticle mass transfer resistances and operating constraints.

The first candidates for purification by PSA were N_2, H_2, O_2, CO and CH_4. Hydrogen purification by PSA became into a scale commercial practice in late 1960s. At that time, Basmadjian and Pogorski[30] introduced the rinse step in an early PSA cycle. During the 1970s interest in alternative separation processes was stimulated by the huge energy costs associated with the rising price of crude oil. Several other patents were filed by Sircar and co-workers in the period 1977–1988, where two or more sets of parallel beds, with different adsorbents, are connected sequentially, in order to isolate different species with as little power input as possible.[31,32] In the late 1980s, the most obvious modes of PSA operation had been discovered and it seemed to mark the passage from drastic innovations to a series of minor, but not less important refinements over the existing cycles. A fully understandable synopsis of PSA patent literature is presented in Ruthven *et al.*,[33] and Tondeur and Wankat.[34] Currently, the search for alternative PSA processes matching the technology to the product specifications continues.

In recent PSA processes, three or more beds are used to synchronize and hold two additional steps to those in the Skarstrom cycle: co-current depressurization or blowdown, and pressure equalization. From the open literature available, there are significant distinctions in the ways in which even the simplest PSA steps can be accomplished. For instance, pressurization with light product is generally superior to pressurization with feed. This is partially done by pressure equalization. Also, simultaneous depressurization with product withdrawal is advantageous over isobaric production or simultaneous pressurization with production. The co-current blowdown step has, likewise, appeared in several forms, either co- or counter-current to the feed step. The rinse step, which follows the feed step and allows the partial admittance of pure heavy product to the bed, shows benefits in terms of recovery by displacing the residual feed that is then recycled before the blowdown.

Typically, a PSA cycle consists of four or more steps, sometimes as many as twelve, which may take from a few seconds to several minutes. Figure 9.2 shows a simple four-step PSA cycle. Through the synchronized cycling process of the composition profiles and flow directions, the heavy species is preferentially adsorbed and separated from the light or less adsorbable one.

The most common steps of a PSA process are:

- Pressurization (PR) with feed or with purified light product obtained from an adsorption step
- High-pressure adsorption step (HPA), where the contaminants are adsorbed and the purified light product is produced and withdrawn
- Blowdown (BD) co/counter-currently to the feed step

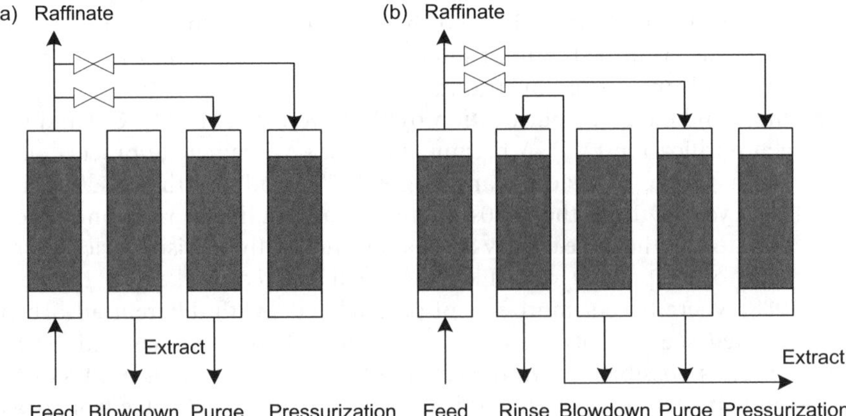

Figure 9.2 Simple (a) four-step and (b) five-step PSA processes. Adapted from Ref. 35.

- Purge (PG) counter-currently to feed step and with purified light product at low pressure, where the desorbed effluent is discharged and the bed is regenerated for the next cycle.

A typical PSA cycle, such as the ones shown in Figure 9.2, presents several general features that explain both the advantages and limitations of this process and determine its suitability for a given application:[35]

- *Product purity and recovery*. They are volume-averaged quantities, since the composition and effluent flow rates from the PSA bed vary with time.
- *Adsorbent productivity*. This is measured by the amount of product or feed mixture processed per unit amount of adsorbent per unit of time. Usually, for a given separation, the product purity is pre-determined, the energy requirement is proportional to the recovery and the bed size is inversely proportional to the adsorbent productivity.
- *Energy requirements*. The separation work of a PSA process compared to its energy consumption is similar to that of processes like distillation. However, PSA uses mechanical energy, more expensive then heat. Thus, the major factor of the PSA operation cost is the power cost. This is greatly minimized if the feed is already available at high pressure, since capital costs, energy requirements, and feed recompression costs are reduced. Even recompressing the product for later cycle usage generates lower costs than feed recompression. Hybrid PSA processes can use these features as an advantage to improve both separation performance and overall costs.
- *Process scaling*. The overall costs increase approximately linearly with feed throughput. The economics tend to favor PSA at low to moderate throughputs and other systems for very large operations.
- *Pressure range*. The process performance is often enhanced for a PSA cycle with desorption at sub-atmospheric pressures, usually called vacuum swing adsorption (VPSA).

When highly pure products are desired and/or a high-energy effort is required for recompression of feed gas streams, the standalone PSA process may be coupled with other separation technologies in order to enhance the performance of the cyclic process.

9.2 Hybrid Membrane/PSA Processes for Gas Separation

Although there is some published work on the development of hybrid membrane/PSA systems for gas separation,[2,3,6,36–38] truly synergistically concepts have only been proposed by Feng *et al.*,[39] Esteves and Mota,[40–43] and Esteves.[44] An important conclusion drawn from these authors is that membrane permeation can be an effective aid in the pressurization and high-pressure adsorption steps of a typical PSA process. The results also indicate the feasibility of incorporating membrane permeation into the blowdown step of the PSA cycle, so that the operating pressure range in the PSA can be used as the driving force for permeation. Therefore, a complete understanding of these hybrid processes for gas separation is crucial, mainly because benefits, such as product quality, plant minimization, environmental impact, and energetic cost reductions, are arising.

Zolandz and Fleming[45] report one example of the two combined technologies for O_2 production from air, for applications on board aircraft, where bulk separation by a membrane is followed by adsorption for further purification. Marcea and Hwang[46] studied O_2 separation from air on a PSA bed with a zeolite followed by O_2 enrichment using a membrane. It was shown that a higher separation degree could be obtained than by either of the two processes alone. Likewise, Baker *et al.*[38] reported the recovery of H_2 and LPG from PSA tail gas, through a combined PSA/membrane process for treating a feed gas containing 90% H_2 and 10% light hydrocarbons (Figure 9.3). More than half of the H_2 formerly lost with the tail gas is recovered and recycled to the feed gas, increasing the overall H_2 recovery of the process from 87% to 95%. The off-gas from the condenser is passed through a silicone-rubber membrane unit that selectively permeates the remaining hydrocarbons.

Baker and Kaaeid[47] applied a membrane to condition gas streams containing H_2 and hydrocarbons, destined for PSA separation. The difficulty that arises from this type of refinery and petrochemical streams is the presence of water vapor, heavier hydrocarbons, and H_2S. These species cause a variety of problems: in PSA they may adsorb preferentially onto the bed and reduce its capacity to adsorb the light hydrocarbons, giving rise to regeneration problems;

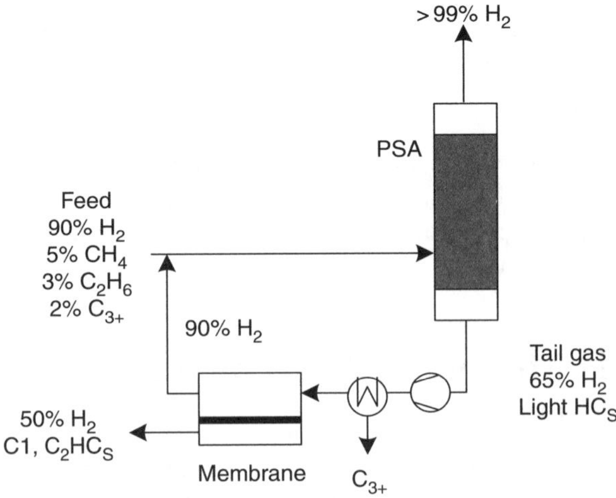

Figure 9.3 Combination of a hydrocarbon-permeable membrane with a PSA unit to increase the recovery of hydrogen from a refinery gas stream from 87 to 95%. Adapted from Ref. 38.

in membranes they may induce its collapse, with flux decreasing and irreversible damages. Usually, for both systems, a pre-treatment of H_2/hydrocarbon mixtures must be carried out to feed only a clean stream to the process. This author's patent[47] describes a process with selective removal of water vapor, H_2S, and C_5–C_8 hydrocarbons using a rubbery membrane (silicone rubber), which preferentially permeates the heavier components, and an adsorption unit capable of selectively remove CH_4 from H_2. Using this scheme, the H_2-enriched stream remains at high pressure on the feed side of the membrane, which means that this stream may be passed directly to the adsorption step of the PSA unit without the recompression that would be needed if the H_2 was in the permeate stream. The latter case is mostly common in industrial separations, through the use of glassy cellulose acetate membranes. The invention is specially developed for mean streams that contain no more than about 80% H_2 and 10% C_5–C_8. Such streams are typically used as fuel gas. A hydrocarbon removal of 86% from the feed mixture was obtained in the membrane unit; the H_2 loss in the permeate stream was about 20%. Nearly 80% of the H_2 from the feed was recovered, with 85% purity, in the residue stream. Thus, this stream was suitable for passing into a PSA process.

Feng *et al.*[39] reported a hybrid, synergistic process that incorporates membrane permeation into a PSA cycle for gas separation. Unlike the traditional membrane processes, where the pressures are kept constant, the pressure in the integrated permeation/adsorption system changes as the permeation proceeds with time. These authors analyzed parametrically permeation processes with varying pressure to study the feasibility of integrating permeation with cyclic PSA processes. They developed two integrated configurations: membrane-assisted feed gas pressurization, and membrane-assisted co-current depressurization. In both cases, the permeation occurs in a cyclic manner and is coupled with the sequential steps of a PSA process. The permeability experiments were performed on an asymmetric cellulose acetate hollow-fiber module, having 20 fibers, 15 cm long and 165 μm OD, with feed into the shell side.

A bench-scale unit was assembled and tested for H_2 purification from a 90/10 wt% H_2/N_2 mixture at 2.17 MPa. The apparatus comprised a membrane module with 341 fibers totaling 778 cm^2 surface area, and two adsorption columns, each packed with 5 g of 5 Å molecular sieve with 350–500 μm particle size. The cyclic operation was synchronized by means of a set of controlled switching valves.

The results show that the integrated process can improve both product purity and recovery, as compared to the standalone PSA unit, especially for the cases where the feed gas mixture contains impurities that are difficult to remove by adsorption. A permeation with varying permeate pressure was investigated. Figure 9.4 presents the schematic diagram of the permeator and receiver system, which would be an adsorber in the integrated membrane/PSA system.

Two cases were considered: one refers to a permeator with constant feed pressure and varying permeate pressure, and another one refers to a permeator

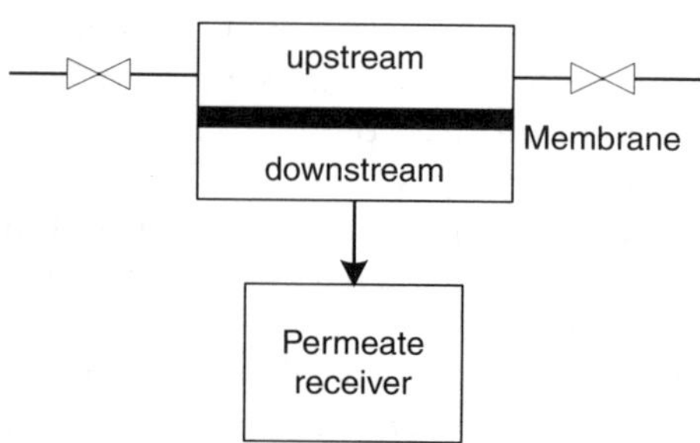

Figure 9.4 Schematic diagram of the permeator and its permeate vessel receiver. Adapted from Ref. 39.

with varying feed and permeate pressures. For the case of a permeator with constant feed pressure and varying permeate pressure, the results show that the pressure in the permeate receiver increases as permeation proceeds with time. The permeation rates determine the rate of pressure built up in the receiver, which in turn affects the permeation rate. The pressure increase is faster initially and gradually slows down. Since the PSA usually operates under fast cycles, this fast pressure built-up is particularly desired for incorporating a membrane into a PSA process. As the permeate pressure builds up, the permeate rate decreases, and since the permeation rate of the fast permeating species is retarded more significantly than the slower one, the permeate concentration decreases with time. Simultaneously, the residue concentration increases with time and approaches the limit of the regular feed concentration when the permeation time is sufficiently long. The effect of the permeate receiver volume is also reported. The smaller the volume, the larger the permeate pressure and the lower the permeate concentration.

Tests with a 50/50 H_2/N_2 mixture at 2.17 MPa in a permeator with about the same size as the permeate receiver result in a 90 wt% H_2 permeate at 0.72 MPa and a residue gas that contains 33 wt% H_2. These results demonstrate that the membrane permeation can be an affective aid to the pressurization step in a typical PSA process. For the case of a permeator with varying feed and permeate pressures, the results indicate the feasibility of incorporating the permeation into the blowdown step of a typical PSA cycle, so that the PSA pressure difference available can be used for permeation operation.

Esteves and Mota,[40–43,48] and Esteves[44] used the contributions from Feng *et al.*[39] for further research and developed innovative schemes for membrane-based hybrid processes for gas separation and waste reduction via adsorption. Figure 9.5 presents the initial configurations developed by Feng *et al.*[39]

Configuration I membrane-assisted feed gas pressurization acts in the feed line of the PSA unit, splitting the mixture into two streams: the permeate, enriched in the fast permeating species and less adsorbed in PSA bed, is fed to column B for pressurization; and the residue, enriched in the slow permeating

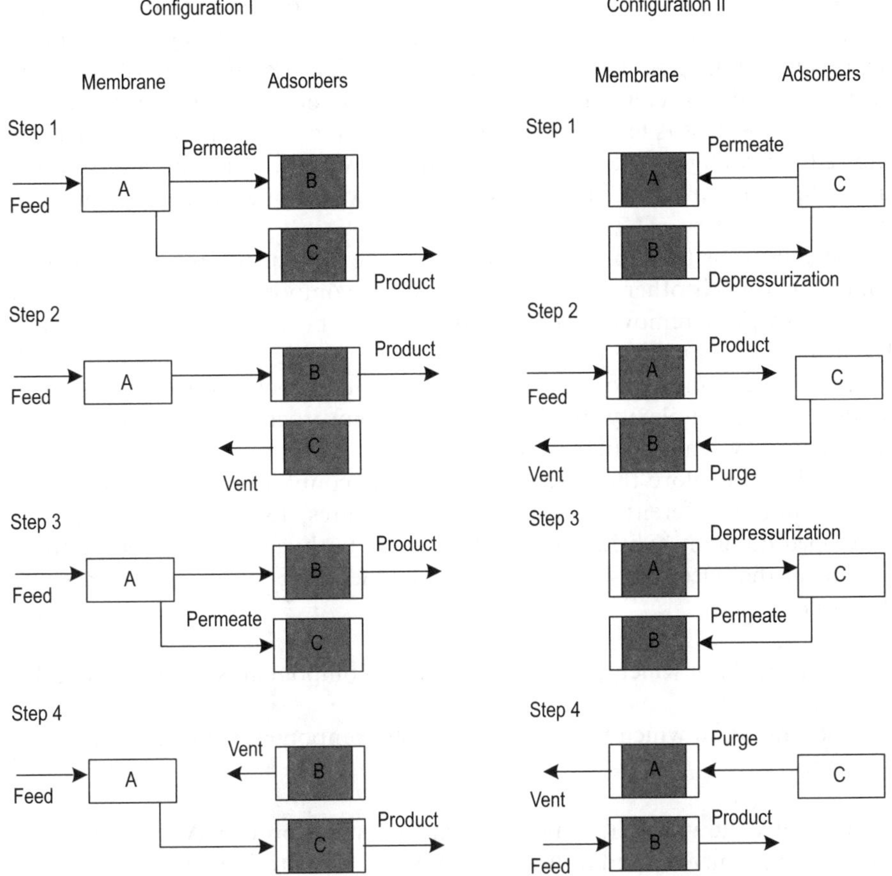

Figure 9.5 Configurations of the coupled membrane/PSA system investigated by Feng and Ghosh.[39] A two-bed PSA system is chosen for simplicity of the principle. I. Operating sequence of a PSA system with membrane-assisted feed gas pressurization. II. Operating sequence of a PSA system with membrane-assisted co-current depressurization. Adapted from Ref. 39.

species, is fed to adsorber C for adsorption (step 1). The increase of the partial pressure of the light species in the bed leads to an increase in the sorption capacity of the adsorber to this species. If the permeate stream is closed (step 2) to restrain permeation, the operation is similar to a typical two-bed process. The purified product is obtained from B and the strong adsorptive impurities, retained in C, are released by venting, which also regenerates the adsorbent for the next adsorption step. The two beds are then reversed. Overall, the system operates in a cyclic fashion.

In configuration II membrane-assisted co-current depressurization the membrane feed inlet is connected to bed B and the permeate outlet is connected to adsorber A that is being pressurized (step 1). Hence, the pressure difference in the PSA is used as the driving force for permeation. Note that the gas used for pressurization is more purified than that produced during the co-current depressurization step. The residue left in the high-pressure side of the membrane is used for counter-current purging the depressurized bed (step 2). Configuration II was tested with a simulated ammonia synthesis purge gas and showed advantages over the standalone PSA unit, since at a given feed throughput, the product purity was enhanced with a marginal reduction in the product recovery.

The concept is therefore straightforward: the two membrane/PSA processes complement each other by removing different components from the mixture. While adsorption removes impurities from the H_2 product, the membrane takes H_2 from the impurities that are retained as residue. Either because of the operational complexity of this type of integrated processes, either due to the mathematical complexity involved in the design procedures, research on feasible hybrid systems is still scarce.

In order to explore the synergy obtained by coupling PSA and membranes for separating different types of gaseous mixtures, Esteves and Mota[40–43,48] developed different hybrid concepts to address both cooperative and opposing regions of the selectivity for the two standalone units; two main schemes were developed:

- Scheme A, in which the more permeable component is the least adsorbed one
- Scheme B, in which the more permeable component is the more adsorbed one.

Figure 9.6 shows a schematic diagram for hybrid process A, consisting of a membrane permeator and a dual-bed PSA unit. The residue is sent directly to the PSA unit but, unlike scheme B, the permeate is temporarily stored in an intermediate tank before being sent to the adsorption process. Also, in scheme A the residue is used to feed the adsorption step instead of the permeate in configuration B.

The cycle for hybrid scheme A starts with an incomplete pressurization (PR_1) of one of the PSA beds using the gas stored in the tank, which is enriched in the least adsorbed species A. The gas stored in the tank corresponds to a permeate stream that was obtained during the previous high-pressure adsorption (HPA) step operating on the other PSA bed. During this step, valve V_1 is kept opened while V_2 and V_3 stay closed until pressure equalization between the tank and the PSA bed is established. To complete the pressurization step (PR_2), the tank outlet is closed by shutting valve V_1, and valve V_3 is opened to pressurize the PSA bed with the residue stream from the membrane unit, which is less rich in species A than the permeate stream employed in step PR_1. During PR_2, the membrane behaves essentially as an empty tube, since both permeate and residue sides are at feed pressure P_h. Thus, during this step, the residue stream essentially has the same composition as the regular feed.

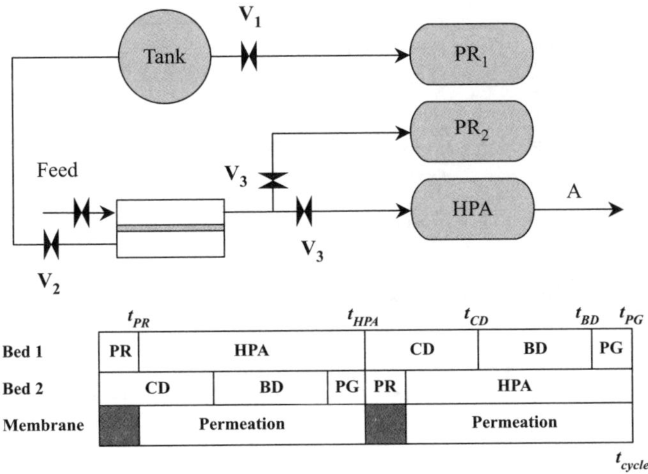

Figure 9.6 Schematic diagram and scheduling chronogram of elementary steps comprising the cyclic operation of hybrid scheme A. The notation is as follows: PR, pressurization; HPA, high-pressure adsorption; CD, co-current blowdown; BD, counter-current blowdown; PG, low-pressure purge. The product from the HPA step is enriched in the less adsorbed species (A). Gray area of chronogram identifies periods during which the membrane is working as an empty tube. Reprinted with permission from Ref. 41.

The cycle then follows with the HPA step, which is initiated by opening valve V_2 while feeding the PSA with the residue stream from the membrane, at a prescribed flow rate. The residue stream is enriched in the strongly adsorbed component B, while the permeate is stored in the intermediate tank to be employed in the next cycle. During the HPA step, the residue pressure is kept constant at the high-pressure value, P_h, whereas the permeate pressure increases with time due to gas build-up in the tank. This happens first through fast equalization between the tank and the permeate side of the membrane and then slowly as more gas is driven through the membrane while both the permeate side of the membrane and the tank together build up pressure at the same rate. The two pressurization steps, PR_1 and PR_2, as well as the HPA step, are illustrated schematically in Figure 9.7.

Ultimately, the PSA cycle proceeds with the following steps: co-current blowdown (CD) to recover the residual amount of A that was pushed to the end of the bed during the HPA step; counter-current blowdown (BD) and purge (PG) to recover species B and to regenerate the bed for the next cycle. During these steps, the membrane module operates with the other PSA bed. Although the operation of each bed is batch-wise, the system as a whole is a continuous one that is operated under cyclic steady-state conditions. Unlike conventional membrane operation, in which the pressures are kept constant, here the permeation occurs cyclically due to its coupling to the PSA cycle.

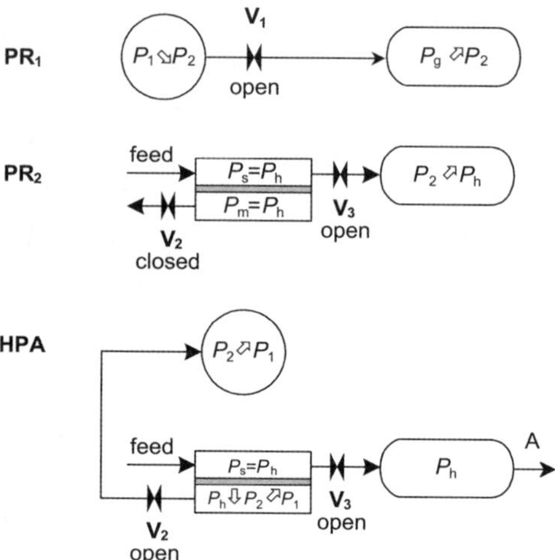

Figure 9.7   Schematic of integrated cycle for scheme A. P_1 is the intermediate pressure in the tank after permeation at the end of the HPA step; P_2 represents the equalization pressure at the end of step PR$_1$. Downward and upward arrows indicate decreasing and increasing pressure changes, respectively. Reprinted with permission from Ref. 41.

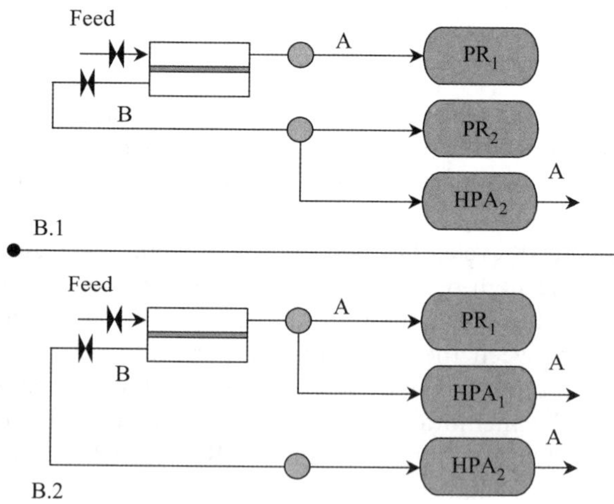

Figure 9.8   Schematic diagrams of the elementary steps comprising the cyclic operation of hybrid scheme B; A and B are the less and more strongly adsorbed species, respectively. Reprinted with permission from Ref. 41.

Figure 9.8 shows the schematic diagram of hybrid process B, which is applicable when the more strongly adsorbed species is also the more permeable one. In this scheme, the pressures on both sides of the membrane are kept constant during permeation. This is another major difference with respect to

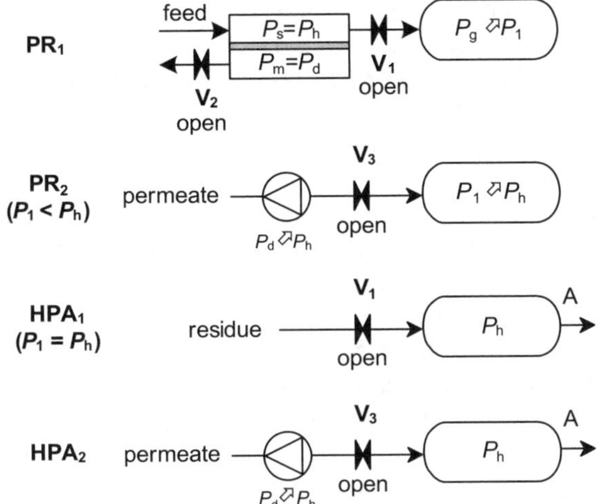

Figure 9.9 Schematic of integrated cycle for scheme B. P_1 is the intermediate pressure in the adsorption bed after permeation at the end of PR$_1$. Upward arrows indicate increasing pressure changes. Reprinted with permission from Ref. 41.

hybrid scheme A. Although the residue stream is sent directly to the PSA, the permeate stream is temporarily stored and is either used to complete the pressurization step or is fed to the bed during the adsorption step. This depends on the total feed amount admitted per cycle. For a more comprehensive understanding of this process, a schematic diagram of its cyclic operating principle is given in Figure 9.9.

The integrated cycle for scheme B starts with a permeation step during which the residue side of the membrane is kept at P_h, where P_h is the pressure value in the adsorption bed during the HPA step, and the permeate side of the membrane is at a lower pressure P_m. The value of P_m must be appropriately selected for each separation because, unlike in process A, the gas coming from the permeate must be repressurized to feed the HPA step. Simultaneously, an initial PR$_1$ takes place with the residue eluent stream from the membrane, enriched in A, until pressure equalization between the membrane and the bed is established. Subsequently, two situations can occur:

- If the pressure in the bed is lower than the adsorption pressure, P_h, the membrane residue is insufficient to complete the pressurization step (case B.1 in Figure 9.8) and the bed is pressurized with permeate gas enriched in the more strongly adsorbed species B (step PR$_2$). After PR$_2$, the membrane is no longer operational for this PSA bed. Then, a high-pressure adsorption step (HPA$_2$) is carried out by feeding the PSA with the rest of the permeate obtained from the membrane at a prescribed flow rate. During this stage, the permeate pressure is kept constant at pressure P_h.
- If the adsorption pressure is attained during permeation and there is residue left from the membrane (case B.2 in Figure 9.8), the cycle proceeds

with a high-pressure adsorption step (HPA_1) using that stream enriched in the lighter component A. After HPA_1, the membrane is no longer operational for this bed. The bed pressure is kept constant at P_h, and the high-pressure adsorption step is then completed using the permeate enriched in component B (step HPA_2).

As in scheme A, the PSA cycle proceeds with co- and counter-current blowdowns, in which the bed is depressurized co-currently to the feed from pressure P_h to an intermediate value P_d and then down to P_g counter-currently to the feed. Finally, a low-pressure ($P = P_g$) purge (PG) with part of the HPA product may occur to recover species B and to regenerate the bed for the next cycle. Hybrid scheme A was detailed and analyzed for H_2/CH_4 separation as a case study, whereas scheme B was studied for CO_2/CH_4 and CO_2/N_2 separations. In order to validate the hybrid concept, experimental studies were performed for CO_2/N_2 system, for both standalone PSA and coupled processes.

For brevity of presentation, we do not reproduce in full detail the model equations and parametric analysis for each scheme developed.[41,44,48,49] Instead, we highlight some of the more representative results that demonstrate the hybrid concept and its enhancements relative to the more conventional units. The dimensions of the laboratory-scale PSA unit parameters, adsorbent properties, membrane properties and main process simulation parameters are listed in Tables 9.1 and 9.2.

Parametric studies and assessment of the performance of the hybrid schemes were implemented in gPROMS, which is a well-known software package for the modeling and simulation of lumped- and distributed-parameter process models with combined discrete and continuous characteristics.[50,51] To simplify the computational modeling, the following assumptions have been considered for the membrane module: isothermal operation and ideal gas law, constant permeances, shell-side feed, and counter-current operation with negligible pressure drop on both sides of the membrane. The dynamics of the PSA unit were modeled using a non-isothermal and variable-velocity axially dispersed plug-flow model, with negligible pressure loss inside the PSA bed. A pore-diffusion model was assumed to govern mass transfer inside the adsorbent particles.[52,53] Transport and thermodynamic properties were taken as constant and temperature independent. Multicomponent adsorption equilibrium was modeled by the Sips isotherm model.

9.2.1 Scheme A: the More Permeable Component is the Least Adsorbed

The performance of scheme A was assessed and compared to that of a standalone PSA unit for the separation of a 50/50% (v/v) H_2/CH_4 mixture. The effects of various operating parameters, such as permeation throughput, total feed amount per cycle, purge-to-feed ratio, and adsorption and blowdown pressures, were analyzed through detailed process simulation. The PSA

Table 9.1   Simulation parameters for H_2/CH_4 separation using hybrid process A

Parameter	Value	Parameter	Value
Physical properties for PSA unit			
Particle diameter, d_p (cm)	0.028	Column radius, R_c (cm)	2.05
Interparticle porosity, ε_p	0.61	Column length, L (cm)	60.0
Particle density, ρ_p (g cm^{-3})	0.850	Wall	
Bulk density, ρ_b (g cm^{-3})	0.498	Thickness, e_w (m)	10^{-3}
Particle tortuosity, τ	10	Density, ρ_w (g cm^{-3})	8.22
Interparticle porosity, ε	0.43	Heat capacity C_{pw} (cal g^{-1} K^{-1})	0.110
Bed heat capacity, C_{pi} (cal g^{-1} K^{-1})	0.250	Conductivity, k_w (W m^{-1} K^{-1})	43.3
Bed conductivity, D°_h (W m^{-1} K^{-1})	0.106	Heat transfer coef., h_w (cal m^{-2} K^{-1} s^{-1})	30
Parameters of Sips adsorption isotherm model			
q_{m,H_2} (mol kg^{-1})	$1.795/T$	q_{m,CH_4} (mol kg^{-1})	$0.308T^{-0.59}$
n_{H_2}	1.00	n_{CH_4}	0.96
b_{0,H_2} (psi^{-n})	3.80×10^{-5}	b_{0,CH_4} (psi^{-n})	4.23×10^{-6}
Q_{H_2} (J mol^{-1})	1.73×10^3	Q_{CH_2} (J mol^{-1})	1.23×10^3
Physical properties for membrane unit			
Tank volume V_v (L)	2.5 (0.25–2.5)	Membrane selectivity, α_{H_2/CH_4}	35
Membrane volumes, $V_a = V_m$	$0.02V_v$	Permeation flow parameter, F_m	17.18 (5.15–51.5)
Feed mixture			
Composition, y_{if} (%)	50/50	Intraparticle diffusivity, D_c (m^2 s^{-1})	2×10^{-8}
Temperature, T_f (K)	293.15	Molecular diffusion, PD_m (m^2 bar s^{-1})	6.85×10^{-5}
Heat capacity, C_{pg} (cal mol^{-1} K^{-1})	7.647		
Operating parameters			
Feed pressure, P_h (bar)	35.5 (25.5–35.5)	Pressurization, Δt_{PR} (s)	30
Purge pressure, P_1 (bar)	1.21	High-pressure adsorption, t_{HPA} (s)	180
P/F ratio	0.09 (0.06–0.22)	Co-current blowdown, t_{CD} (s)	90
Blowdown pressure, P_d (bar)	12.73 (4.73–14.73)	Counter-current blowdown, t_{BD} (s)	90
Feed amount per cycle (L STP)	47.7 (47.7–80)	Purge, t_{PG} (s)	
(STP: 0 °C, 1 atm)		Cycle time, t_{cyc} (min)	7.0

Numbers in parentheses represent ranges spanned in parametric study.

employs activated carbon as the adsorbent,[52] whereas the membrane permeances are those commonly expected for a polysulfone membrane with a 0.1 mm thick separation layer.[54] It has a typical gas selectivity α_{H_2/CH_4} of 35; the

Table 9.2 Parameters for CO_2/N_2 separation using hybrid process B

Parameter	Value	Parameter	Value
Adsorbent and PSA column			
Intra-particle porosity, ε_p	0.656	Wall	
Intra-particle porosity, ε	0.476	Thickness, e_w (mm)	9.5
Particle density, ρ_p (g cm^{-3})	0.763	Density, ρ_w (g/cm^3)	8.221
Particle hydraulic radius, r_p (mm)	1.668	Heat capacity, C_{pw} (cal g^{-1} K^{-1})	0.110
Mean pore radius, r_μ (Å)	12.82	Conductivity, k_w (W m^{-1} K^{-1})	43.30
Bed length, L_b	40–60	Heat transfer coef., h_w (W m^{-1} K^{-1})	63.2
Bed radius, R_b (cm)			

Sips adsorption isotherm model[48]
($r^2 > 0.999, 0.011$ FSE(N_2), 0.013 FSE CO_2), Residuals < 0.04 mol kg^{-1})

Parameter	Value	Parameter	Value
Loading at saturation: q_{m,N_2} (mol kg^{-1})	11.282	q_{m,CO_2} (mol kg^{-1})	20.608
Heterogeneity parameter: n_{0,N_2}	1.140	n_{0,CO_2}	1.252
Exponent parameter: α_{N_2}	0.383	α_{CO_2}	0.126
Affinity coefficient at T_Q: b_{0,N_2} (bar^{-n})	0.018	b_{0,CO_2} (bar^{-1})	0.036
Interaction parameter: η_{N_2}	1.04	η_{CO_2}	0.76
Reference temperature: T_{0,N_2} (K)	288.2	T_{0,CO_2} (K)	299.1
Heat of adsorption: Q_{N_2} (kJ mol^{-1})	8.974	Q_{CO_2} (J mol^{-1})	19.583

Membrane unit

Parameter	Value	Parameter	Value
Cross section relation	$V_k = 0.1 V_{PSA}$	Membrane selectivity, α_{CO_2/N_2}	34.4
Permeation area, A_{mem} (m^2)	0.14–0.82	CO_2 permeance, ρ_{CO_2} (GPU)	30.3

Operating parameters range

Parameter	Value	Parameter	Value
Feed pressure, P_h (bar)	6–12	Feed temperature, T_f (K)	299–319
Purge pressure, P_1 (bar)	0.34–0.44	Feed composition, y_f (%)	30; 50; 80
P/F ratio	0.0–0.8	Feed amount per cycle (L STP)	20.6–180.6
HBD pressure, P_{HBD} (bar)	2.5–5.5	LBD pressure, P_{LBD} (bar)	1.0–2.0
Feed amount per cycle (L STP)	20.6–180.6	Dispersion parameters, $\gamma_1, \gamma_2, \gamma_3$	0.7, 0.5, 0.75

H_2 and CH_4 permeances are assumed to be 100 GPU and 2.86 GPU, respectively. Process and operating parameters are listed in Table 9.1. The values of the main operating parameters for the reference case are: $V_v = 2.5$ dm^3, $F_m = 17.2$, $P_h = 35:5$ bar, $P_d = 12.7$ bar, $P_1 = 1.21$ bar, $P/F = 0.09$, and $F = 47.7$ dm^3 (STP), which were then changed in order to study the influence of each operating parameter.

The hybrid scheme required 11 cycles of operation to attain CSS. Unlike conventional membrane operation where the two pressures P_s and P_m are kept constant, here P_m swings cyclically due to its coupling to the PSA unit. As expected, the larger the volume of intermediate tank, the higher its equalization pressure and the lower the value of P_m at the end of the HPA step.

During the HPA step, permeation occurs driven by the pressure difference, $P_s - P_v$, between the residue side of the membrane and the storage tank. Since the permeation rate decreases as P_m increases, because of gas build up in the tank, the mole fraction of H_2 in the permeate stream decreases with time, approaching its value in the feed gas. This effect is more pronounced for smaller storage tanks.

Figures 9.10 and 9.11 show the effect of total feed amount admitted per cycle on product purity and recovery for both the standalone PSA unit and the integrated system. The results obtained for two different values of P_h, 25.5 bar and 35.5 bar, show that for the same separation performance, the hybrid system withstands a higher feed throughput than the PSA operating alone. Increasing the feed flow rate results in longer bed coverage at the end of the HPA step, thus improving both purity and recovery. If the total feed amount admitted per cycle exceeds 65 L (STP), the H_2 product is diluted with CH_4

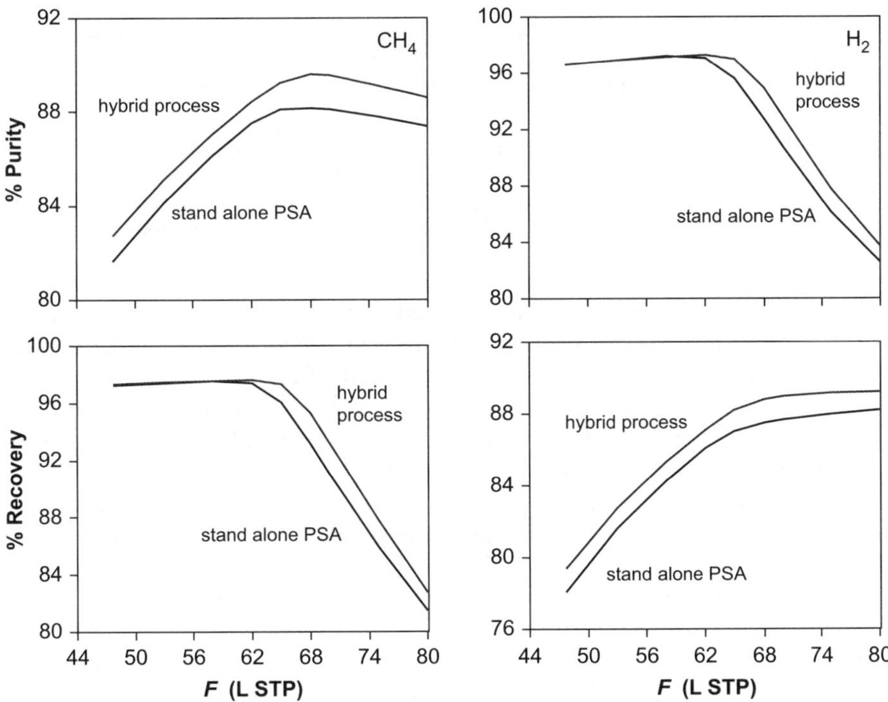

Figure 9.10   Impact of total feed amount per cycle, F, on CH_4 and H_2 purities and recoveries, for both standalone PSA and integrated process, at $P_h = 25.5$ bar. All other operating parameters are fixed at their reference values. Lines represent simulated results. Reprinted with permission from Ref. 41.

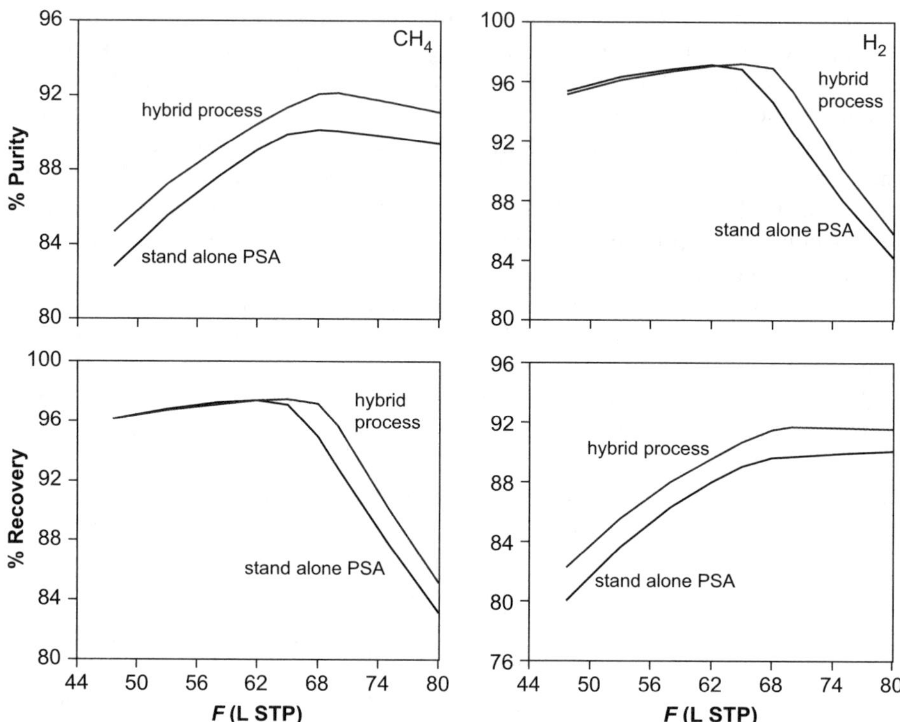

Figure 9.11 Impact of total feed amount per cycle, F, on CH_4 and H_2 purities and recoveries, for both stand-alone PSA and integrated process, at $P_h = 35.5$ bar. All other operating parameters are fixed at their reference values. Lines represent simulated results. Reprinted with permission from Ref. 41.

and both H_2 purity and CH_4 recovery deteriorate significantly. Nevertheless, in this region, it is clearly seen that the hybrid process performs better than the conventional PSA unit for the same amount of feed.

The purge step regenerates the bed for the next cycle and recovers the product stream enriched in CH_4 by striping it out of the bed. Increasing the purge-to-feed ratio, P/F, steadily increases H_2 purity and CH_4 recovery, but with loss in H_2 recovery and dilution of the CH_4 product stream (Figure 9.12). It is worth noting that the purge step is performed at the expense of partial consumption of H_2 product obtained during the HPA step. It is also observed that the H_2 recovery and CH_4 purity for the integrated system are higher than those for the conventional PSA unit operating alone.

The effect of the intermediate blowdown pressure, P_d, on separation performance of the integrated system was also analyzed. As expected, increasing P_d, which corresponds to the bed pressure at the end of the first depressurization step, the purity of the less adsorbed species, H_2, increases significantly, but less product is recovered. The opposite trend occurs for CH_4, where the product is more diluted as P_d is increased.

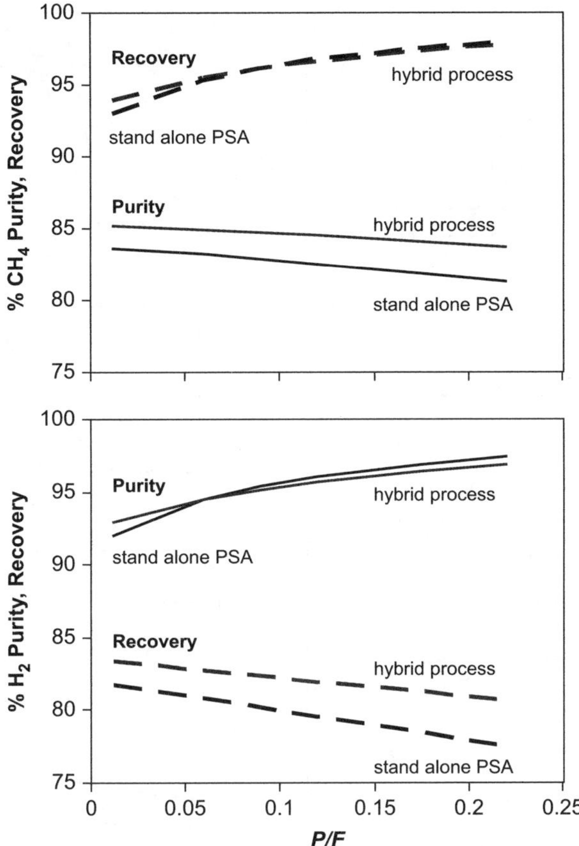

Figure 9.12 Impact of P/F ratio on product purity and recovery, for both standalone PSA and integrated process. Operating parameters are fixed at their reference values. Lines represent simulated results. Reprinted with permission from Ref. 41.

Finally, the simulated H_2 and CH_4 concentration profiles along the packed bed for the standalone PSA unit were compared with those for scheme A, considering two different totals feed amounts: $F = 47.7$ and $F = 70.0$ L (STP). This is illustrated in Figures 9.13 and 9.14. According to Figure 9.11, these amounts correspond to the lowest feed throughput considered in the parametric study, as well as the feed throughput that exhibits the maximum process performance for the given operating conditions. The comparison between the profiles for the simple PSA and the ones for the hybrid process is carried out at the end of each cycle step. It is seen that the concentration profiles for scheme A are sharper than those for the PSA alone. Once again, this results in an enhanced separation performance for the hybrid process. Table 9.3 lists the average composition of the various gaseous streams fed to the PSA during the PR and HPA steps for the reference case analyzed in this work.

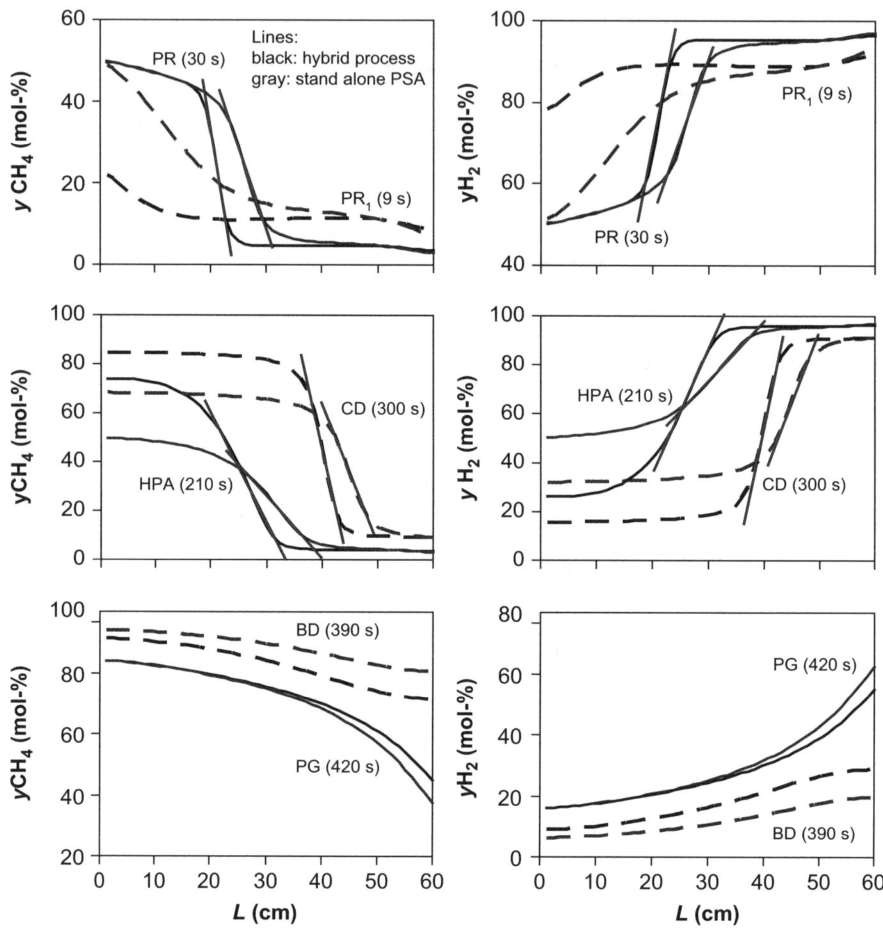

Figure 9.13   Simulated CH$_4$ (left) and H$_2$ (right) axial composition profiles in gas phase along the packed bed at the end of each PSA step, under CSS conditions, for both standalone PSA and integrated process. Operating parameters are fixed at their reference values, except for the feed amount which is $F = 47.7$ L STP. Reprinted with permission from Ref. 41.

Unlike in a conventional PSA process, the adsorbent bed in the integrated system is fed with a varying-composition gas stream, initially rich in the more permeable but less adsorbed component, which is progressively enriched in the other component having opposite behavior. The membrane performs a pre-bulk separation of the feed, and simultaneously works with the intrinsically dynamic periodic operation of the PSA, to enhance separation performance when compared to the two individual units. This sequence of progressively heavier feed steps has the net effect of sharpening the composition wave fronts inside the adsorbent bed, giving rise to a decrease of band broadening and higher product purity. This effect is more pronounced in the high-feed throughput region.

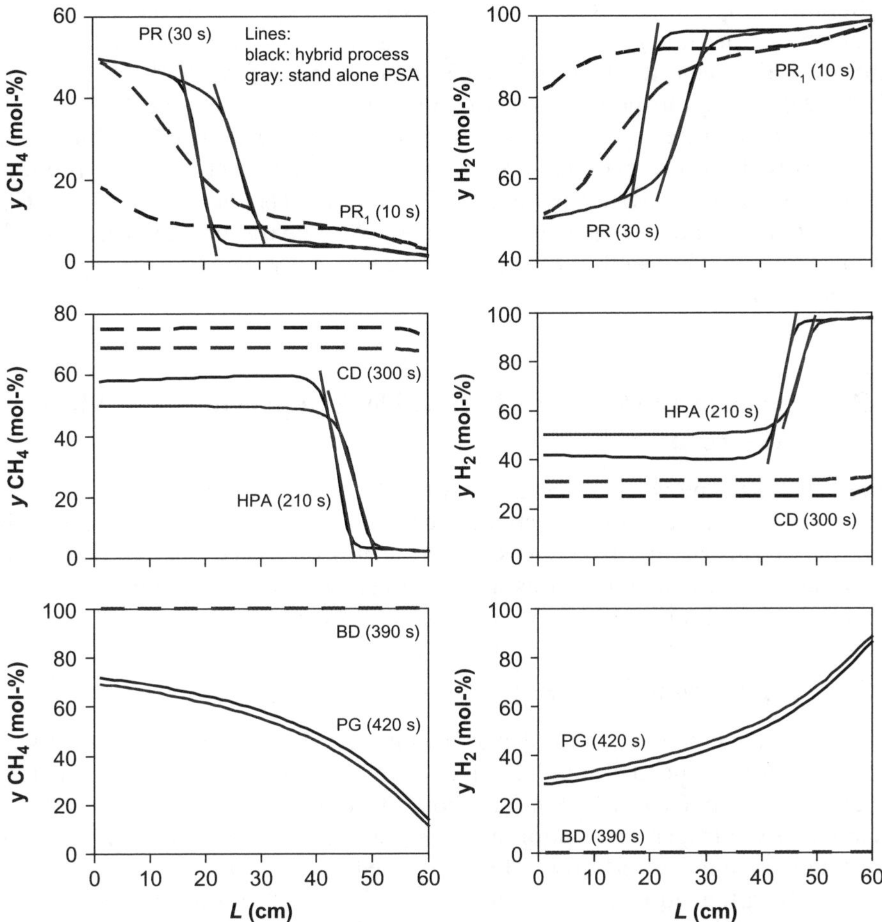

Figure 9.14 Simulated CH_4 (left) and H_2 (right) axial composition profiles in gas phase along the packed bed at the end of each PSA step, under CSS conditions, for both standalone PSA and integrated system. Operating parameters are fixed at their reference values, except for the feed amount which is $F = 70.0$ L STP. Reprinted with permission from Ref. 41.

Table 9.3 Average composition and recovery, when applicable, of the various streams fed to the PSA during the PR and HPA steps for the reference case analyzed in this work

Stream	Step	% CH_4 (v/v)	% H_2 (v/v)	Product recovery
Feed	PR_2	50.0	50.0	—
Residue	HPA	57.9	42.1	92.1
Permeate	PR_1	18.5	81.5	32.4

Summarizing, the assessment of hybrid scheme A shows that the inclusion of a membrane module into the cyclic adsorption process in a synergistic mode, improves the separation performance when compared with the standalone PSA. Depending on the values of the operating parameters, and for an equimolar feed mixture at 35 bar, a 7 min integrated cycle gives H_2 and CH_4 products with purities within 83–97% and 81–99%, respectively. The product recoveries obtained are in the range of 77–99% for H_2 and 81–98% for CH_4. Whether the obtained improvements for a given separation are large enough to overcome capital and extra operating costs, must be assessed on a case per case basis.

9.2.2 Scheme B: the More Permeable Component is the More Adsorbed

The hybrid scheme B was successfully applied to the bulk separation of CO_2/N_2 and CO_2/CH_4 mixtures over activated carbon.[44] The process performance is reported in this chapter for CO_2/N_2 in terms of product recovery and purity at CSS. The numerical simulations are validated with experimental work on a composite membrane and a laboratory scale PSA unit.

9.2.2.1 Experimental Unit

The adsorbent employed in the experimental work is a coal based, high-activity (109% CTC), extruded carbon (2 mm diameter pellets), courtesy from Sutcliffe Speakman Carbons Ltd (UK). The pore structure of the carbon was characterized by nitrogen adsorption at 77 K and mercury porosimetry.[48] A few structural parameters for this carbon are listed in Table 9.2.

The adsorption equilibria measurements of N_2 and CO_2 on activated carbon were performed using a standard static gravimetric method. Further details of these measurements are reported elsewhere.[44,48] The Sips isotherm extended to multi-component adsorption was adopted to fit the experimental equilibrium data (Table 9.2). This model has the Langmuirian form applied to non-uniform surfaces and it has been extensively used to model gas adsorption on microporous adsorbents and PSA systems.[23,33] Optimal parameters were found for the adsorption isotherm model, by fitting simultaneously all the data at multiple temperatures. A global isotherm was obtained for each species as illustrated in Figure 9.15.[48]

Figure 9.16 shows the PSA unit built in-house; it is a single-column set-up operated automatically. Further details are described elsewhere.[49] A five-step PSA cycle was developed for separation of CO_2/N_2 mixtures: pressurization (PR) with feed, high-pressure adsorption (HPA), co-current blowdown (HBD), counter-current blowdown (LBD), and low-pressure purge (LPG). During HPA and HBD, N_2 product is obtained, while during LBD and LPG, CO_2 is produced. The PSA model was validated against experimental runs and subsequently employed to study process performance in terms of product purity

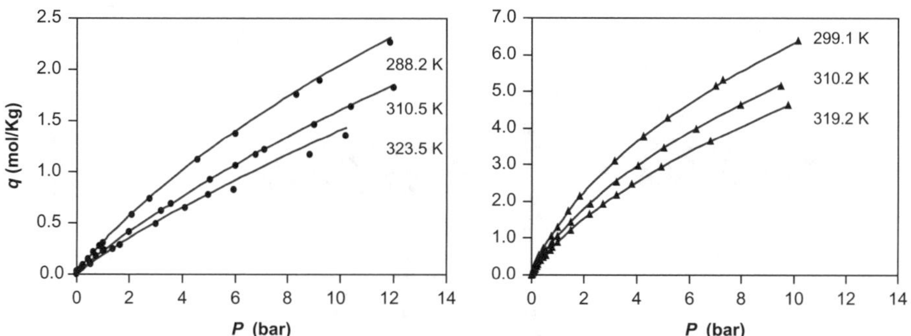

Figure 9.15 Fitting of the experimental adsorption data for N_2 (left) and CO_2 (right) by the Sips isotherm model, whose parameters are listed in Table 9.2.[42] Symbols represent experimental data and lines are the isotherm model. Adapted from Ref. 48.

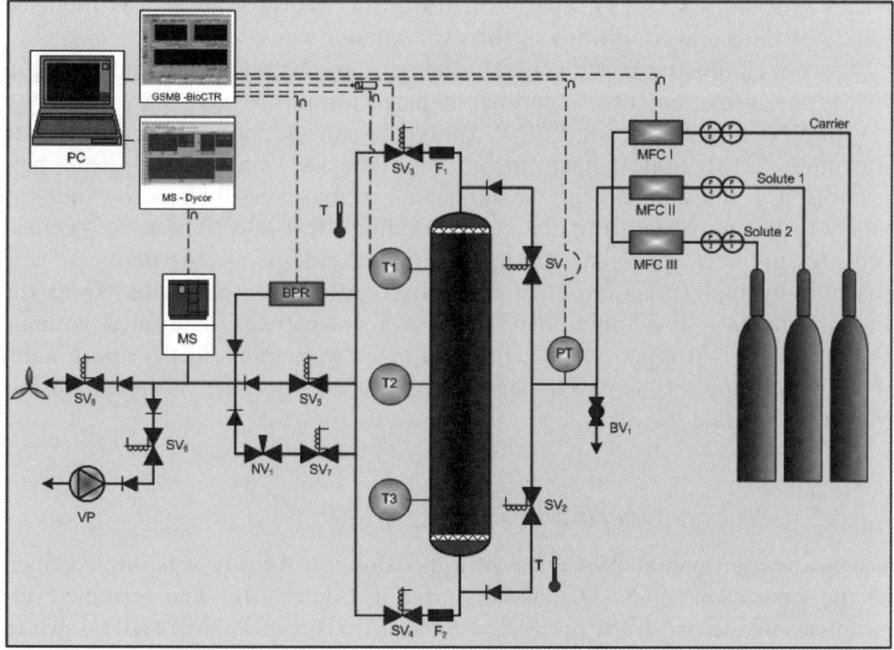

Figure 9.16 Schematic of the one-column PSA experimental apparatus. Symbols: BPR, back-pressure regulator; MFC, mass-flow controller; SV, solenoid valve; BV, ball valve; F, filter; VP, vacuum pump; MS, mass spectrometer.

and recovery. Temperature, gas- and adsorbed-phase concentration profiles and velocity profiles along the PSA bed were analyzed, as well as their dynamic behavior histories of those variables during the PSA cycle and until the periodic steady-state was attained. In order to confirm the enhanced performance of the

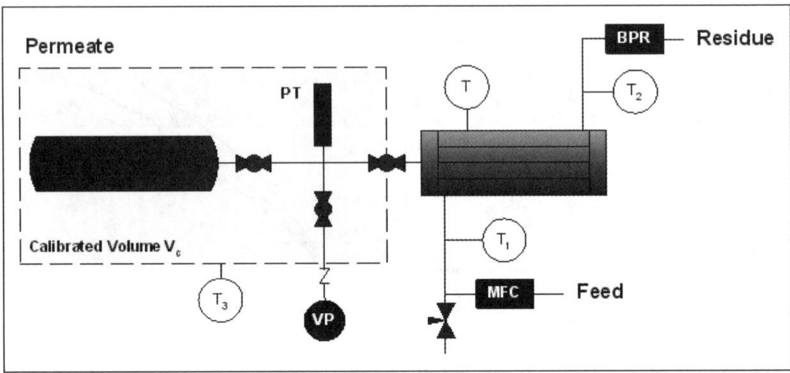

Figure 9.17 General schematic of the membrane unit set-up used in the permeation measurements of pure components.

hybrid process, a CO_2/N_2 separation using the hybrid scheme was modeled exactly at the same conditions of the experimental runs.

Figure 9.17 illustrates the general schematic of the membrane unit used in permeation measurements. Experimental permeation runs were performed with CO_2 and N_2 in a composite PES/PI (polyethersulfone Sumikaexcel/polyimide Matrimid 5218) hollow-fiber membrane.[55,56] The single-component permeances were measured using a standard variable-pressure method and the binary ratios converted into the corresponding ideal selectivities. A specified feed pressure was applied to the upstream shell-side, while the permeate side was initially under vacuum. The permeation rates were calculated from the pressure increase as a function of time, in a downstream calibrated volume. The permeation stopped when equalization between pressures on both sides of the membrane was achieved. The main physical properties of the membrane are listed in Table 9.2.

9.2.2.2 Process Operation

A typical experimental PSA cycle of approximately 14 min was implemented for the separation of CO_2/N_2, as illustrated in Figure 9.18. The periodic cycle comprises five steps: (i) PR up to $P_h = 6$ bar, (ii) HPA at P_h, (iii) HBD down to $P_M = 3.2$ bar, (iv) LBD down to $P_l = 0.3$ bar, and (v) LPG at P_l. During HPA and HBD, N_2 product is obtained, while during LBD and LPG, CO_2 is produced. In order to confirm the performance enhancement of the hybrid process, this system was modeled both as a conventional standalone PSA process and as an equivalent hybrid unit at exactly the same experimental operating conditions. Both models required 11 cycles to attain CSS regime; this suggests that adding the membrane module does not slow down the convergence to cyclic steady state. The results show that for the simple PSA and depending on the feed composition and amount of feed admitted per cycle, the CO_2 purities and

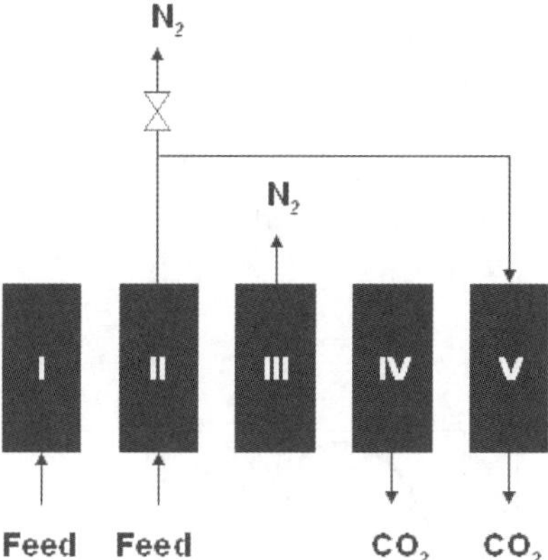

Figure 9.18 Schematic diagram of the experimental PSA cycle implemented for the CO_2/N_2 separation. (I) PR from $P_1 = 0.3$ bar to $P_h = 6.0$ bar; (II) HPA at P_h with N_2 recovery; (III) HBD from P_h to $P_d^{CD} = 3.2$ bar with N_2 recovery; (IV) LBD from P_d^{CD} to P_1 with CO_2 withdraw; (V) LPG at P_1 through N_2 product and with CO_2 withdraw.

recoveries are generally in the range of 54–92% and 83–99%, respectively; for N_2, the purity and recovery ranges are 51–99% and 66–69%, respectively. The good agreements between the predicted and experimental pressure profiles and the predicted and experimental CO_2 effluent composition histories, obtained for a typical PSA run under CSS, validates the standalone PSA model.

In order to assess scheme B, two experimental runs, one for the standalone PSA (Run A) and another for the integrated process (Run B), were performed for CO_2/N_2 separation.[49] Figure 9.19 shows both experimental and predicted pressures, CO2 gas-phase composition, and temperature histories under CSS, for the two illustrative runs. The simulated results successfully reproduce the experimental temporal profiles obtained, predicting the cyclic behavior of both processes.

Instead of the regular 30/70% (v/v) mixture fed to the standalone PSA, in the hybrid process an average feed stream with only 16.5% CO_2 is initially fed to the PSA. This fact is highly advantageous, since both the PR and HPA_1 steps are performed with a feed stream enriched in the lighter component (N_2), which is the first to be withdrawn from the column as high-pressure product. This first product is then obtained in Run B at higher purity (94.1% against 84.5% in the traditional PSA). Subsequently, the HPA_2 step occurs with a feed stream enriched in the more adsorbed component, while high-purity N_2 (91.7%) is simultaneously recovered. The HBD step produces a purer N_2 stream than that obtained in the standalone PSA process. Subsequently, the CO_2 product is collected in the LBD step, were an increase of 23% in the CO_2 average purity is obtained relative to Run A. Note that these enhancements observed in the hybrid scheme have to be balanced with the need to compress the membrane permeate stream in order to feed it to the PSA unit.

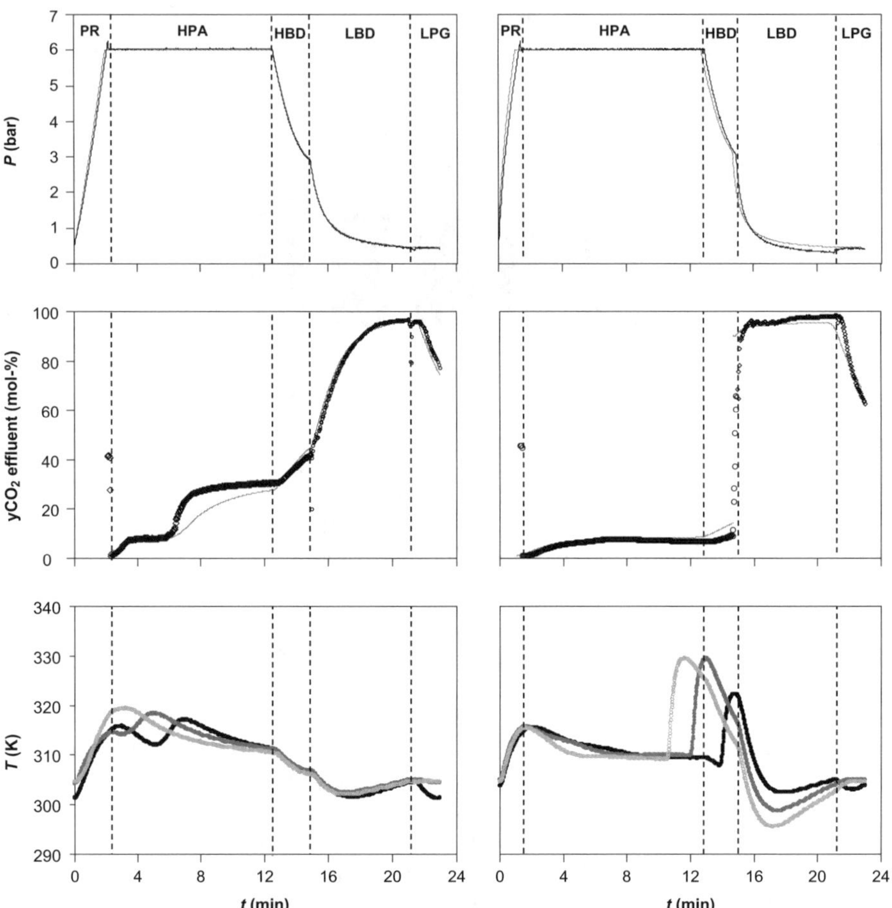

Figure 9.19   Experimental (symbols) and predicted (lines) pressure, CO_2 gas-phase composition and temperature (from left to right: bottom, middle and top of the column) histories for Run A (simple PSA, left) and Run B (hybrid process, right) at CSS. CO_2/N_2 30/70% (v/v) feed mixture at 309 K, $P_h = 6$ bar, $P_l = 0.4$ bar, total feed flow rate of 3.0 slpm, total feed/cycle of 40 L STP (STP: 0 °C, 1 atm), $\alpha_{CO_2/N_2} = 34.4$, and total surface area available for permeation $A_{mem} = 0.82$ m^2. Reprinted with permission from Ref. 49.

9.3   Concluding Remarks

Future research and development in hybrid membrane/adsorption-based technologies will certainly persist in several areas of intervention, either due to the greenhouse effect, to meet environmental legislation, to minimize costs, or simply to reduce waste and save resources. This will include mainly landfill gas (CO_2/CH_4); flue gases (CO_2/N_2); natural gas (N_2/CH_4); synthesis gas ($CO/CO_2/H_2/CH_4$); moisture, O_2 and N_2 from air; H_2 from reformer off/gases; CO_2

sequestration; olefin/paraffin separation; VOC vapor recovery; and automotive fuel emissions control.

Production and purification of raw materials and products using new reversible and resistant adsorbents, and permeating materials are still needed in the industrial field. Fully developed processes, such as PSA and membranes, should be not only synergistically integrated and optimized to improve energy efficiency (*i.e.* higher recoveries at given purities and differential pressures), but also the turning point to search for innovative technologies and feasible schemes. Moving-bed and simulating moving-bed technologies applied to gas-phase, incorporating a complete eluent closed-loop recycle and minimized two-column schemes are examples very recently explored by the scientific community.[57,58] Moreover, the use of molecular simulation, process modeling and optimization strategies are still required to fulfill the gap concerning both the interaction between organic molecules and materials,[59,60] and their impact on process design. Note that this should be encouraged by factors that boost the innovation, such as the capital intensity of the new processes, their amortization period and technical superiority relative to the more traditional ones.

References

1. P. P. Radecki, J. C. Crittenden, D. R. Shonnard and J. L. Bullock (ed.). *Emerging Separation and Separative Reaction Technologies for Process Waste Reduction, CWRT-AIChE*, New York, 1999.
2. R. W. Baker, *Ind. Eng. Chem. Res.*, 2002, **41**, 1393.
3. E. Drioli and M. Romano, *Ind. Eng. Chem. Res.*, 2001, **40**, 1277.
4. H. Strathmann, *AIChE J.*, 2001, **47**, 1077.
5. S. A. Stern, *J. Membr. Sci.*, 1994, **94**, 1.
6. W. J. Koros and G. K. Fleming, *J. Membr. Sci.*, 1993, **83**, 1.
7. W. J. Koros, *J. Membr. Sci.*, 1999, **155**, 145.
8. J. D. Wind, C. Staudt-Bickel, D. R. Paul and W. J. Koros, *Ind. Eng. Chem. Res.*, 2002, **41**, 6139.
9. J. D. Wind, D. R. Paul and W. J. Koros, *J. Membr. Sci.*, 2004, **228**, 227.
10. E. K. Lee and W. J. Koros, *Encyclopedia of Polymer Science and Technology*, Academic Press, New York, 2002, p. 279.
11. R. Mahajan, D. Q. Vu and W. J. Koros, *J. Chin. Inst. Chem. Eng.*, 2002, **33**, 77.
12. K. Ghosal and B. D. Freeman, *Polym. Adv. Technol.*, 1994, **5**, 673.
13. L. M. Robeson, *J. Membr. Sci.*, 1991, **62**, 165.
14. R. Mahajan and W. J. Koros, *Ind. Eng. Chem. Res.*, 2000, **39**, 2692.
15. R. Mahajan and W. J. Koros, *Polym. Eng. Sci.*, 2002, **42**, 1432.
16. D. Q. Vu, W. J. Koros and S. J. Miller, *J. Membr. Sci.*, 2003, **211**, 311.
17. C. Hibshman, C. J. Cornelius and E. Marand, *J. Membr. Sci.*, 2003, **211**, 25.
18. J. D. Wind, *et al.*, *Macromolecules*, 2003, **36**, 1882.
19. C. Cao, *et al.*, *J. Membr. Sci.*, 2003, **216**, 257.

20. M. E. Rezac, E. T. Sorensen and H. W. Beckham, *J. Membr. Sci.*, 1997, **136**, 249.
21. S. N. Ushakov and I. Szanto, *Acta Chim. Hung.*, 1960, **24**, 343.
22. R. T. Yang, *Gas Separation by Adsorption Processes*, Butterworth Publishers, Stoneham, MA, 1987.
23. D. Finlayson and A. J. Sharp, *British Patent* 365 092, 1930.
24. R. L. Hasche and W. N. Dargan, *US Patent* 1 794 377, 1931.
25. H. Kahle, *Chem.-Ing.-Tech.*, 1953, 23.
26. H. Kahle, *Chem.-Ing.-Tech.*, 1954, 26.
27. G. A. Perley, *US Patent* 1 896 916, 1933.
28. C. W. Skarstrom, *US Patent* 2 944 627, 1958.
29. P. Guerin de Montgareuil and D. Domine, *French Patent* 1 223 261, 1957; *US Patent* 3 155 468, 1964.
30. D. Basmadjian and L. A. Pogorski, *US Patent* 3 279 153, 1966.
31. S. Sircar, *et al.*, *US Patents* 4 077 779, 1978; 4 171 206, 1979; 4 171 207, 1979; 4 249 915, 1981; 4 264 340, 1981; 4 756 723, 1988.
32. T. R. White, R. Kumar, S. Sircar and E. G. Greskovitch, *US Patent* 4 477 265, 1984.
33. D. M. Ruthven, S. Farooq and K. S. Kneabel, *Pressure Swing Adsorption*, VCH Publishers Inc., New York, 1994.
34. D. Tondeur and P. C. Wankat, *Sep. Purif. Methods*, 1985, **14**, 157.
35. R. H. Perry and D. W. Green*Perry's Chemical Engineers*, McGraw-Hill, 1999.
36. R. W. Baker and K. A. Lokhandwala, *US Patent* 6 011 192, 2000.
37. R. W. Baker, *US Patent* 5 785 739, 1998.
38. R. W. Baker, K. A. Lokhandwala, Z. He and I. Pinnau, *US Patent* 6 183 628, 2001.
39. X. Feng, C. Y. Pan, J. Ivory and D. Ghosh, *Chem. Eng. Sci.*, 1998, **53**, 1689.
40. I. A. A. C. Esteves and J. P. B. Mota, *Desalination*, 2002, **148**, 275.
41. I. A. A. C. Esteves and J. P. B. Mota, *Ind. Eng. Chem. Res.*, 2007, **46**, 5723.
42. I. A. A. C. Esteves and J. P. B. Mota, in *CHEMPOR 2001*, ed. F. Ramôa Ribeiro and J. J.C. Cruz Pinto, Aveiro, Portugal, 2001, vol. 1, p. 525.
43. I. A. A. C. Esteves and J. P. B. Mota, in *Fundamentals of Adsorption 7 (FOA7)*, ed. K. Kaneko, H. Kanoh and Y. Hanzawa, Int. Adsorption Society (IK Int. Pub.), 2002, vol. 7, p. 918.
44. I. A. A. C. Esteves, *Gas Separation Processes by Integrated Adsorption and Permeation Technologies*, Ph.D. Dissertation, Universidade Nova de Lisboa (FCT/UNL), Lisbon, 2005.
45. R. R. Zolandz and G. K. Fleming, Gas permeation, in *Membrane Handbook*, ed. W. S. W. Ho and K. K. Sirkar, Kluwer Academic Publishers, Boston, 1992.
46. P. V. Marcea and S. T. Hwang, *J. Membr. Sci.*, 1994, **88**, 131.
47. R. W. Baker and K. A. Kaaeid, *US Patent* 6 011 192, 2000.
48. I. A. A. C. Esteves and J. P. B. Mota, *Sep. Purif. Technol.*, 2008, **62**, 281.
49. I. A. A. C. Esteves and J. P. B. Mota, *Adsorpt. Sci. Technol.*, 2007, **25**, 693.

50. P. I. Barton and C. C. Pantelides, *AIChE J.*, 1994, **40**, 966.
51. M. Oh and C. C. Pantelides, *Comput. Chem. Eng.*, 1996, **20**, 611.
52. R. T. Yang and S. J. Doong, *AIChE J.*, 1985, **31**, 1829.
53. R. T. Yang and S. J. Doong, *AIChE J.*, 1986, **32**, 397.
54. D. T. Coker, B. D. Freeman and G. K. Fleming, *AIChE J.*, 1998, **44**, 1289.
55. G. C. Kapantaidakis, S. P. Kaldis, X. S. Dabou and G. P. Sakellaropoulos, *J. Membr. Sci.*, 1996, **110**, 239.
56. G. C. Kapantaidakis, G. H. Koop and M. Wessling, *Desalination*, 2002, **145**, 353.
57. J. P. B. Mota and I. A. A. C. Esteves, *Ind. Eng. Chem. Res.*, 2007, **46**, 6978.
58. J. P. B. Mota, I. A. A. C. Esteves and M. F. J. Eusébio, *AIChE J.*, 2007, **53**, 1192.
59. F. J. A. L. Cruz, I. A. A. C. Esteves, J. P. B. Mota, S. Agnihotri and E. A. Muller, *J. Nanosci. Nanotechnol.*, 2010, **10**, 2537.
60. I. A. A. C. Esteves, F. J. A. L. Cruz, E. A. Muller, S Agnihotri and J. P. B. Mota, *Carbon*, 2009, **47**, 948.

Subject Index

Page numbers in *italics* refer to entries in Membrane Engineering for the Treatment of Gases, Volume 2.